Natural Resource and Environmental Economics

Natural Resource and Environmental Economics

TONY PRATO

Iowa State University Press / Ames

Tony Prato is professor of resource economics and management, and director of the Center for Agricultural, Resource and Environmental Systems at the University of Missouri–Columbia. He has written extensively on the assessment and management of natural resources.

© 1998 Iowa State University Press, Ames, Iowa 50014
All rights reserved

Authorization to photocopy items for internal or personal use, or the internal or personal use of specific clients, is granted by Iowa State University Press, provided that the base fee of $.10 per copy is paid directly to the Copyright Clearance Center, 27 Congress Street, Salem, MA 01970. For those organizations that have been granted a photocopy license by CCC, a separate system of payments has been arranged. The fee code for users of the Transactional Reporting Service is 0-8138-2938-0/98 $.10.

∞ Printed on acid-free paper in the United States of America

First edition, 1998

International Standard Book Number: 0-8138-2938-0

Library of Congress Cataloging-in-Publication Data

Prato, Tony
Natural resource and environmental economics / Tony Prato.—1st ed.
p.   cm.
Includes bibliographical references and index.
ISBN 0-8138-2938-0
1. Natural resources—Economic aspects. 2. Environmental economics.
I. Title.
HC21.P73   1998
333.7—dc21   97-25983

Last digit is the print number: 9 8 7 6 5 4 3 2 1

Contents

Preface xi
Acknowledgments, xiii

1. **Importance of Natural Resources and Environment** 3

    Nature and Importance of Problems 3
    Global Warming 4
    Ozone Depletion 5
    Acid Deposition 7
    Conservation of Biological Diversity 8

    Temporal, Scientific and Policy Aspects 10

    Contributing Factors 11
    Population 11
    Per Capita Resource Use 13
    Environmental Damages 14
    Technology 15

    Approaches to Resource and Environmental Issues 17

    Role of Economics 19
    Reductionist versus Holistic Science 20

    Summary 21

    Questions for Discussion 22

    Further Readings 23

    Notes 23

2. **Economic and Financial Concepts in Resource Management** 27

    Consumption and Demand Theory 27
    Utility Function and Indifference Map 28
    Constraints 29
    Household Equilibrium 31

    Production and Supply Theory 33
    Production Function and Isoquant Map 34
    Constraints 35
    Efficient Input Use 37
    Firm's Input Demand Curve 38
    Efficient Output 40

    Market Equilibrium 43
    Market Demand and Supply 44
    Example of Market Equilibrium 48

Present Value 49

Summary 51

Questions for Discussion 52

Further Readings 53

3. **Historical Views of Natural and Environmental Resource Capacity 55**

   Economic Views 55
   Classical Economics 55
   Neoclassical Economics 59

   Conservationism 60

   Contemporary Views 62
   Limits to Growth 62

   Other Factors 63

   Indicators of Resource Capacity 65
   Physical Measures 65
   Economic Measures 67
   Environmental Resources 68

   Summary 68

   Questions for Discussion 70

   Further Readings 70

   Notes 71

4. **Economy and Environment 73**

   Circular Flow Economy 73

   Material Balances 75

   Ecological Economics 78

   Sustainable Development 80
   Sustainable Resource Use 83

   Summary 86

   Questions for Discussion 87

   Further Readings 87

   Notes 87

5. **Property Rights and Externalities 91**

   Markets and Efficient Property Rights 91
   Markets 91
   Efficient Property Rights 93

   Transaction Costs 95

   Market Failures 96

Common Property and Open Access Resources   97
Public Goods   100

Externalities   100
Types of Externalities   101
Examples of Externalities   102

Externalities and Property Rights   104
Assignment of Property Rights   105
Government Intervention   107

Summary   108

Questions for Discussion   109

Further Readings   110

Notes   110

6. **Natural Resource Decisions**   111

   Natural Resource Management   111

   Types of Resources   115
   Exhaustible Resources   115
   Renewable Resources   116

   Static Efficiency   118
   Pure Competition   118
   Imperfect Competition   122

   Summary   123

   Questions for Discussion   123

   Further Readings   124

   Note   124

7. **Exhaustible Resource Use**   125

   Market Dynamics   125

   Net Social Benefit   127

   Two-Period Dynamic Efficiency   128
   Case 1: Two-Period Efficiency with Constant Oil Demand   129
   Case 2: Two-Period Efficiency with Variable Oil Demand   133

   Multiple-Period Efficiency   139
   Equilibrium Conditions   139
   Time Path of Prices and Extraction   140

   Underlying Factors   143
   Technological Progress   143
   Imperfect Competition   143
   External Costs   144
   Discount Rate   147
   Recycling   149
   Exploration and Development   153

Summary 153

Questions for Discussion 154

Further Readings 154

Notes 155

8. Renewable Resource Management 157

Simplifying Assumptions 158

Natural Growth 159

Static Efficiency 162
Private Property 163
Examples 164
Common Property 167
Multiple Objective Management 171

Dynamic Efficiency 173
Objectives and Constraints 174
Necessary Conditions 175

Renewable Resource Policies 182
Tax on Harvest 182
Access Fees, Effort Restrictions and Other Approaches 184

Summary 185

Questions for Discussion 186

Further Readings 187

Notes 187

9. Economics of Environmental Pollution 189

Residual Emissions, Pollution and Pollution Damages 190
Expanded Material Balances Model 191

Efficient Reduction in Environmental Pollution 194
Static Efficiency with Pure Competition 195
Static Efficiency with Imperfect Competition 206
Dynamic Efficiency 207

Establishing Property Rights for Environmental Resources 210

Pollution Abatement Policies 211
Point Source Pollution 211
Nonpoint Source Pollution 229

Summary 236

Questions for Discussion 238

Further Readings 239

Notes 240

10. Natural and Environmental Resource Accounting   243

Accounting Deficiencies   244
Defensive Expenditures   245
Resource Capacity   246
Residual Pollution Damages   249

Resource Accounting Methods   250
Physical Accounts   250
Monetary Accounts   252
Satellite Accounts   257

Resource-Specific Accounting   257

Implications for Sustainable Development   259

Summary   260

Questions for Discussion   261

Further Readings   262

Notes   262

11. Benefit-Cost Analysis of Resource Investments   265

Socially Efficient Investment   266
Net Social Benefit   266
Timber Harvesting Example   267

Special Topics in Investment Analysis   270
Continuous versus Discrete Time   271
Investment Evaluation Period   271
Efficiency and Equity Implications of the Discount Rate   272
Selection of Discount Rate   273
Capital and Operating Costs   279
Economic versus Financial Feasibility   279
Local versus Global Efficiency   280
Independent versus Interdependent Investments   280
Capital Rationing   281
Primary and Secondary Benefits and Costs   281
Risk and Uncertainty   282

Alternative Investment Evaluation Criteria   283
Payback Period   283
Average Rate of Return   284
Net Present Value   285
Annual Net Benefit   286
Benefit-Cost Ratio   286
Internal Rate of Return   287
Example   287

Evaluation of Independent and Interdependent Investments   290
Independent Investments   290

Interdependent Investments   291
Evaluation of Multiple Resource Investments   294

Summary   296

Questions for Discussion   297

Further Readings   298

Notes   299

12. Nonmarket Valuation of Natural and Environmental Resources   301

Importance of Nonmarket Valuation   301
Efficient Use of Exhaustible Resources   302
Natural and Environmental Resource Accounting   304
Resource Protection Policies   304
Efficient Resource Investment   304

Theoretical Basis for Nonmarket Valuation   305
Willingness to Pay and Accept Compensation   305
Valuing Changes in Resource Price   306
Valuing Changes in Resource Quality   309
Inequality between Willingness to Pay and Willingness to Accept Compensation   310

Use and Nonuse Values   311
Use Value   311
Option Value   312
Existence Value   314
Bequest Value   314

Estimation of Nonmarket Values   314
Indirect Market Methods   315
Direct Valuation Methods   322

Summary   329

Questions for Discussion   331

Further Readings   331

Notes   332

Index   335

Preface

Newspapers, books, reports and scientific journals are replete with articles and research findings dealing with a wide array of natural and environmental resource issues, such as depletion of fossil fuels and ozone, global warming, deforestation, and conservation of biodiversity. Considering the abundance of literature on these issues, it is appropriate to ask whether we really need another book on natural resource and environmental economics. At the time this book was conceived, there were several introductory books on the subject; however, the most recent upper-division undergraduate- and graduate-level text was 10 years old. Hence, the motivation for this book.

Natural Resource and Environmental Economics is a contemporary upper-division undergraduate- and graduate-level book that addresses a broad range of natural resource and environmental issues from an economic perspective. It covers topics in natural resource and environmental economics with greater analytical rigor than that found in introductory textbooks. The book is designed for students with a background in introductory microeconomics and algebra.

Natural Resource and Environmental Economics emphasizes the links between the economy and the ecosystem. It adopts a conceptual framework that views the economy as a subsystem of a finite and nonexpanding ecosystem that imposes biophysical limits on economic growth. This framework is consistent with the concept of sustainable development and sustainable resource use. Addressing the interconnections between the economic and ecological spheres requires that natural and environmental resource issues be approached from a multidisciplinary perspective. Natural resource and environmental economics is one of the disciplines in this nexus.

The initial chapters of the book provide background on the importance and relevance of natural resources and the environment (Chapter 1), basic economic and financial concepts in resource management (Chapter 2), and historical perspectives on natural and environmental resource capacity and its relationship to economic growth (Chapter 3).

The text then examines four paradigms for understanding the relationships between the economy and the environment or ecosystem (Chapter 4); effects of property rights and externalities on markets and resource allocation (Chapter 5); attributes of natural resource decisions and types of natural resources (Chapter 6);

principles of exhaustible resource use (Chapter 7) and renewable resource management (Chapter 8); privately and socially efficient levels of environmental pollution and policies for reducing pollution (Chapter 9); alternative methods of accounting for changes in natural and environmental resource capacity (Chapter 10); principles and application of benefit–cost analysis to resource investments (Chapter 11); and theory and methods of nonmarket valuation of natural and environmental resources (Chapter 12).

Natural Resource and Environmental Economics contains more material than can be covered in one semester. Specific chapters may, however, be selected for a course in natural resource economics or environmental economics or one that combines both areas. Chapters 1 through 8 are appropriate for courses in natural resource economics, and Chapters 1 through 6 and 9 and 12 are suitable for courses in environmental economics. Chapters 10 and 11 cover topics that are appropriate for courses that combine both areas.

A few sections of the book deal with topics that are better suited for students familiar with the interpretation of derivatives and integrals. These topics include efficient intertemporal use of exhaustible resources in Chapter 7 and renewable resources in Chapter 8, and dynamic evaluation of environmental pollution in Chapter 9. These topics can be excluded from upper-division (undergraduate) courses and included in graduate-level courses.

Acknowledgments

My understanding and appreciation of the socioeconomic, cultural and ecological aspects of managing natural and environmental resources have been enriched by the writings of the authors cited in the book. I want to thank colleagues and associates who, knowingly or unknowingly, have inspired and influenced my thinking and stimulated my professional interest in natural and environmental resource issues. These include Max Langham of the University of Florida, Bob Young and Dick Walsh of Colorado State University, Bob Kalter of Cornell University, Wally Tyner of Purdue University, Bob Caldwell and Bill Daly of the Natural Resources Conservation Service, Merlyn Brusven and Dave Walker of the University of Idaho, and Ted Napier and the late Joseph Havlicek of The Ohio State University.

I have been especially influenced by the writings and philosophies of Edward B. Barbier, Kenneth E. Boulding, Lester R. Brown, Robert Costanza, Herman E. Daly, Salah El Serafy, Anthony Fisher, Nicholas Georgescu-Roegen, John Krutilla, Aldo Leopold, Lewis and Clark, John McPhee, John Muir, Margaret E. and Olaus J. Murie, Richard B. Norgaard, David Pearce, E. F. Schumacher, Henry David Thoreau and others.

The preparation of this book was greatly enhanced by the computer graphics done by Zeyuan Qiu and Nicki Love. I appreciate the staff of the Center for Agricultural, Resource and Environmental Systems for tolerating my frequent absences from the office while writing the book. I sincerely appreciate the Department of Agricultural Economics, Social Sciences Unit and College of Agricultural, Food and Natural Resources, University of Missouri–Columbia, for providing an academic setting that encourages scholarly pursuits and professional development.

I deeply appreciate my parents, Jean and the late Alfred Prato, for supporting and encouraging my educational and professional pursuits, and my brothers, Gary and the late Richard Prato, for the wonderful backpacking trips together that cultivated my love of wild places. I am especially grateful to Karla, my wife and best friend, and to my children for their support and love.

Finally, I am especially grateful to the Lord Jesus Christ for sustaining my every breath and thought, generously pouring out His wonderful grace and love on my life, and continually reminding me of His power, presence and majesty through His creation.

Natural Resource and Environmental Economics

CHAPTER 1

Importance of Natural Resources and Environment

If current predictions of population growth prove accurate and patterns of human activity on the planet remain unchanged, science and technology may not be able to prevent either irreversible degradation of the environment or continued poverty for much of the world.

—U.S. NATIONAL ACADEMY OF SCIENCES AND ROYAL SOCIETY OF LONDON, 1993

While the history of environmental legislation qualifies the 1970s as the decade of the environment in the United States, the 1990s deserves to be billed as the decade for global environmental action. Is this because environmental problems have become more severe? Has affluence increased the demand for environmental quality? Are the rate of resource depletion and the magnitude of environmental damages arousing public concern? Are rapid improvements in worldwide communications and the democratization of countries in the former Soviet Union simply increasing our awareness of past environmental degradation?

The evidence suggests affirmative answers to all of these questions. Regardless of how you answered, you are likely to agree that resource and environmental issues are complex, widespread and deserving of attention. This chapter discusses five elements of resource and environmental problems: *nature and importance; temporal, scientific and policy aspects; contributing factors; alternative approaches; and the role of economics.*

Nature and Importance of Problems

Examples of resource and environmental problems include global warming; pollution of soil, water and air resources; food contamination; acid deposition; overexploitation of renewable resources (forests, fish and wildlife); depletion of exhaustible resources (oil, natural gas and coal); disposal of human and toxic wastes; losses in biodiversity; and degradation of ecosystems. Most natural and environmental resource problems involve three elements: depletion of exhaustible resources; overexploitation of renewable resources; and environmental pollution. The remainder of this section provides a brief overview of the nature and importance of four major natural and environmental resource problems: *global warming; ozone depletion; acid precipitation; and conservation of biological diversity.*

GLOBAL WARMING. Global warming refers to changes in climate (rainfall and temperature) caused by increased concentrations of carbon dioxide and other greenhouse gases (nitrogen oxide, methane and chlorofluorocarbons [CFCs]) in the upper atmosphere. Accumulation of greenhouse gases in the upper atmosphere traps the heat reflected from the earth's surface, which increases surface temperatures. In the last two centuries, atmospheric concentrations of greenhouse gases have increased 25 percent. Concentrations of greenhouse gases can double by the middle of the next century if no action is taken to reduce emissions of greenhouse gases.

Negative impacts of global warming include a reduction in agricultural production and forest biomass and extensive property damages from coastal flooding due to a rise in sea level. Production of agricultural and forest products is expected to be negatively impacted by an increase in global mean temperature of between 3.6 and 4.1 degrees Fahrenheit (2 and 2.3 degrees Celsius) by the end of the next century. While agricultural production in some countries would be enhanced by global warming, the net effect on agricultural production is expected to be negative. Sea-level rises from global warming are expected to be between 11.8 and 39.4 inches (30 and 100 centimeters).[1] Hence, global warming is a form of environmental pollution.

The root of the global warming problem is found in worldwide changes in fossil fuel consumption. When world population and energy use per person were low, wood was the major source of energy for cooking, heating and transportation. Because wood is a renewable resource, its use can continue indefinitely provided the rate at which trees are harvested is no greater than the rate of tree growth. As population and personal income increased, per capita demand for energy increased and coal replaced wood as a primary energy source. In the United States, coal had almost completely replaced wood by 1910. Coal is still the major source of energy for many Asian countries.

There are three problems associated with coal. First, unlike wood, it is an exhaustible resource, which means that increased use of coal decreases coal reserves. Second, coal mining, especially strip mining, causes major land disturbance and, in humid climates, water pollution. Third, underground mining of coal is hazardous to health. Underground mining accidents have also claimed many lives. In an effort to improve air quality, reduce land disturbance and water pollution, and satisfy increases in the demand for energy to power motor vehicles, petroleum (oil and natural gas) began to replace coal as a source of energy. By 1957, petroleum, which is an exhaustible fossil fuel, had replaced coal in many uses.

After two major shifts in energy use (wood to coal and coal to petroleum), the United States economy became heavily dependent on exhaustible fossil fuels. The share of total energy use in the United States that is supplied by exhaustible energy sources increased from 9 percent in 1850 to 95 percent in 1981.[2] The environmental consequences of this dependence are further magnified by the rise in the per capita use of processed energy that has occurred in response to the use of more energy-intensive technologies. For example, in developed countries, there has been a major shift in transportation modes from trains and buses to automobiles and airplanes. In the meantime, development of renewable energy sources, such as solar energy, has been slow because renewable sources have been relatively more expensive than exhaustible sources of energy. In addition, energy policies and programs have generally not supported the development and use of renewable energy sources, such as solar energy.

Heavy dependence on imported oil has geopolitical implications as well. Many developed countries import a significant share of their crude oil from Middle Eastern countries. The implications of importing oil from a politically unstable part of the world were dramatically demonstrated during the Arab oil embargo in 1973 and the Iraqi conflict in 1992.

Enter the global warming problem. As a result of significant growth in petroleum use in developed countries and steady growth in coal use in developing countries, concentrations of carbon dioxide in the upper atmosphere have increased from 260 parts per million in 1860 to 346 parts per million in the mid-1980s.[3] The trend in global warming is exacerbated by fossil fuel prices that are significantly below the full social and environmental costs of extraction, processing and distribution. The underpricing of fossil fuels results in overconsumption of fossil fuels and inadequate development of renewable resource substitutes, such as solar and geothermal energy.

Global warming has an important spatial dimension. Because developed countries burn proportionately more fossil fuels than do developing countries, developed countries account for the bulk of the carbon dioxide (CO_2) emissions worldwide. Country-by-country emissions of carbon dioxide are shown in Figure 1.1. While developed countries account for 70 percent of the fossil fuel–based carbon dioxide emitted since 1950 (69 percent in 1990), recent trends suggest that emissions from developing countries will increase from 1.8 billion tons in 1990 to 5.5 billion tons in 2025.[4] Hence, major reductions in worldwide carbon dioxide emissions will require reducing the current use of fossil fuels in developed countries and restraining growth in fossil fuel consumption in developing countries.

Uncertainty about the magnitude of, and damages from, global warming has hampered the willingness of many countries to respond. For example, the cost of restricting carbon dioxide emissions in the United States to their 1990 levels has been estimated to be between $8 billion and $36 billion per year in 1990 dollars. In comparison, the benefits of restricting greenhouse gas emissions to twice their preindustrial levels have been estimated to be $8 and $66 per ton of emissions.[5] Unilateral action by one country to reduce greenhouse gas emissions is not going to solve the global warming problem. It will take a concerted, worldwide effort to reduce the upward trend in carbon dioxide concentrations in the upper atmosphere. International coordination in the area of global warming was advanced when the United States and 154 other countries signed the United Nations Framework Convention on Climate Change at the United Nations Conference on Environment and Development in 1992. This convention has the goal of stabilizing emissions of greenhouse gases at their 1990 levels by the year 2000.[6]

OZONE DEPLETION. The stratosphere contains an ozone layer that protects the earth from the sun's harmful ultraviolet radiation known as UV-B. Increased

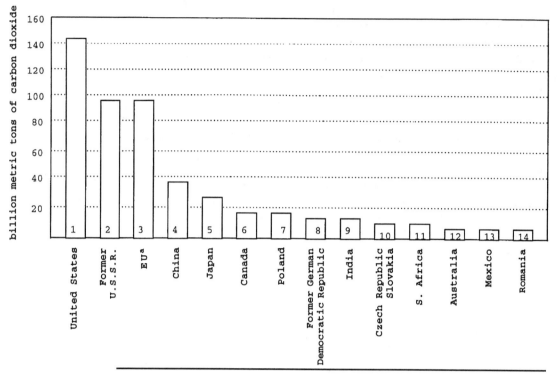

Figure 1.1. Cumulative emissions of CO$_2$ from fossil fuels, 1959–1989.

[a]The European Union (EU) comprises 12 countries: Belgium, Denmark, France, Germany, Greece, Ireland, Italy, Luxembourg, the Netherlands, Portugal, Spain, and the United Kingdom.

SOURCE: Carbon Dioxide Information Analysis Center (CDIAC), Oak Ridge National Laboratory, unpublished data (CDIAC Oak Ridge, Tennessee, August 1989).

UV-B radiation at the earth's surface could increase the incidence of skin cancer and eye disease, reduce agricultural productivity, and increase the loss of terrestrial and marine biomass. Accumulation in the stratosphere of industrial compounds containing chlorine and bromine have the capacity to deplete the ozone layer over a long period of time. Compounds containing these chemicals include CFCs, halons and chlorine-based solvents. For more than 30 years, these chemicals were used as aerosol propellants, coolants in refrigerators and air conditioners, cleaning and foam-blowing agents and fire extinguishers. Hence, ozone depletion is a form of environmental pollution.

In 1974, it was discovered that CFCs used as aerosol propellants could deplete stratospheric ozone. In response to this discovery, the United States Environmental Protection Agency (U.S. EPA) banned this use of CFCs. In 1985, a hole was discovered in the ozone layer above Antarctica. Two years later, in 1987, the Montreal Protocol on Substances that Deplete the Ozone Layer (Montreal Protocol) was signed by 24 countries, including the United States. The goal of the Montreal Protocol was for each country to achieve a 50 percent reduction in emissions of CFCs by 1996. In 1988, DuPont, the primary manufacturer of CFCs, announced its decision to phase out production of CFCs by the year 2000. In 1992, the Montreal Protocol was amended to set a target of banning CFCs by 1996.[7]

Actions to reduce ozone depletion appear to be paying off. Research by scientists with the National Oceanic and Atmospheric Administration reported in mid-1996 indicates that ground-level readings on the total amount of ozone-depleting chemicals taken on three continents and two Pacific Ocean islands have peaked and are starting to decline. This turnabout is expected to lead to a recovery in the ozone layer around the year 2005 or 2010.[8]

The apparent success of the Montreal Protocol and other events in protecting the ozone layer stands in stark contrast to the global warming problem. This is not surprising given the high economic stakes of reducing global warming versus ozone depletion. In the case of ozone depletion, the causes and effects of ozone depletion were indisputable and the development of substitutes for ozone-depleting chemicals was relatively easy. In comparison, the benefits of reducing global warming are uncertain and the costs are high because it will require reducing greenhouse gas emissions and, hence, fossil consumption.

ACID DEPOSITION. Acid deposition is a process whereby part of the sulfur and nitrogen gases released to the atmosphere by the burning of fossil fuels is converted to acids that can be carried to the earth's surface by rain, snow and dry deposits. Acid deposition can damage biological resources (lakes and trees) and human health and structures (buildings, bridges and statues) and can reduce visibility. Adverse biological effects of acid deposition were discovered in the mid-1960s by fishers from Sweden, Norway, Canada and New York, who noticed that the productivity of mountain lakes was declining.[9] In New York's Adirondack Mountains, 14 percent of the lakes became so acidified that they lost their ability to support many species of fish.[10] In the late 1970s, German foresters observed major declines (up to 50 percent) in the health of forest ecosystems. Stunted tree growth was also reported in Canada, the northeastern United States and southeastern California.[11]

The Office of Technology Assessment reported that sulfate pollution accounts for up to 50,000 premature deaths per year in the United States.[12] Structural damages occur as acid rain erodes the surfaces of buildings, bridges and statues. Sulfate particles in the atmosphere cause light to scatter, which increases haze and reduces visibility. Hence, acid precipitation is a form of environmental pollution caused by the burning of fossil fuels.

Much of the acid precipitation in the United States is caused by conditions in the Midwest. Air masses in this region transport air pollutants (primarily sulfur dioxide from coal-burning power plants) up the Mississippi and Ohio River valleys. While in transit, the pollutants slowly form acids and sulfate particles, which become attached to rain, snow and dry deposits. Because the physical and chemical reactions that lead to acid deposition are very complex, there is not necessarily a 1:1 relationship between air emissions from power plants and rainfall acidity. Even the evidence regarding rainfall acidity is mixed. Gene Likens and associates at Cornell University published maps showing substantial increases in rainfall acidity from the mid-1950s to the mid-1970s. Data collected from the Hubbard Brook Experimental Forest in New Hampshire (which maintains the longest record on rainfall acidity) indicates, however, that the acidity of precipitation has remained constant or declined slightly since 1964. While sulfuric acid decreased, nitric acid increased.[13]

Two general approaches can be taken to reduce acid deposition: reducing emissions of substances that contribute to acid rain and ameliorating the impacts of acid deposition in environmentally sensitive areas. The first approach involves curbing emissions of sulfur dioxide or nitrogen oxide at the source by installing sulfur dioxide scrubbers in smokestacks. The second approach involves adding lime to sensitive lakes and other areas. Lime helps to neutralize sulfuric and nitric acids.[14]

Despite uncertainty regarding the extent to which reductions in sulfur dioxide and nitrogen oxide emissions will reduce acid deposition and the geographic extent and severity of ecological damages caused by acid deposition (especially damages to terrestrial ecosystems), political pressure was exerted by Canada and the New England states on the United States Congress to pass legislation that would reduce emissions of these substances. In response to this pressure, the United States Congress included a provision in the 1990 Clean Air Act to control acid deposition. The goal of this provision is to reduce sulfur dioxide emissions by 10 million tons per year relative to their 1980 levels and nitrogen oxide emissions by 2.5 million tons per year by the year 2000. These emission reductions are deemed sufficient to protect aquatic ecosystems. Economic analysis indicates that this provision has a favorable net economic benefit, with annual benefits of $2 billion to $9 billion and annual costs of $4 billion.[15] A unique feature of the acid precipitation provision is that the target reduction in emissions is being achieved using tradable emission permits, which are discussed in Chapter 9.

CONSERVATION OF BIOLOGICAL DIVERSITY. Accelerated losses in biological diversity have become a major economic and environmental issue and the principal motivation for achieving sustainable use of land and water resources. Noss and Cooperrider[16] define biological diversity as follows: "Biological diversity is the variety of life and its processes. It includes the variety of living organisms, the genetic differences among them, the communities and ecosystems in which they occur, and the ecological and evolutionary processes that keep them functioning, yet ever changing and adapting." Losses in biodiversity occur directly through exploitation of species such as Africa's black rhinoceros and indirectly through modification of habitats and ecosystems, as when mangrove forests in Asia are cleared to make way for shrimp ponds.

Losses in biodiversity occur at three levels of biological organization: genetic diversity within species, species diversity, and diversity of communities and ecosystems. Genetic diversity has declined for domesticated (food and fiber) and wild species of plants and animals. The success of the Green Revolution in developing high-yielding varieties of cereal was achieved at the expense of fewer traditional varieties adapted to local environmental conditions. If current trends continue, three-fourths of India's rice fields are likely to be planted to only 10 varieties by 2005. In the United States, 71 percent of the corn is planted using just six varieties and half of the wheat land is planted to nine varieties. Fewer varieties make crop production more vulnerable to pest and disease outbreaks. Diversity within wild species of fish have also declined. In the Pacific Northwest region of the United States, at least 106 major populations of salmon and steelhead have been lost and 214 anadromous species are at risk due to modifications of the Columbia River.[17]

Extinction of species of plants and animals is the most familiar loss in biodiversity. Most of the losses in species occur among invertebrates in tropical forests. Wilson estimates that a minimum of 50,000 invertebrate species becomes extinct each year due to destruction of their rain forest habitat.[18] Peter Raven of the Missouri Botanical Gardens stated that one-fourth of all tropical plants are likely to be lost in the next 30 years.[19] American oysters in Chesapeake Bay have decreased 99 percent since 1870. In Soviet Asia, certain rivers and lakes have lost more than 90 percent of commercial fish species.[20] More than half of the fish that existed before commercial logging operations in Malaysia have been lost and more than 3,000 species of native plants are in danger of extinction.[21]

Losses in ecosystem biodiversity have been especially severe. Nearly half of the original area in tropical rain forest has been reduced due to logging and agricultural or urban development. The United Nations Food and Agricultural Organization[22] estimated that tropical deforestation proceeded at the rate of 6,564 miles^2 (170,000 kilometers^2) during the 1980s, which is equivalent to destroying about 0.9 percent of the total forested area in 1980. Losses in biodiversity due to tropical deforestation are especially significant because tropical rain forests contain the majority of terrestrial species. Less than 10 percent of the old-growth forests of the Pacific Northwest remain. Wetland ecosystems, which are among the world's most productive ecosystems, have been especially hard hit due to drainage of wetlands for agricultural production and aquacultural ponds. Half of the wetlands in the United States that existed in precolonial times have been destroyed. Wetland losses not only reduce fish and wildlife habitat, but they increase runoff and pollution of water bodies by sediment, nutrients and chemicals. Other examples of losses in biodiversity include losses in the North American tall grass prairie, cedar groves in Lebanon and old-growth hardwood forests in Europe.[23]

Why conserve biodiversity? Moral, ethical and economic reasons have been used to justify the conservation of biodiversity. Barbier et al.[24] state that "Conservation of biological diversity is of vital importance to humankind because some level of biodiversity is essential to the functioning of ecosystems on which not only human consumption and production but also existence depends." The International Union for the Conservation of Nature, the United Nations Environment Program and the World Wildlife Fund point out that "Biological diversity should be conserved as a matter of principle, because all species deserve respect regardless of their use to humanity . . . and because they are all components of our life support system. Biodiversity also provides us economic benefits and adds greatly to the quality of our lives."[25]

In the United States, the primary legislation for conserving biodiversity is the Endangered Species Act of 1973. The purpose of this act is "to protect the ecosystems upon which endangered species and threatened species depend." The act provided a legal basis for protecting the habitat of the northern spotted owl, which afforded protection for old growth forests in the Pacific Northwest. The importance of conserving biodiversity has been recognized in several world forums including the First World Conservation Strategy in 1980, the Second World Conservation Strategy in 1991 and the 1992 United Nations Conference on Environment and Development (UNCED). At the UNCED conference, 154 nations signed the Convention on Biological Diversity, which attempts to conserve biodiversity on a global scale.

Temporal, Scientific and Policy Aspects

Public attitudes toward resource and environmental issues have changed over time. Prior to the 1970s, there was little public concern for resource depletion and environmental degradation. Most of the emphasis was on resource development and use. A notable exception was the concern over soil erosion in the Dustbowl (U.S. prairie states) that occurred in the 1930s. Stirred by Rachel Carson's 1962 book, *Silent Spring,* on the human and environmental impacts of pesticides,[26] smog problems in major cities, and the Santa Barbara oil spill of 1969, the United States became more concerned about environmental issues in the late 1960s and early 1970s. During this period, there was a burst of environmental legislation and activity, including the passage of the National Environmental Protection Act in 1969 and creation of the U.S. EPA in 1972.[27] In recent times, there has been growing public concern over the protection of aesthetic or amenity values such as solitude and scenic beauty.

Concern about environmental degradation has been influenced by scientific studies of physical, biological and social processes. It is not surprising, for example, that public interest in groundwater protection increased when agencies such as the U.S. Geological Survey and the U.S. EPA became involved in testing or regulating the quality of drinking water. The finding that babies and pregnant women are especially vulnerable to pesticide residues in food heightened public and government interest in food safety.

Solutions to resource and environmental problems are complicated by the interaction of complex biological, hydrological, geological, chemical, social, economic and institutional processes from which these problems derive. Contamination of rivers and streams by agricultural pesticides illustrates this point. Timely application of pesticides to crops is important in reducing yield losses due to pests. Because crop yields directly affect farm profits, use of pesticides is motivated by economic reasons. Farmers usually apply pesticides according to a schedule. When significant rainfall occurs shortly after pesticide application, there is a high likelihood that runoff from that field contains pesticides. Rain and runoff are hydrological events. The volume of runoff is affected by the amount of crop residue left on the land, which varies by crop, the type of soil in which the crop is grown, the way the soil is prepared for planting and the steepness of the land. Soil type and steepness of the land are determined by geological conditions. Concentration of pesticides in runoff depends both on the rate and method of pesticide application to the crop and on the volume of runoff. The choice of crop rotations and farming practices is influenced by socioeconomic conditions and public policy.

Biological effects of pesticides on streams and rivers depend on the volume of water in the stream or river, as well as the potency and persistence of the pesticide. Potency and persistence are chemical properties of pesticides that influence their effects on plants and animals. Pesticides that are potentially harmful to plants and animals and degrade slowly pose the greatest threat to biological activity. If the water in the stream or river is used as a source of drinking water and the concentration of the pesticide exceeds the maximum contaminant level established for drinking water, then the pesticide is likely to pose a risk to human health.

On the social side, numerous private individuals, organizations and governmental agencies influence the contamination of streams and rivers by pesticides.

These groups are referred to as stakeholders. They all have a stake in the use and/or effects of pesticides. The manufacturer of the pesticide determines the chemical composition of the pesticide, which affects its potency and persistence and the recommended rate of application under different soil and climatic conditions. The pesticide distributor makes recommendations to the farmer regarding the frequency and timing of application. In some cases, the pesticide is custom-applied by private pesticide applicators hired by the farmer. Pesticide applicators typically have the highest health risk from pesticides. Several organizations, such as the Natural Resources Conservation Service and the Extension Service, assist the farmer in determining farming methods that reduce the risk of pesticide contamination. National, state and local environmental groups want to reduce the adverse impacts of pesticides on plants and animals.

Agricultural and environmental policies affect the use of pesticides. National farm policy, which is updated every five years, directly affects the profitability of growing different crops. In the past, agricultural policy has favored the production of crops such as corn, cotton and rice, which use considerable pesticides. Finally, the general public has become increasingly concerned about the adverse health and environmental risks associated with pesticide residues in food and water. This concern has led to stricter national environmental policies, such as the Safe Drinking Water Act and the Endangered Species Act. Under the authority of the Endangered Species Act, the U.S. EPA has the authority to restrict or ban the use of pesticides in areas containing endangered plants and/or animals.

There are many success stories of efforts to curb natural resource degradation and environmental pollution. The banning of DDT in the 1960s has contributed to the resurgence of raptors, including the bald eagle and the peregrine falcon. Development and application of no-till farming methods have substantially reduced erosion on cropland where this method has been adopted by farmers. Legislation to reduce water pollution from point sources, such as factories and sewage treatment plants, has been very successful in reducing water pollution from these sources. Implementation of safe drinking standards has reduced the risk of contamination of public drinking water supplies.

Contributing Factors

Four major factors govern resource depletion and environmental pollution: namely, population, per capita resource use, damages per unit of resource use and technology, as shown in Figure 1.2. In its simplest form, population times per capita resource use equals total resource use. Total resource use times environmental damages per unit of resource use equals total environmental damages. Technology influences population, resource use and environmental damages.

POPULATION. World population is growing at an annual rate of 92 million people.[28] For the period 1950 to 1990, world population grew at an exponential rate, as shown in Figure 1.3. Population growth is the major force behind population dy-

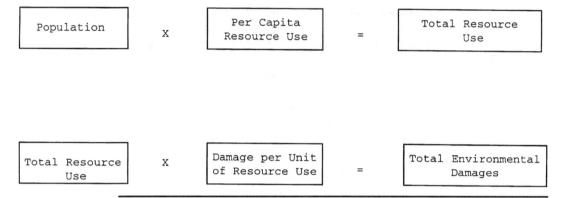

Figure 1.2. Determinants of resource use and environmental damages.

namics. The implications of exponential population growth can be seen by examining the number of years it takes to add one billion people to the world population, as shown in Table 1.1.

Since current world population is 5.5 billion, the last two rows in Table 1.1 are based on a population growth rate equal to the current rate of 1.7 percent. Projections are that the world population will reach 10 billion to 14 billion by the year 2100.

Resource and environmental impacts are further complicated by the uneven distribution of worldwide population growth. Figure 1.3 shows that growth in world population from 1950 to 1990 is substantially greater for developing countries than it is for developed countries. For example, the United States has a current population growth of 1 percent per year, whereas the population of many African countries is growing in excess of 3 percent per year. At these growth rates, population in the United States doubles every 70 years and the populations of African countries double every 23 years.

Differences between population growth in developed and developing countries are attributed to the higher fertility rates existing in developing countries. On a worldwide basis, total fertility rates in 1990 were 1.7 children per woman in high-income countries versus 3.8 children per woman in low- and middle-income countries. Fertility rates (births per childbearing woman) generally decrease as per capita income and educational level increases. One study showed that the average woman

Table 1.1. Years required to achieve successive one billion increments in population

Number of Years	Increments
2-5 million	1st
130	2nd
30	3rd
15	4th
12	5th
10	6th
10	7th

Source: Summarized from G. Tyler Miller, Jr., *Resource Conservation and Management* (Belmont, California: Wadsworth Publishing Co., 1990), p. 4.

1. **Importance of Natural Resources and Environment** 13

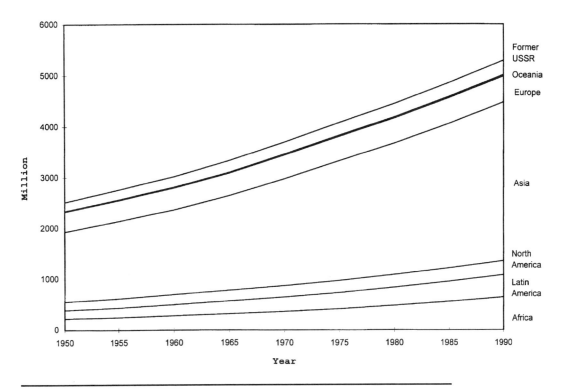

Figure 1.3. World population by region from 1950 to 1990 in 5-year intervals.

SOURCE: United Nations, 1990 Demographic Yearbook (United Nations, New York, 1992).

in developing countries has seven children when none of the women have a secondary education and only three children when 40 percent have a secondary education.[29] Rapid population growth contributes directly to resource depletion. For example, future growth in population over the next 30 years is expected to increase total energy use in developed countries by 70 percent, even with current levels of per capita energy use.[30]

PER CAPITA RESOURCE USE. Per capita resource use varies over time and space. Worldwide use of fossil fuels on a per capita basis increased from 0.625 tons of coal equivalent in 1900 to 2.40 tons in 1986, a 280 percent increase.[31] Energy consumption per capita is significantly greater in developed than in developing countries, as shown in Table 1.2.

Relationships between population and per capita resource use have important implications for international attempts to resolve resource and environmental problems. A disproportionate 80 percent of the world's processed exhaustible energy and mineral resources are consumed by the 23 percent of the world's population living in developed countries, as shown in Table 1.3. A major factor contributing to this distortion is the higher per capita national income in developed versus developing countries ($10,700 versus $640). Part of the income distortion is explained by the

Table 1.2. Per capita energy consumption in selected countries, 1987

Country	Energy Use (gigajoules)
Developed Countries	
United States	280
Soviet Union	194
West Germany	165
Japan	110
Average	187
Developing Countries	
Mexico	50
Turkey	29
Brazil	22
China	22
Indonesia	8
India	8
Nigeria	5
Average	18

Source: U.S. Bureau of the Census, *Statistical Abstracts of the United States: 1990* (Washington, DC: U.S. Government Printing Office, 1990); World Resources Institute, *World Resources, 1990-91* (New York: Oxford University Press, 1990).

Table 1.3. Comparison of population and resource use in developed and developing countries, 1988

Characteristic	Developed Countries	Developing Countries
Number	33	142
Location	North America, Europe, Japan, Australia, New Zealand	Africa, Asia, Latin America
Population (billion)	1.2	4
% World's population	23	77
% World's energy and mineral resources	80	20
Average GNP per person	$10,700	$640
Population doubling time (years at indicated annual growth rate)	117 @ 0.6%	33 @ 2.1%
Environmental features	Temperate latitudes with more favorable climate and soils	Tropical latitudes with less favorable climate and soils

Source: Summarized from G. Tyler Miller, Jr., *Resource Conservation and Management* (Belmont, California: Wadsworth Publishing Co., 1990), p. 5 and Figure 1-4, p. 8. GNP = gross national product.

more favorable climate and soils and lower population generally present in developed countries.

ENVIRONMENTAL DAMAGES. Total environmental damages equal total resource use times environmental damages per unit of resource use. Evidence is mounting that environmental damages are substantial. The International Institute of Applied Systems Analysis estimates that losses in forest productivity from sulfur dioxide emissions from automobiles and power plants amounts to $30.4 billion annually.[32] While not quantified in dollar terms, tropical deforestation (excess of harvest rate over natural regrowth and tree planting) amounts to 43 million acres (17.4 million hectares) per year.[33] In the past decade, deforestation has claimed an area equivalent to the combined size of Malaysia, the Philippines, Ghana, the Congo,

Ecuador, El Salvador and Nicaragua.[34] Excessive harvesting of forests is usually the result of land clearing for agricultural production, export timber sales, shelter wood and fuel wood.

The United Nations reports that worldwide degradation in irrigated and rainfed cropland and rangeland resulted in crop and livestock production losses of $42 billion in 1990. About 70 percent of the losses occurred in Africa and Asia.[35] Land degradation results from soil erosion, overgrazing, water logging, salinity and chemical use. Cropland losses from soil erosion in the United States amount to $3.5 billion per year.[36] Annual offsite damages from soil erosion in the United States are approximately $10 billion.[37]

One study indicates that a doubling of greenhouse gases (global warming) by 2025 would cause agricultural losses of $18 billion, increase air conditioning costs by $11 billion, and increase the cost of mitigating adverse impacts of rises in the sea level by $7 billion, plus other losses, for a total loss of $58 billion annually.[38] Air pollution in the United States has been estimated to cost the nation $40 billion per year in the form of increased health care costs and losses in worker productivity.[39]

There are many other examples of natural resource depletion and environmental degradation from around the world:

- Overfishing occurs in four out of the world's 17 fishing areas.[40]
- Irrigation diversions of water from the river feeding the Aral Sea have increased salinity to the point at which fish can no longer exist.[41]
- Water pollution in the Chesapeake Bay, a major world estuary, has reduced oyster production from eight million to less than one million bushels per year.[42]
- Compared with those living elsewhere, Bulgarian people living near heavy industries have asthma rates that are nine times higher, skin diseases seven times more frequent, liver diseases four times greater, and nervous system disorders three times higher.[43]
- The head of the Russian Academy of Sciences admitted that chemicals and organic toxins have taken a major toll on human health. Eleven percent of the children are born with defects, half of the drinking water and a tenth of the food supply are contaminated, 55 percent of the school age children have health problems, and rates of illness and early death of individuals in the 25–40 age bracket have increased.[44]
- Skin cancer fatalities in the United States are expected to increase by 200,000 per year as a result of a recent revision in the rate of ozone depletion.[45]
- Cleanup of hazardous waste sites in the United States is expected to cost $750 billion.[46]
- At least 50,000 invertebrate species per year become extinct as a result of the despoliation of tropical rain forests.[47]

TECHNOLOGY. Technological change directly influences population growth, use of natural resources and environmental damages. Rapid changes in agricultural technology during the 1950s and 1960s increased the productivity of agricultural labor and caused a massive off-farm migration of labor as illustrated in Figure 1.4. Rapid growth in the nonagricultural labor force, combined with major advancements in industrial technology, stimulated production and consumption of industrial products and increased per capita income. Developing countries have not experienced the same technology-driven increase in agricultural productivity and the associated growth in industrial production and per capita income as have developed countries.

Technology has benefited society. For example, technology has been instrumental in reducing food prices, increasing the abundance and diversity of food and fiber products, improving human health and nutrition, increasing labor productivity, and improving transportation and communications. Technological advancements have also enhanced the efficiency of exploring for petroleum resources; extracting and using energy, mineral, forest and fishery resources; capturing, storing and utilizing solar energy; and in recycling products derived from natural resources. Technological advancement has significantly increased use of certain exhaustible resources and degraded environmental quality. Agriculture is a prime example. In the 1950s, farmers throughout the world were essentially self-sufficient in energy. Livestock wastes were the major source of fertilizer, and draft animals were the dominant power source for farming operations. From 1950 to 1985, worldwide use of energy in agriculture increased from 276 million to 1,903 million barrels of oil equivalent, a sevenfold increase, and world use of fertilizer in agriculture increased

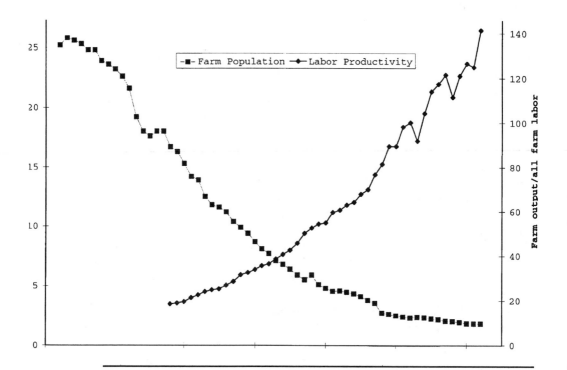

Figure 1.4. Percentage of U.S. population on farms and farm labor productivity, 1932–1993, United States.

SOURCE: Economic Research Service, U.S. Department of Agriculture, *Agriculture Resources and Environmental Indicators,* update no. 6 (Washington, D.C., July 6, 1996); and U.S. Census Bureau, *Current Population Reports* (Washington, D.C., 1994).

NOTE: 1982 = 100%.

ninefold.[48] In addition, the use of farm tractors quadrupled and the area under irrigation tripled.[49] Intensity of energy use in agriculture also increased. From 1950 to 1985, energy use per ton of grain produced increased from 0.44 to 1.14 barrels of oil equivalent.[50]

Technology-driven growth in agricultural productivity has increased use and depletion of other natural resources. From 1950 to 1985, world acreage devoted to cereal grains increased from 1,210 to 1,766 million acres (490 to 715 million hectares).[51] During the same period, expansion of dryland and irrigated acreage in the United States caused the plowing up of grasslands and the drainage of wetlands in the Great Plains. In the 1970s, fencerow-to-fencerow farming was encouraged and practiced. These and other land use changes have increased wind and water erosion; mining of groundwater for irrigation; pollution of surface and groundwater by fertilizers, pesticides and wastes; and loss of wildlife habitat. Mining of groundwater occurs when the rate of pumping exceeds the natural rate of recharge. The water table for the Ogallala Aquifer in the central United States has fallen substantially over the past several years due to groundwater mining.

Technological optimists and proponents of the cornucopian view argue that the potential for technological change, information management and substitution of less abundant for more abundant resources removes any natural limits to economic growth and development. Cornucopians often dispute the claim that there is an environmental crisis.[52] Others argue that degradation in natural and environmental resources is a direct result of unbridled economic growth, which is not sustainable, and that immediate action is needed to restore balance between humans and the natural environment.[53]

Other factors besides population growth, per capita resource use, environmental damages per unit of resource use, and technology influence the occurrence and persistence of resource and environmental problems. These other factors include religion, culture, tradition, customs, poverty, hunger, public policy, resource endowments and climate.

Approaches to Resource and Environmental Issues

Resource and environmental issues can be analyzed several ways. Two extreme approaches are discussed here: the reductionist method and the holistic method. Most analyses use an approach that is somewhere between the two extremes. The reductionist method applies scientific concepts from a particular discipline to a narrowly defined issue. The issue is examined in terms of a set of hypotheses supported by the knowledge base in that discipline. Hypotheses are tested by applying statistical techniques to data obtained from controlled experiments or surveys. The reductionist approach has a long history of use and acceptance in the scientific community.

The reductionist approach is illustrated for the issue of whether or not petroleum resources are becoming more scarce over time. An economist is likely to address this issue by examining the trend over time in relative prices of oil and natural gas. If relative prices of petroleum decrease over time, the economist concludes that petroleum is becoming less scarce. This conclusion was reached by Barnett and Morse.[54]

A petroleum geologist is likely to approach scarcity in terms of the trend in world petroleum reserves, which reflects use and additions to reserves resulting from new discoveries. For the geologist, scarcity increases as world reserves decrease. An energy analyst might judge scarcity in terms of the number of years it takes to use up remaining reserves at current or expected rates of use (reserves-to-use ratio) or by the percent of use supplied by domestic production. The energy analyst interprets an increasing reserves-to-use ratio as a reduction in scarcity and a decreasing ratio or greater reliance on imported oil as an increase in scarcity.

Scarcity indicators chosen by the economist, geologist and energy analyst are not always consistent with one another. Because energy markets are subject to government intervention, political upheavals, wars and monopolistic power, relative petroleum prices are not necessarily a good indicator of petroleum scarcity. In 1973, crude oil prices quadrupled in a relatively short period of time because the Organization of Petroleum Exporting Countries (OPEC) cartel effectively controlled a significant share of the world's oil supply. When OPEC's grip on the world oil market weakened, energy prices declined. Moreover, relative energy prices do not necessarily move in the same direction as the reserves-to-use ratio and/or dependence on imported oil. There have been periods when United States dependence on imported oil, and oil prices, have both increased. For these reasons, the reductionist approach to energy scarcity in particular, and resource or environmental issues in general, has its limitations.

The holistic approach addresses resource and environmental issues by synthesizing and integrating concepts, data and results from several disciplines. Systems analysis, which is the hallmark of the holistic approach, attempts to consider the relevant dimensions of an issue. The International Institute of Applied Systems Analysis[55] captured the essence of the systems approach in the following remark: "The existence of many linkages among the elements of a complex system implies that a change in some part may reverberate through the system, eventually triggering changes in ways that are not immediately obvious, especially if one is examining, because of a narrow jurisdictional or scientific specialty, only part of the system."

A systems analyst recognizes that resource and environmental problems are inherently complex and involve multiple interactions among physical, biological, social, economic, political and legal processes. A systems approach tries to understand the relationships among the elements in the system. Systems analysts usually take the position that resource and environmental issues, such as global warming, acid precipitation, land degradation and deforestation, involve multiple processes and multiple political entities (states and nations) that cannot be understood and resolved using a narrow disciplinary approach.

The systems approach has its share of problems. First, it runs counter to the way generations of scientists have been trained. The reductionist approach is the cornerstone of scientific inquiry. Second, as a holistic or systems approach to problem solving is more complex, it requires considerable interaction between the practitioners of several disciplines. This is a difficult task because of differences in theory, methods and data. For example, classical ecology views humans as just another species within a resource-limited system. Human activities are generally ignored in classical ecology except to the extent they alter the natural environment. In contrast, classical economics treats human satisfaction as the ultimate purpose of economic activity. The resource base is viewed as essentially unlimited due to technological change and resource substitution. Despite these problems, there is growing support for the holistic approach to resource and environmental issues.

Role of Economics

The primary objective of economics is to determine the efficient allocation of scarce resources among alternative uses. Several subdisciplines of economics deal with natural and environmental resource use and the interactions between the economy and the natural environment. These subdisciplines include *resource economics, environmental economics* and *ecological economics.* All three subdisciplines rely on the concepts of supply, demand and market equilibrium, which are discussed in Chapter 2.

Resource economics provides an analytical framework for determining an efficient allocation of natural and environmental resources over space and time. *Efficiency* is defined as maximizing one or more objectives subject to technical and physical limitations. While the objective of maximizing profit is key to conventional economic theory, resource economics emphasizes the objective of maximizing benefits to society. The production technology available for extracting or harvesting a natural resource is a technical limitation, whereas the amount of the resource available in different locations and time periods is a physical limitation. Efficiency can be also be defined in terms of minimizing the cost of achieving one or more objectives subject to the same technical and physical limitations.

Resource economics deals with two broad categories of resources: exhaustible resources, such as petroleum and minerals, and renewable resources, such as forests, fish and wildlife. Exhaustible resources are finite. Use of an exhaustible resources implies depletion. Economics of exhaustible resource use explains where and how rapidly exploration and development should occur and how much of the developed resource should be used in the production of different consumer products.

Renewable resources have the capacity to regenerate over time. If growth and use are in balance, then the stock of a renewable resource is maintained indefinitely, apart from natural disasters and events. The economics of renewable resource use describes the efficient rates of harvest in different locations and time periods. It provides a basis for evaluating whether society is better off by increasing commercial harvest rates to meet the expanding demand for products made from renewable resources or by reducing harvest rates to protect biological diversity. Finally, resource economics supplies a rational framework for determining the economic benefits and costs of investing in technologies to develop substitutes for exhaustible resources and for determining the economic benefits and costs of expanding the boundaries of a national park or a national wildlife refuge.

Environmental economics has three primary thrusts. First, the material balances approach in environmental economics adds an environmental sector to the traditional circular flow model of conventional economics. The environmental sector explains the origin and fate of wastes generated by production and consumption activities and the potential impacts of wastes on environmental quality. Second, environmental economics affords a framework for evaluating alternative technologies and public policies for reducing environmental pollution. Third, environmental economics provides analytical methods for estimating the economic value of improving environmental quality. These methods are especially important when markets do not exist or are inadequate for determining the value of improving environmental quality.

Ecological economics goes a step further than the material balances approach of environmental economics by recognizing the full range of interrelationships between the economic system and the ecological system. The economy is recognized as a subsystem of a finite and nonexpanding ecosystem. Ecological economics argues that depletion of exhaustible resources, overexploitation of renewable resources, and environmental pollution constitutes natural limits to economic growth.[56] Natural limits are incorporated in conventional national income accounts through a process known as natural resource and environmental accounting.

REDUCTIONIST VERSUS HOLISTIC SCIENCE. The traditional way of advancing knowledge in economics and other disciplines is through specialization, which implies compartmentalization. This mode of scientific inquiry supports the reductionist method, which breaks down a subject into its elementary parts. Each part is studied from a narrow disciplinary perspective. Specialization and reductionism have allowed science to make significant contributions to understanding a wide range of phenomena.

An alternative view is that the value of economics or of any single discipline in understanding and resolving complex resource and environmental problems is maximized through its integration with other disciplines. In the holistic approach, economists do not blindly cling to the assumptions, prescriptions and predictions of economic theory; rather, there is willingness to view economic concepts in light of physical and biological principles. The goal of the holistic approach is much broader than advancing the knowledge base of individual disciplines. It is to develop theoretical and practical knowledge that is useful in achieving greater harmony between humans and the environment.

Early support for a holistic approach is reflected in the writings of Schumacher, who was an economist. He makes a strong plea for practicing metaeconomics, which he defines as having the "aims and objectives from a study of man, and . . . at least a large part of its methodology from a study of nature." Conventional economics derives much of its methodology from quantitative sciences such as physics, not from the study of nature. Schumacher[57] is critical of this quantitative orientation, noting that "the great majority of economists is still pursuing the absurd ideal of making their 'science' as scientific and precise as physics, as if there were no qualitative differences between mindless atoms and men made in the image of God." This same criticism of conventional economic methodology is echoed in the writings of Herman Daly.

The implications of the holistic approach for the study of economics can be illustrated with regard to a pivotal assumption of economic theory, namely, that humans are motivated by selfishness. This assumption underlies the theory of household behavior and the theory of the firm, which is central to microeconomics. Daly and Cobb[58] criticize the assumption that households maximize utility and firms maximize profit oblivious to social community and biophysical interdependence, stating: "What is neglected is the effect of one person's welfare on that of others through bonds of sympathy and human community, and the physical effects of one person's production and consumption activities on others through bonds of biophysical community."

Another consequence of taking a holistic approach to resource and environmental issues is that it raises serious concerns about pursuing unbridled economic growth. Until recently, growth has been the undisputed goal of economic development. While this goal maximizes living standards in terms of material wealth, there is strong evidence that it contributes to natural resource and environmental degradation, which decreases human welfare and the quality of life.

What are the implications for economics of a holistic approach to natural resource and environmental problems? It requires the student of economics to become familiar with physical and biological principles governing the natural world and to integrate these principles with economic ones in addressing resource and environmental problems. In this framework, natural resource and environmental economics is viewed not so much as a self-contained body of knowledge but rather as a set of concepts that in combination with other scientific principles enhances society's understanding of resource and environmental issues. This view has evolved through the contributions of many contemporary economists including Kenneth Building, Nicholas Georgescu-Roegen, Richard Norgaard, Herman Daly and others.

Summary

Economic development has improved income and living standards, accelerated the use of natural resources, and increased environmental pollution. Four major resource and environmental problems associated with economic development include global warming, ozone depletion, acid precipitation and loss of biological diversity. These and other resource and environmental problems can be viewed in terms of three elements: depletion of exhaustible resources, overexploitation of renewable resources, and environmental pollution.

Resource use and environmental pollution have important spatial and temporal dimensions. In the spatial dimension, developed countries account for only 23 percent of the world's population but use 80 percent of the world's energy and mineral resources and generate 70 percent of the fossil fuel–based carbon dioxide emitted since 1950. Important temporal aspects of resource use are the replacement of renewable resources (wood) with exhaustible resources (coal and petroleum) and the slow development and adoption of solar energy.

Total resource use is the product of population and per capita resource use. Total environmental damage equals total resource use times environmental damage per unit of resource use. While per capita resource use is higher in developed countries than in developing countries, population growth is lower. Technological progress influences population growth, per capita resource use and environmental damages per unit of resource use.

Resource and environmental problems can be addressed from a reductionist or holistic perspective. *Reductionism* applies scientific concepts from a single discipline to a narrowly defined problem or issue. *Holism* synthesizes and integrates theory, methods and data from several disciplines. Systems analysis is the hallmark of the holistic approach. Conventional economics typically takes a reductionist approach to natural resource use that ignores the relationships between the economy and the natural environment. The subdisciplines of resource, environmental and ecological economics focus on one or more aspects of this relationship. There is a growing trend toward addressing resource and environmental issues from a holistic perspective.

Questions for Discussion

1. Many developing countries desire to achieve the same standard of living as developed countries. Several scientists and economists argue that there are not enough natural resources in the world for developing countries to achieve the same per capita use of resources as that of developed countries. Discuss ways in which this apparent conflict can be resolved.

2. Many developing countries believe that developed countries should assist them in applying technologies that reduce carbon dioxide emissions, which are the primary cause of global warming. They claim that developed countries generated most of the carbon dioxide emissions and should pay the cost of helping developing countries reduce emissions. What is your opinion of this viewpoint?

3. Advocates claim that increased use of solar energy would reduce dependence on exhaustible energy resources, the incidence of environmental pollution, and vulnerability to disruptions in crude oil production. What are some of the advantages and disadvantages of increased dependence on solar energy?

4. In the spotted owl controversy, economic development interests argued that income and employment losses from reduced harvesting of old growth forests in the Pacific Northwest were too high a price to pay to conserve this ecosystem. To what extent should economic impacts of conserving biodiversity be considered?

5. In an effort to control population growth, the People's Republic of China severely penalizes couples who have more than one child. Some view this policy as draconian. Others believe population control is the only effective way to reduce adverse consequences of rapid population growth in developing countries. Do you think that China's population policy is too drastic?

6. Cornucopians believe that changes in technology and substitution of manufactured capital for natural resources are sufficient to offset the depletion of natural resources and curb environmental pollution. Neomalthusians argue that technological change has contributed to increased total and per capita resource use and environmental pollution and that such impacts will continue unless immediate action in taken. What is your opinion of both positions?

Further Readings

Building, K. E. 1981. *Evolutionary Economics.* Beverly Hills, California: Sage Publications.

Bormann, F. Herbert, and Stephen R. Kellert, eds. 1991. *Ecology, Economics, Ethics: The Broken Circle.* New Haven, Connecticut: Yale University Press.

Field, Donald R., and William R. Burch, Jr. 1988. *Rural Sociology and the Environment.* New York: Greenwood Press.

Flavin, Christopher. 1996. "Facing Up to the Risks of Climate Change." In *State of the World, 1996.* New York: W.W. Norton & Co.

Norgaard, Richard B. 1989. "The Case for Methodological Pluralism." *Ecological Economics,* 1:37–57.

Portney, Paul R. "Acid Rain, Making Sensible Policy." *Resources,* Winter 1994, pp. 9–12.

Notes

1. Joel Darmstadter, "The U.S. Climate Change Action Plan: Challenges and Prospects," *Resources,* Winter 1995, pp. 19–23.

2. Nicholas Lenssen, "Providing Energy in Developing Countries," in *State of the World, 1993* (New York: W.W. Norton & Co., 1993), p. 106.

3. Lester R. Brown and Sandra Postel, "Thresholds of Change," in *State of the World, 1987* (New York: W.W. Norton & Co., 1987), p. 9.

4. John Gever et al., *Beyond Oil: The Threat of Food and Fuel in the Coming Decades,* 3rd ed. (Niwot, Colorado: University Press of Colorado, 1991), p. 43.

5. William D. Nordhaus, "To Slow or Not to Slow: The Economics of the Greenhouse Effect," *The Economic Journal* 101(1991):920–937; and A.S. Manne and R.G. Richels, "CO_2 Emission Limits: An Economic Cost Analysis of the USA," *The Energy Journal* 11(1990):51–74.

6. Darmstadter (1995).

7. Peter M. Morrisette, "Negotiating Agreements on Global Change," *Resources,* Spring 1990, pp. 8–11.

8. *Columbia (Missouri) Tribune,* "Study: Ozone Layer May Begin to Close," 31 May, 1996.

9. Hillary French, "Clearing the Air," in *State of the World, 1990,* Lester R. Brown, ed. (New York: W.W. Norton & Co., Inc., 1990).

10. NAPAP, *National Acid Precipitation Program, 1990 Integrated Assessment Report* (Washington, D.C.: National Acid Precipitation Program, 1991).

11. French (1990).

12. Office of Technology Assessment, United States. Congress, *Acid Rain and Transported Air Pollutants: Implications for Public Policy* (Washington, D.C.: United States Government Printing Office, 1984), p. 47.

13. Glen E. Gordon, "Acid Rain, What is It?" *Resources,* Winter 1994, pp. 6–8.

14. Winston Harrington, "Breaking the Deadlock on Acid Rain Control," *Resources,* Fall 1988, pp. 1–4.

15. Paul Portney, "Policy Watch: Economics and the Clean Air Act," *Journal of Economic Perspectives* 4(1989):173–182.

16. Reed F. Noss and Allen Y. Cooperrider, *Saving Nature's Legacy: Protecting and Restoring Biodiversity* (Washington, D.C.: Island Press, 1994), p. 5.

17. John C. Ryan, 1992, "Conserving Biological Diversity," in *State of the World, 1992* (New York: W.W. Norton & Co., 1992), pp. 9–26.

18. Edward O. Wilson, *The Diversity of Life* (Cambridge, Massachusetts: Harvard University Press, 1992).

19. Peter H. Raven, "Biology in an Age of Extinction: What is Our Responsibility?" (Plenary Address, Fourth International Congress of Systematic and Evolutionary Biology, College Park, Maryland, July 1–4, 1990).

20. Carl Safina and Ken Hinman, "Stemming the Tide: Conservation of Coastal Fish Habitat in the United States" (Summary of a National Symposium on Coastal Fish Habitat Conservation, Baltimore, Maryland, March 7–9, 1991).

21. Jared M. Diamond, "The Present, Past and Future of Human-Caused Extinctions," in *Philosophical Transactions of the Royal Society of London,* vol. B325 (1980); and Philip Shabecoff, "Plant Lovers' Ambitious Goal Is No More Extinctions," *New York Times,* November 13, 1990.

22. Food and Agricultural Organization, *Forest Resources Assessment: Tropical Countries,* Forestry Paper No. 112, Food and Agriculture Organization of the United Nations, Rome, 1993.

23. Ryan (1992).

24. Edward B. Barbier, James C. Burgess, and Carl Folke, *Paradise Lost? The Ecological Economics of Biodiversity* (London: Earthscan Publications, Ltd., 1994), p. 3.

25. IUCN/UNEP/WWF, *Caring for the Earth: A Strategy for Sustainable Living* (Gland, Switzerland: IUCN, 1991).

26. Rachel Carson, *Silent Spring* (New York: Houghton-Mifflin, 1962).

27. Legislation passed during the 1970s to protect land, air and water resources included the Clean Air Acts of 1970 and 1977, the Federal Water Pollution Control Act of 1972, the Endangered Species Act of 1973, the Toxic Substances Control Act of 1976, the Resource Recovery and Conservation Act of 1976 and the National Energy Act of 1978. Most of these acts have since been amended.

28. Population Reference Bureau, *1991 Population Data Sheet* (Washington, D.C.: 1991), U.S. Bureau of the Census.

29. World Bank, *World Development Report 1992: Development and the Environment* (New York: Oxford University Press, Inc., 1992), p. 29.

30. Nicholas Lenssen, "Providing Energy in Developing Countries," in *State of the World, 1993* (New York: W.W. Norton & Co., 1993), p. 106.

31. Brown and Postel (1987), p. 5.

32. International Institute of Applied Systems Analysis, "The Price of Pollution," *Options,* September 1990.

33. United Nations Food and Agriculture Organization as cited in World Resources Institute, *World Resources 1992–1993* (New York: Oxford University Press, 1992), pp. 118–119.

34. Lester R. Brown, "A New Era Unfolds," in *State of the World, 1987* (New York: W.W. Norton & Co., 1987), pp. 5–6.

35. H. Dregne et al., "A New Assessment of the World Status of Desertification," *Desertification Control Bulletin,* No. 20, 1991.

36. E. H. Clark, III, J. A. Haverkamp, and W. Chapman, *Eroding Soils: The Off-Farm Impacts* (Washington, D.C.: The Conservation Foundation, 1985).

37. Marc Ribaudo, *Water Quality Benefits from the Conservation Reserve Program* (Washington, D.C.: U.S. Department of Agriculture, Economic Research Service, February 1989).

38. William Cline, *Global Warming: The Economic Status* (Washington, D.C.: Institute for International Economics, 1992).

39. From Thomas Crocker, University of Wyoming, as described in James S. Cannon,

The Health Costs of Air Pollution (New York: American Lung Association, 1985).

40. United Nations Food and Agriculture Organization as cited in World Resources Institute, *World Resources 1992–1993* (New York: Oxford University Press, 1992), p. 179.

41. Lester Brown, "The Aral Sea: Going, Going," *World Watch* (Washington, D.C.: Worldwatch Institute, January/February 1991), pp. 20–27.

42. Tom Horton and William M. Eichbaum, *Turning the Tide: Saving the Chesapeake Bay* (Washington, D.C.: Island Press, 1991).

43. Josh Friedman, "Bulgaria's Deadly Secret," *Newsday,* April 21, 1990.

44. Chrystia Freeland, "Russians Doomed for Next 25 Years," *Financial Times,* October 8, 1992.

45. William K. Reilly, *Statement on Ozone Depletion* (Washington, D.C.: April 4, 1991).

46. Milton Russell et al., "The U.S. Hazardous Waste Legacy," *Environment,* July/August 1992.

47. Wilson (1992).

48. Lester R. Brown, "Sustaining World Agriculture," in *State of the World, 1987* (New York: W.W. Norton & Co., 1987), pp. 128, 131.

49. U.S. Department of Agriculture, *Agricultural Statistics* (Washington, D.C.: U.S. Government Printing Office, various years); Gordon Sloggett, *Energy and U.S. Agriculture: Irrigation Pumping, 1974–83* (Washington, D.C.: U.S. Government Printing Office, 1985).

50. Brown and Postel (1987), pp. 10–11.

51. U.S. Department of Agriculture, Economic Research Service, *World Indices of Agricultural and Food Production, 1950–85* (Unpublished printout, Washington, D.C., 1986).

52. Julian Simon, *The Ultimate Resource* (Princeton: Princeton University Press, 1981); and Julian L. Simon and Herman Kahn, eds., *The Resourceful Earth* (New York: Basil Blackwell, 1984).

53. F.H. Borman, "Unlimited Growth: Growing, Growing, Gone?" *BioScience* 22, no. 12 (1972):706–709; and Herman E. Daly, *Steady-State Economics,* 2nd ed. (Washington, D.C.: Island Press, 1992).

54. Harold J. Barnett and Chandler Morse, *Scarcity and Growth: The Economics of Natural Resource Availability* (Baltimore, Maryland: Johns Hopkins University Press, 1963).

55. International Institute of Applied Systems Analysis, *Options,* December 1991, p. 8.

56. Robert Costanza, Herman E. Daly, and Joy A. Bartholomew, "Goals, Agenda, and Policy Recommendations for Ecological Economics," in *Ecological Economics: The Science and Management of Sustainability,* Robert Costanza, ed. (New York: Columbia University Press, 1991), pp. 3–4.

57. E. F. Schumacher, *Small Is Beautiful: Economics as if People Mattered* (New York: Harper & Row, Publishers, 1973), pp. 47, 49.

58. Herman E. Daly and John B. Cobb, Jr., *For the Common Good: Redirecting the Economy Toward Community, the Environment, and a Sustainable Future* (Boston, Massachusetts: Beacon Press, 1989), p. 37.

CHAPTER 2

Economic and Financial Concepts in Resource Management

The goal of the economist is not merely to train a new generation in his arcane mystery, it is to understand this economic world in which we live.

—GEORGE J. STIGLER, 1987

Natural and environmental resources are utilized in the production and consumption of goods and services (commodities). Households, firms and governments produce commodities from natural or environmental resources and consume the amenity services provided by these resources. Production and consumption generate economic value in the form of income and employment and generate residuals that are either recycled or released to the environment. Releasing residuals to the environment can degrade the quality of natural and environmental resources and reduce the amenity services that they provide. Economic theory of natural or environmental resources utilizes several key concepts: namely, a) consumption and demand, b) production and supply, c) market equilibrium and d) present value. This chapter explains each of these concepts in narrative, graphical and mathematical terms. The concepts are used in later chapters.

Consumption and Demand Theory

Households indirectly consume natural and environmental resources when they purchase a house containing wood, plastic and metal products and use electricity, oil and natural gas to heat, cool and light the house. Households directly consume natural and environmental resources when they breathe air, drink water, or use a forest for outdoor recreation. Consumption generates residuals that can degrade natural and environmental resources. The income that households use to purchase commodities is earned by selling natural and human resources, such as land and labor to firms and the government. Household savings are a source of capital that firms and governments use to finance production activities.

Consumption and demand theory explains how an individual decides what to consume, how much to consume, and how consumption varies with commodity prices, household income and other socioeconomic characteristics of the household. Four key assumptions underlie household demand theory. First, commodities provide utility or satisfaction to households. Second, the objective of households is to select commodities that maximize their utility or satisfaction. Third, total household expenditure on commodities is limited by income and commodity prices. Fourth, household consumption does not affect the prices paid for commodities (households are price takers).

UTILITY FUNCTION AND INDIFFERENCE MAP. Household preferences for commodities are represented by a *utility function*. This utility function ranks the household's preferences for all commodities. Consider a household that receives utility or satisfaction from consuming two commodities, X and Y. The utility function for the household is:

$$U = F(X, Y).$$

This function states that total utility (U) is a function (F) of the quantities of X and Y consumed. Changes in preferences for commodities alters the utility function.

The utility function has four properties. First, it is *single valued,* which means that specific values of X and Y result in a unique level of utility. Second, utility is an *increasing function* of X and Y, which means U increases (decreases) as more (less) X and/or Y are consumed. Third, the utility function exhibits *diminishing marginal utility* for X and Y. Diminishing marginal utility requires that the increments in total utility become progressively smaller as equal increments of X are consumed and consumption of Y is held constant. Likewise, increments in total utility become progressively smaller as equal increments of Y are consumed and consumption of X is held constant. Fourth, the utility function is *ordinal.* An ordinal utility function ranks different commodity bundles according to the household's preferences. A *commodity bundle* is a particular combination of X and Y, such as four units of X and six units of Y. Ordinality requires that if commodity bundle c is more (less) preferred than commodity bundle a, then c provides greater (less) utility than does a.

The utility function is represented graphically by an *indifference map* like the one shown in Figure 2.1. In this figure, the quantity consumed of X is measured on the horizontal axis and the quantity consumed of Y is measured on the vertical axis. Every point between the Y axis and X axis represents a commodity bundle. Commodity bundle a consists of X_a units of X and Y_a units of Y and commodity bundle b contains X_b units of X and Y_b units of Y.

Suppose commodity bundles a and b provide the same level of utility. All commodity bundles that provide the same utility as bundles a and b lie on the *indifference curve* U_1. Total utility is constant along an indifference curve. Moving up (down) an indifference curve shows consumption of Y increasing (decreasing) and consumption of X decreasing (increasing). A collection of indifference curves from the same utility function is called an *indifference map.* An indifference map contains a very large number of indifference curves. Three indifference curves (U_1, U_2 and U_3) are shown in Figure 2.1.

Indifference curves have three properties. First, an indifference curve is *convex* (U) to the origin because of diminishing marginal utility for X and Y. Diminishing marginal utility implies that increasing increments of Y are needed to offset equal decrements in X when moving up the indifference curve (a to b). Conversely, increasing increments of X are needed to offset equal decrements in Y when moving down the indifference curve (b to a). Second, indifference curves are *nonintersecting* because the utility function is single valued. If indifference curves intersected, then the same commodity bundle would provide two different levels of utility, which is not possible when the utility function is single valued. Third, indifference curves are *monotonically increasing,* which means utility increases by moving to a higher indifference curve. This property follows from the assumption that the utility function is increasing in X and Y. In Figure 2.1, bundle c on U_2 is preferred to bundles a and b on U_1 because $U_2 > U_1$. Similarly, bundles on U_3 are preferred to bundles on U_2 and U_1 because $U_3 > U_2 > U_1$.

CONSTRAINTS. Household consumption of X and Y is constrained by household income and market prices of X and Y. Apart from borrowing, household expenditure on X and Y cannot exceed household income. In addition, the household

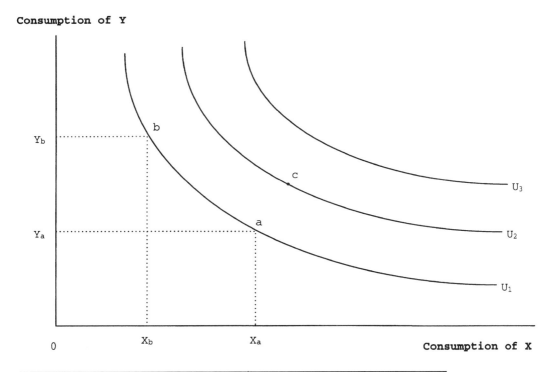

Figure 2.1. Indifference map for commodities X and Y. U = utility.

must pay the market prices for X and Y. Constraints are imposed on the household's purchases of X and Y by income and prices as illustrated by the *budget set* and *budget line* in Figure 2.2. The budget set contains all commodity bundles, such as d, e and f, that the household is able to purchase with a given income and market prices. The budget line, labeled B_2, contains all combinations of X and Y that exhaust the household's income. For example, commodity bundles e and f contain different amounts of X and Y, but they cost the same amount and utilize all of the household's income.

B_2 corresponds to a household income of $100, a price of X of $10, and a price of Y of $20. With this budget constraint, the household can purchase a maximum of 10 units of X ($100/$10) and a maximum of five units of Y ($100/$20). Therefore, the X intercept of the budget line is 10 units and the Y intercept is five units. Because a household cannot influence market prices, the budget line is linear.

The most efficient commodity bundles lie on the budget line. Why? Consider bundle d. This bundle is in the budget set but below the budget line. If the household chooses bundle d, then there is unused income because bundle d is below the budget line. When utility is an increasing function of X and Y, unused income implies that the household can increase utility by consuming more X and/or Y. Therefore, moving from bundle d to any bundle on B_2 increases utility and exhausts the household's income.

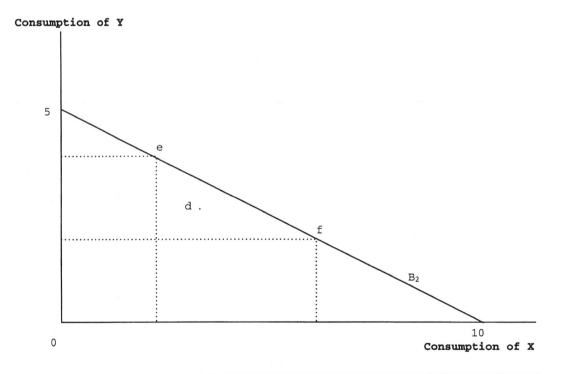

Figure 2.2. Budget set and budget line (B_2) for commodities X and Y when household income is $100; price of X is $10 and price of Y is $20.

2. Economic and Financial Concepts

HOUSEHOLD EQUILIBRIUM. A household achieves equilibrium by choosing a commodity bundle on the budget line that maximizes utility. Stated differently, equilibrium is achieved by selecting the bundle on the budget line that is on the highest indifference curve. The equilibrium bundle is found at the point of tangency between the indifference curve and the budget line, namely, bundle f in Figure 2.3. For bundle f, $X = 6$, $Y = 2$, utility is U_2 and household expenditure is B_2. The household is in equilibrium with bundle f because utility is maximized and all income is spent.

Household Demand Curve. The *household demand curve* for X (D_x) is defined as the relationship between the equilibrium quantities and corresponding prices of X when the price of Y, income and the utility function (preferences for commodities) are held constant. D_x is illustrated in Figure 2.4. This figure shows p_x on the vertical axis and consumption of X on the horizontal axis. For bundle f in Figure 2.3, $X = 6$ when $p_x = \$10$, $p_Y = \$20$ and household income is $100. The quantity–price combination $X = 6$ and $p_x = \$10$ becomes point g on D_x.

Let p_x increase from $10 to $20, as shown in Figure 2.5. The higher price reduces the X intercept of the budget line from 10 ($100/$10) to 5 ($100/$20), which causes the budget line to rotate clockwise from B_2 to B_1. Household equilibrium is re-established at bundle i, where budget line B_1 is tangent to indifference curve U_1.

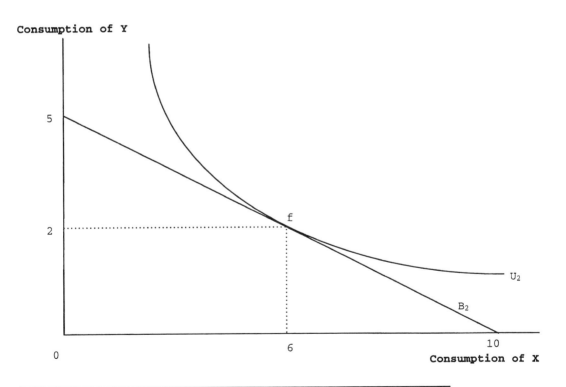

Figure 2.3. Household equilibrium when household income is $100; price of X is $10 and price of Y is $20. B_2 = budget line; and U_2 = utility.

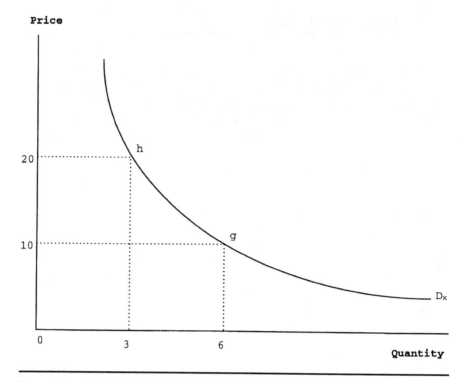

Figure 2.4. Household demand curve for X. D = demand.

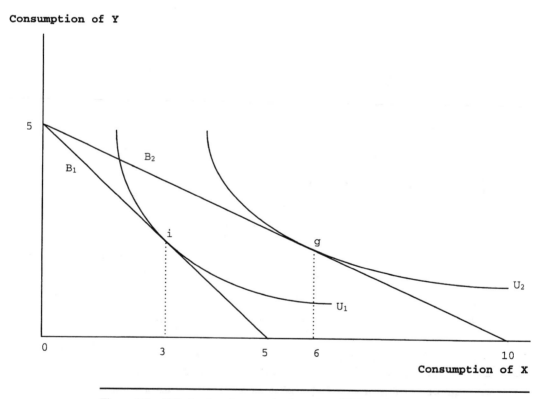

Figure 2.5. Effects of an increase in the price of X from $10 to $20 on household equilibrium. B = budget line; and U = utility.

Notice that the new equilibrium occurs on a lower indifference curve ($U_1 < U_2$). At the new equilibrium, $X = 3$ and $p_x = \$20$, which gives point h on D_x. Connecting points g and h gives the household demand curve for X. The demand curve shown in Figure 2.4 is called a *Marshallian demand curve*. Along a Marshallian demand curve, household income and prices of all other commodities are constant.

D_x is downward sloping because of diminishing marginal utility for X. Therefore, along the demand curve, an increase (decrease) in p_x results in a decrease (increase) in consumption of X. Hence, quantity demanded and price of X are inversely related. The demand curve for Y is derived in a similar manner by connecting the equilibrium quantities associated with various prices of Y, holding the price of X and household income constant.

Elasticity of demand for a commodity measures the percentage change in quantity demanded corresponding to a 1 percent change in the price of the commodity, holding other prices and income constant. Because the demand curve is usually negatively sloped, the elasticity of demand is negative. When the elasticity of demand is greater (less) than 1 in absolute value, demand is said to be *elastic (inelastic)*. Increasing the price of a commodity whose demand is elastic (inelastic) causes household expenditure on that commodity (price times quantity purchased) to decrease (increase). Conversely, when demand is elastic (inelastic), lowering the price causes household expenditure to increase (decrease). Commodities considered to be necessities, such as basic food and clothing, have inelastic demands. Less essential commodities, such as luxury homes and fancy sports cars, have elastic demands.

The relationship between quantity demanded and household income when all prices are held constant determines whether a commodity is a *normal good* or an *inferior good*. The demand curve for a normal (inferior) good shifts upward (downward) from D_{x2} to D_{x3} (D_{x2} to D_{x1}) when household income increases, as shown in Figure 2.6.

If X and Y are *substitutes* (butter and margarine), then the demand curve for X shifts to the right (left) when the price of Y increases (decreases), holding price of X and household income constant. Conversely, if X and Y are *complements* (bread and butter), then the demand curve for X shifts to the left (right) when the price of Y increases (decreases), holding price of X and household income constant.

Production and Supply Theory

Firms use natural and environmental resources to produce commodities. Crude oil is extracted from underground reservoirs and refined into petroleum products, such as heating oil, gasoline and jet fuel. Expenditures made by firms on natural or environmental resources are a source of income to those, such as households and governments, who own the resources. Production activities generate residuals that, when released to the environment, can degrade natural and environmental resources. An oil spill caused by a ruptured pipeline or tanker is a threat to ecosystems. The burning of fossil fuels such as oil, natural gas and coal produces carbon dioxide, which is the primary cause of global warming, and sulfur dioxide, which contributes to acid rain.

The theory of production and supply explains how individual firms determine

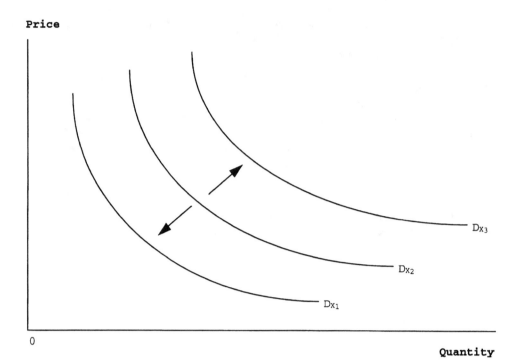

Figure 2.6. Effects of increase in household income on demand (D) for a normal good (D_{X2} to D_{X3}) and on demand for an inferior good (D_{X2} to D_{X1}).

the most efficient quantities of inputs and outputs in production and how changes in prices and technology influence input and output levels. An *input* refers to a resource used in production, and an *output* refers to the commodity that results from production. Three key assumptions underlie the economic theory of production and supply. First, firms attempt to maximize their own profits. Second, a firm's ability to convert inputs into outputs is determined by technology, which, along with market prices of inputs and outputs and the firm's cost budget, determines the most economically efficient levels of inputs and outputs. Third, under pure competition, the firm cannot influence the prices paid for inputs or the prices received for outputs.

PRODUCTION FUNCTION AND ISOQUANT MAP. The production technologies available to a firm determine the output obtained from different combinations of inputs. Consider a firm that utilizes two inputs, Z_1 and Z_2, to produce one output, Y. The *production function* shows the relationship between maximum production of Y and use of Z_1 and Z_2:

$$Y = G(Z_1, Z_2).$$

Maximum production for any given combination of Z_1 and Z_2 is achieved by employing the most efficient technology.

A production function has three properties. First, it is *single valued,* which means that specific values of Z_1 and Z_2 result in a unique level of production. Second, Y is a *monotonically increasing function* of Z_1 and Z_2. This means that Y increases (decreases) as use of Z_1 and Z_2 increases (decreases). Third, the production function exhibits *diminishing marginal returns* for inputs Z_1 and Z_2. When production is subject to diminishing marginal returns, equal increments of Z_1 result in progressively smaller increments in Y, holding Z_2 and the production function (technology) constant. Likewise, equal increments of Z_2 result in progressively smaller increments in Y, holding Z_1 and the production function constant.

Fourth, a production function exhibits *constant, increasing* and/or *decreasing returns to scale.* Returns to scale refers to the relationship between changes in production and simultaneous and proportional changes in all inputs. When production and inputs change in the same proportion, there are constant returns to scale. When changes in production are proportionately greater (less) than the changes in inputs, there are increasing (decreasing) returns to scale. For example, if all inputs are doubled at the same time, then there are constant returns to scale when production just doubles, increasing returns to scale when production more than doubles, and decreasing returns to scale when production less than doubles. A production function can exhibit more than one type of returns to scale, such as increasing returns to scale followed by decreasing returns to scale. Fifth, the production function is *cardinal.* Cardinality means that specific amounts of Z_1 and Z_2 result in a measurable output.

The production function is represented graphically by an *isoquant map* like the one depicted in Figure 2.7. The isoquant map consists of individual *isoquants* Y_1, Y_2 and Y_3. Each isoquant contains all input bundles that provide the same level of output. An *input bundle* is a specific combination of Z_1 and Z_2. For example, bundle a contains Z_{1a} of Z_1 and Z_{2a} of Z_2, whereas bundle b contains Z_{1b} of Z_1 and Z_{2b} of Z_2. Both bundles give the same output, namely, five units. Changes in production technology alter the production function and the isoquant map.

Isoquants have three properties. First, isoquants are *convex* (∪) to the origin because of diminishing marginal returns to inputs. The latter implies that progressively larger increments of Z_1 are required to offset equal decrements in Z_2 when moving up an isoquant (a to b). Conversely, progressively larger increments of Z_2 are needed to offset equal decrements in Z_1 when moving down an isoquant (b to a). For example, if increasing Z_1 from one to two units, two to three units and three to four units causes Y to increase by six, four and three units, respectively, then there are diminishing marginal returns to Z_1.

Second, isoquants are *nonintersecting* because the production function is single valued. If isoquants intersected, then one input bundle results in two different outputs. This is not possible when the production function is single valued. Third, isoquants are *monotonically increasing,* which means that output increases when moving from lower to higher isoquants. In Figure 2.7, input bundle c on isoquant Y_2 results in greater output than input bundles a and b on isoquant Y_1 ($Y_2 > Y_1$). Similarly, input bundles on isoquant Y_3 result in greater output than input bundles on isoquants Y_2 and Y_1 ($Y_3 > Y_2 > Y_1$).

CONSTRAINTS. The firm's selection of efficient inputs and output is influenced by the production function, the cost budget for inputs and prices of inputs and out-

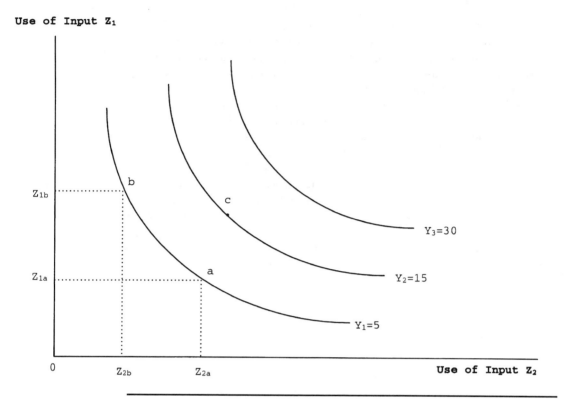

Figure 2.7. Isoquant map for inputs Z_1 and Z_2. Y = output.

puts. Input prices and the firm's cost budget are combined into the firm's *cost set*, as shown in Figure 2.8. The cost set contains all input bundles, such as d and e, which the firm can purchase with a given cost budget at given input prices.

The maximum combinations of Z_1 and Z_2 that the firm can purchase lie on the *isocost line* labeled C_2. If the firm's budget is $80, the price of Z_1 is $16, and the price of Z_2 is $10, then the firm can purchase a maximum of five units of Z_1 ($80/$16) and a maximum of eight units of Z_2 ($80/$10). Therefore, the Z_1 intercept of the isocost line is 5 and the Z_2 intercept is 8. When the firm cannot influence input prices, the isocost line is a straight line.

Changes in the firm's cost budget and/or input prices alter the position of the isocost line. Increasing the firm's cost budget while holding input prices constant results in an outward parallel movement in the isocost line. A decrease (increase) in the price of Z_1 causes the Z_1 intercept of the isocost line to increase (decrease), holding constant the cost budget and the price of Z_2. Similarly, a decrease (increase) in the price of Z_2 causes the Z_2 intercept of the isocost line to increase (decrease), holding constant the cost budget and the price of Z_1.

All efficient input bundles lie on the isocost line. A bundle in the interior of the cost set, such as bundle d, is not efficient because production can be increased without exceeding the cost budget.

2. Economic and Financial Concepts

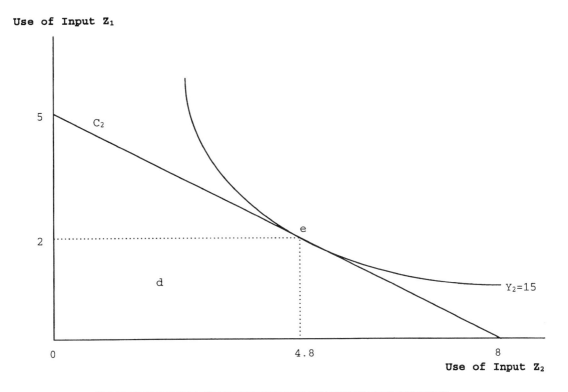

Figure 2.8. Cost set, isocost line (C_2), isoquant (Y_2) and firm equilibrium (e) when the cost budget is $80; price of Z_1 is $16; and price of Z_2 is $10.

EFFICIENT INPUT USE. The efficient input bundle for producing a given level of output is the input bundle that minimizes the cost of producing that level of output. Suppose the firm wants to determine the most efficient input bundle for producing Y_2 (15 units). The efficient input bundle is given by the point of tangency between isocost line C_2 and isoquant Y_2, which occurs at bundle e where $Z_1 = 2$ and $Z_2 = 4.8$. Employing any other combination of Z_1 and Z_2 on the isocost line results in less production, which is inefficient. Hence, $Z_1 = 2$ and $Z_2 = 4.8$ is the most efficient input bundle for producing Y_2. Also, Y_2 is the maximum output the firm can produce when the cost budget is $80, the price of Z_1 is $16, and the price of Z_2 is $10.

The efficient input bundles for producing different levels of output lie along the *expansion path* shown in Figure 2.9. The expansion path is derived by connecting the points of tangency between the isocost lines and the isoquants when input prices and the production function are held constant. When input prices are constant, increasing (decreasing) the firm's cost budget from C_1 to C_2 to C_3 causes the isocost line to shift outward (inward) in a parallel fashion. Therefore, moving up the expansion path requires successively larger cost budgets. The efficient input bundles are $Z^{(1)}_1$ and $Z^{(1)}_2$ for five units of Y, $Z^{(2)}_1$ and $Z^{(2)}_2$ for 15 units of Y, and $Z^{(3)}_1$ and $Z^{(3)}_2$ for 30 units of Y.

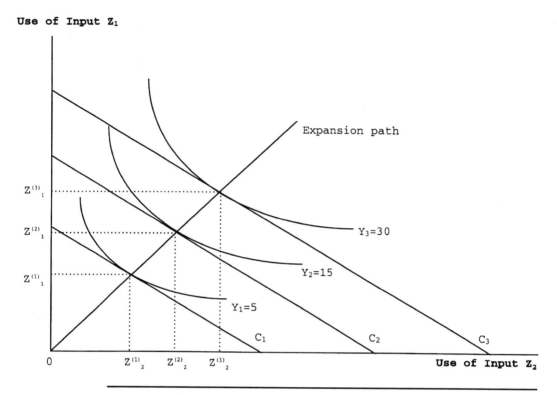

Figure 2.9. Expansion path for Y. C = isocost line; and Y = output.

FIRM'S INPUT DEMAND CURVE. The firm's demand curve for an input shows the quantity demanded of an input at various prices when the production function, cost budget, and prices of other inputs are held constant. Efficient use of Z_2 for a cost budget of $80, a price of Z_1 of $16, and a price of Z_2 of $10 is given by point e in Figure 2.10 (which is the same as point e in Figure 2.8). At e, efficient use of Z_2 is 4.8 and output is 15. Let the price of Z_2 increase from $10 to $16. The higher price reduces the Z_2 intercept of the isocost line from 8 ($80/$10) to 5 ($80/$16), which makes the isocost line rotate clockwise from C_2 to C_1. The new equilibrium occurs at point f, where isocost line C_1 is tangent to isoquant Y_1. At the new equilibrium, $Z_2 = 2$ and $Y = 5$. Hence, a higher price for Z_2 reduces both the efficient use of Z_2 and output.

The demand curve for Z_2 (D_{Z2}) is determined by plotting efficient use of Z_2 against the price of Z_2 when the production function, cost budget and price of Z_1 are held constant, as shown in Figure 2.11. Points g and h on D_{Z2} correspond to points e and f, respectively, in Figure 2.10. Connecting g and h gives D_{Z2}. Therefore, efficient input use is inversely related to input price. *Elasticity of demand for an input* measures the responsiveness of quantity demanded of the input to a change in the price of the input when all other input prices and the cost budget are held constant. The demand function for an input shifts to the right (left) when the productivity of the input is increased (decreased) by a change in technology.

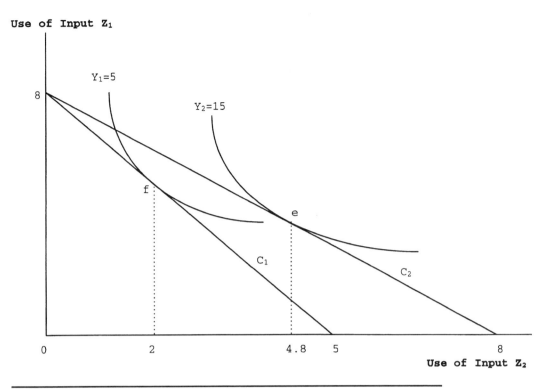

Figure 2.10. Effects of increase in price of Z_2 from $10 to $16 on efficient use of Z_2. C = isocost line; and Y = output.

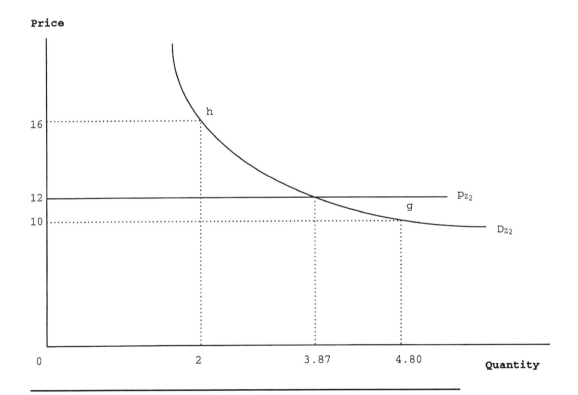

Figure 2.11. Demand curve for input Z_2 holding the production function, cost budget and price of Z_1 constant.

Efficient input use by a firm occurs where the price of the input (p_{Z2}) intersects the demand curve for that input. For example, when the price of Z_2 is \$12, efficient use of Z_2 is 3.87, as shown in Figure 2.11.

EFFICIENT OUTPUT. The traditional criterion for determining the efficient output for a firm is to maximize profit subject to the production function, the prices of outputs and inputs, and the cost budget. An important distinction is made between short-run and long-run profit maximization. Short-run (long-run) profit maximization is based on the short-run (long-run) production function. The short-run production function includes both *variable inputs* and *fixed inputs*. Variable inputs can be changed, and fixed inputs cannot be changed, in the short run. In the short run, a mining company can increase the amount of coal that is mined by increasing the number of laborers working the mine, but it cannot vary the equipment and structures in the mine. Labor is a variable input, and equipment and structures are fixed inputs, in the short run. The long-run production function includes only variable inputs. In the long run, the mining company can change its use of labor, equipment, structures and all other inputs.

Short-Run Profit Maximization. Let Z_1 be the fixed input and Z_2 the variable input. The amount of Z_1 available to the firm is Z_{1f} in the short run. Then the short-run production function is:

$$Y = G(Z_1, Z_2) \text{ where } Z_1 \leq Z_{1f}.$$

If an output of Y_f requires Z_{1f} units of Z_1, then the fixed input Z_1 is not a constraint on production as long as $Y \leq Y_f$.

Short-run marginal cost (SMC) is the change in total cost of production associated with a change in output when there is at least one fixed input. For $Y \leq Y_f$, SMC is determined by the returns to scale in the production function and the prices of Z_1 and Z_2. If the production function exhibits constant, increasing or decreasing returns to scale, then SMC is constant, decreasing or increasing, respectively, as shown by the segment of SMC for $Y \leq Y_f$ in Figure 2.12.

For $Y > Y_f$, the fixed input Z_1 is being fully utilized and increases in output can only be achieved by increasing Z_2. When Z_2 is increased while holding Z_1 constant at Z_{1f}, production is subject to diminishing marginal returns. Hence, for $Y > Y_f$, increasing output by equal increments requires progressively greater increments in Z_2. When this occurs, SMC increases with output as shown in Figure 2.12.

The equilibrium condition for *short-run profit maximization* is to equate the price of output to short-run marginal cost ($p_Y = $ SMC) on the increasing portion of the SMC curve. Short-run profit maximization occurs at Y^*_s in Figure 2.12. For $Y < Y^*_s$, $p_Y > $ SMC, which implies that raising output causes total revenue to increase more than total cost. Profit increases by raising output. Conversely, when $Y > Y^*_s$, $p_Y < $ SMC, which implies that a lower output reduces total cost more than it reduces total revenue. Profit increases by lowering output. Hence, Y^*_s is the most profitable level of output in the short run.

The firm's *short-run supply curve* gives the quantities of the commodity that the firm is willing to supply at various prices in the short run. As mentioned earlier,

Value per Unit

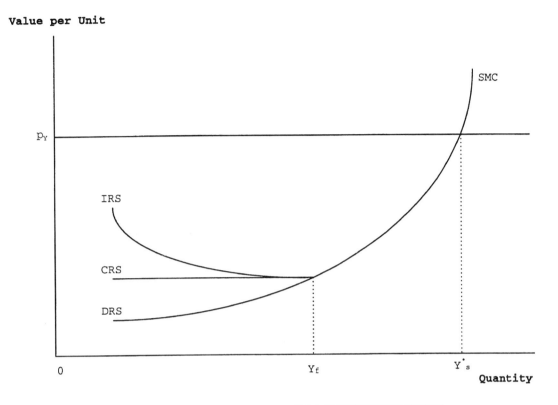

Figure 2.12. Short-run marginal cost (SMC) curve for Y with fixed input. IRS = increasing returns to scale; CRS = constant returns to scale; DRS = decreasing returns to scale; and p_Y = price of Y.

there are two kinds of costs in the short run, variable cost and fixed cost. *Total variable cost* is the amount of the variable input used by the firm times its price. *Total fixed cost* is the cost of the fixed input. *Average fixed cost* (AFC) is total fixed cost divided by output. *Average variable cost* (AVC) is total variable cost divided by output. *Average total cost* (ATC) is the sum of AFC and AVC. AFC, AVC, ATC and SMC are shown in Figure 2.13. Because total fixed cost is independent of the level of output, AFC is a decreasing function of output. When there is increasing marginal returns followed by decreasing marginal returns to the variable input, the AVC and ATC curves are U-shaped. ATC reaches a minimum at a higher level of output than AVC because AFC decreases with respect to output.

The short-run supply curve is the SMC curve above minimum AVC. Suppose output price is below minimum AVC (p_{Y1} < min AVC), as shown in Figure 2.13. When production is zero, losses to the firm equal total fixed cost. When production is nonzero, however, losses to the firm equal total fixed cost plus total variable cost minus total revenue. Because total variable cost exceeds total revenue when p_Y < min AVC, the loss with production exceeds the loss without production. Therefore, the firm minimizes losses when p_Y < min AVC by producing zero output.

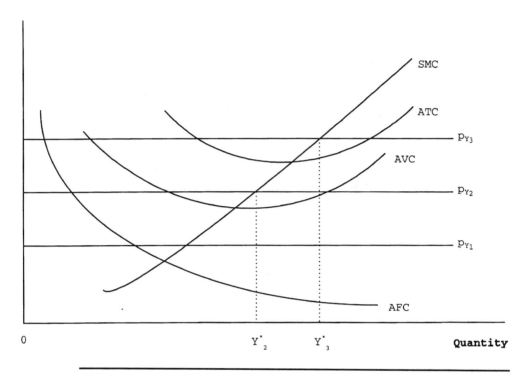

Figure 2.13. Short-run cost curves and supply curve. AFC = average fixed cost; AVC = average variable cost; ATC = average total cost; SMC = short-run marginal cost; and p_Y = price of Y.

Suppose p_Y is between minimum AVC and minimum ATC (min AVC < p_{Y2} < min ATC). In this case, if the firm produces Y^*_2 where p_{Y2} = SMC, then total revenue exceeds total variable cost. The excess of total revenue over total variable cost can be used to pay some (but not all) of the fixed cost. In this case, losses are minimized by producing Y^*_2.

Finally, suppose p_Y exceeds minimum ATC (p_{Y3} > min ATC). If the firm produces Y^*_3 where p_{Y3} = SMC, then maximum profit is earned. In summary, the firm minimizes losses or maximizes profits in the short run by selecting the level of output for which p_Y = SMC, provided $p_Y \leq$ min AVC. If $p_Y <$ min AVC, then the firm minimizes its losses by producing zero output. Therefore, the short-run supply curve is the SMC curve above minimum AVC.

Changes in technology and input prices cause the firm's supply curve to shift to the left or right. Improvements in technology, which increase the productivity of the variable input, and lower input prices cause the short-run supply curve to shift downward. Increases in input prices cause the short-run supply curve to shift upward.

Long-Run Profit Maximization. Because all inputs are variable in the long run, the long-run production function is:

$Y = G(Z_1, Z_2)$,

with no restrictions on Z_1 and Z_2. *Long-run marginal cost* (LMC) is the change in total cost of production with respect to a change in output when all inputs are variable. The shape of the LMC curve is determined by the returns to scale in the long-run production function.

As shown in Figure 2.14, LMC is: a) a decreasing function of output for increasing returns to scale (Figure 2.14a); b) a constant function of output for constant returns to scale (Figure 2.14b); and c) an increasing function of output for decreasing returns to scale (Figure 2.14c). The long-run production function can exhibit a combination of returns to scale. For example, when the long-run production function exhibits increasing returns to scale followed by decreasing returns to scale, the LMC curve is U-shaped.

The equilibrium condition for *long-run profit maximization* is equality between price of output and long-run marginal cost (p_Y = LMC) on the increasing portion of the LMC curve. Equilibrium is not achieved at p_Y = LMC under increasing returns to scale, as demonstrated in Figure 2.14a, because profit increases as output rises. When there is constant returns to scale and p_Y < LMC, as with LMC_2 in Figure 2.14b, no level of output is profitable. When p_Y > LMC, as with LMC_1, the firm has an incentive to increase output because it increases profit. There is not a unique output that maximizes profit when p_Y = LMC under constant returns to scale. Finally, under decreasing returns to scale, profit is maximized at Y^*_L where p_Y = LMC, as shown in Figure 2.14c.

Long-Run Supply Curve. As already shown, long-run equilibrium does not exist, and, therefore, the long-run supply curve is not defined for increasing and constant returns to scale. When the long-run production function exhibits decreasing returns to scale, long-run profit maximization occurs where p_Y = LMC. Because all inputs are variable in the long run, points along the LMC curve give the lowest average cost of producing any given level of output. Therefore, under decreasing returns to scale, the long-run supply curve equals LMC.

The firm's long-run supply curve is affected by changes in technology and input prices. Improvements in technology that increase the productivity of one or more inputs and reductions in input prices cause the long-run supply curve to shift downward. The long-run supply curve shifts upward when input prices increase.

Market Equilibrium

The preceding sections on consumption and demand and production and supply describe the economic theory underlying a household's demand curve for a commodity and a firm's demand curve for an input and supply curve for a commodity. Market equilibrium is determined by the *market demand curve* and *market supply curve* for a commodity. This section explains the market demand and supply curves and explains market equilibrium.

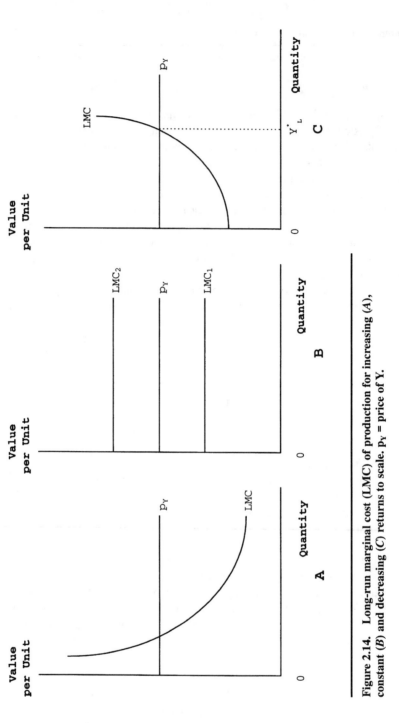

Figure 2.14. Long-run marginal cost (LMC) of production for increasing (A), constant (B) and decreasing (C) returns to scale. p_Y = price of Y.

MARKET DEMAND AND SUPPLY. The *market demand curve* for a commodity is determined by summing the household demand curves for that commodity, as illustrated in Figure 2.15. When there are n households and the price of Y is p_{Y1}, the quantity demanded is $q^{(1)}_{Y1}$ for household 1, $q^{(2)}_{Y1}$ for household 2 and $q^{(n)}_{Y1}$ for household n. Summing the quantities demanded by all households at p_{Y1} gives a market quantity demanded of Q_{Y1} that corresponds to point a. Similarly, when the price of Y is p_{Y2}, quantity demanded is zero for household 1, $q^{(2)}_{Y2}$ for household 2

2. Economic and Financial Concepts

and $q^{(n)}_{Y2}$ for household n. Adding together the quantities demanded by all households at p_{Y2} gives a market quantity demanded of Q_{Y2} that corresponds to point b. Connecting points a and b gives the market demand curve for Y, namely, D_Y. Because the household demand curves are negatively sloped, D_Y is negatively sloped.

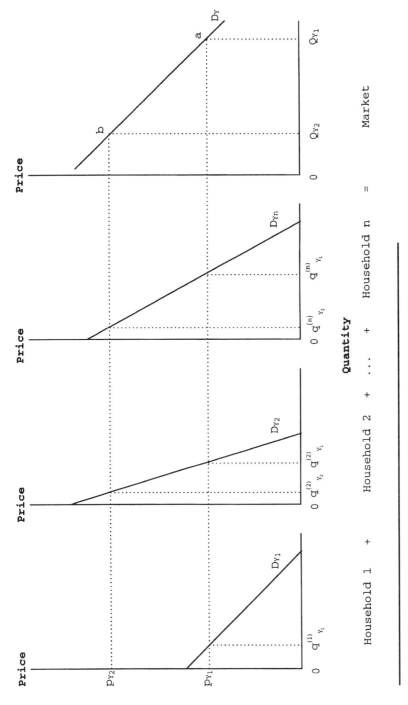

Figure 2.15. Derivation of market demand (D) curve for Y. p_Y = price of Y.

The *market supply curve* is derived by summing the supply curves for individual firms, as shown in Figure 2.16. When m firms produce Y and the price of Y is p_{Y1}, the quantity supplied is $q^{(1)}_{Y1}$ for firm 1, $q^{(2)}_{Y1}$ for firm 2 and $q^{(m)}_{Y1}$ for firm m. Summing the quantities supplied by all firms at p_{Y1} gives a market quantity supplied of Q_{Y1} that corresponds to point c. Similarly, when the price of Y is p_{Y2}, quantity

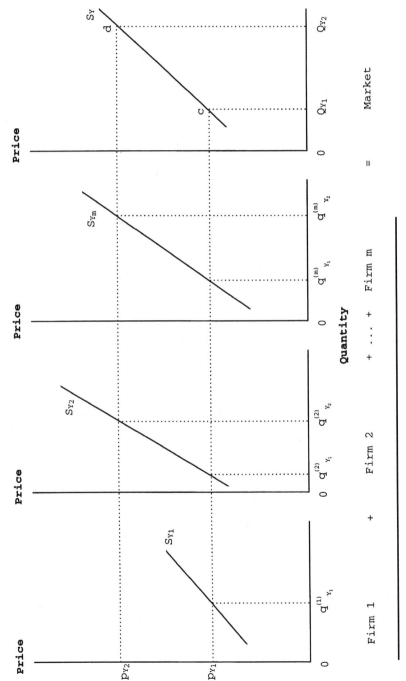

Figure 2.16. Derivation of market supply (S) curve for Y. p_Y = price of Y.

supplied is zero for firm 1, $q^{(2)}_{Y2}$ for firm 2 and $q^{(m)}_{Y2}$ for firm m. Adding together the quantities supplied by all firms at p_{Y2} gives a market quantity supplied of Q_{Y2} that corresponds to point d. The market supply curve for Y, S_Y, is obtained by connecting points c and d. S_Y is positively sloped because the individual firm supply curves are positively sloped.

Combining the market demand curve from Figure 2.15 and the market supply curve from Figure 2.16 gives Figure 2.17. The *market equilibrium price* and *market equilibrium quantity* of Y occur at point e, where quantity demanded equals quantity supplied. Why is point e an equilibrium? When the price of Y is p_{Y1}, quantity supplied is Q_{sY1} and quantity demanded is Q_{dY1}, which results in a *deficit* of $Q_{dY1} - Q_{sY1}$. To eliminate the deficit, price must increase to p^*_Y. When the price of Y is p_{Y2}, quantity supplied is Q_{sY2} and quantity demanded is Q_{dY2}, which results in a *surplus* of $Q_{sY2} - Q_{dY2}$. Elimination of the surplus requires the price to decrease to p^*_Y. At p^*_Y, $Q_{dY} = Q_{sY} = Q^*_Y$. As long as the market demand and supply curves remain unchanged, the equilibrium stays at p^*_Y and Q^*_Y. A similar procedure is used to determine market demand and supply curves and market equilibrium for X.

Factors that shift household demand curves also shift the market demand curve. Increases (decreases) in household incomes and prices of substitutes and decreases (increases) in the prices of complements cause the market demand curve to shift upward (downward). In addition, increases (decreases) in the number of households cause the market demand curve to shift upward (downward). An upward (downward) shift in the demand curve is known as an increase (decrease) in demand.

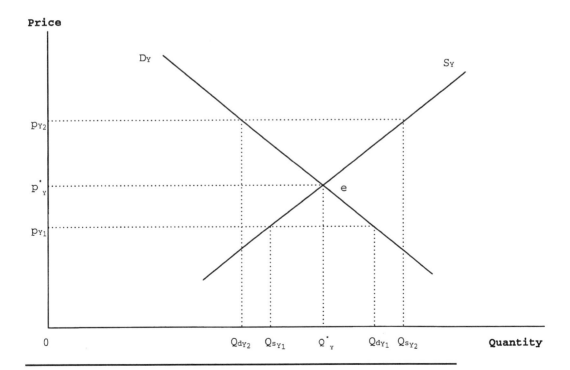

Figure 2.17. Market equilibrium for Y. p_Y = price of Y.

Likewise, the factors that shift firm supply curves also shift the market supply curve. Improvements in technology and decreases in the prices of inputs cause the market supply curve to shift downward. Increases in the price of inputs cause the market supply curve to shift upward. Additionally, expansion (contraction) in the number of firms producing the commodity causes the market supply curve to shift downward (upward). An upward (downward) shift in the supply curve is known as an increase (decrease) in supply.

When the market demand curve increases and/or the market supply curve decreases, equilibrium market price increases. Conversely, when the market demand curve decreases and/or the market supply curve increases, the equilibrium market price decreases. Increases in demand and/or supply cause the equilibrium quantity to increase. Conversely, decreases in demand and/or supply cause the equilibrium quantity to decrease. Increases in demand coupled with decreases in supply cause equilibrium price to increase but, in general, have an indeterminate effect on equilibrium quantity. Likewise, decreases in demand coupled with increases in supply cause equilibrium price to decrease but, in general, have an indeterminate effect on equilibrium quantity.

EXAMPLE OF MARKET EQUILIBRIUM. Market equilibrium price and quantity can be determined for particular market demand and supply curves. For example, suppose the market demand and supply functions for Y are as follows:

$p_d = 50 - 0.5Q_d$ Demand

$p_s = 5 + 0.5Q_s$ Supply

The constant term (50 in the demand function and 5 in the supply function) is the intercept on the price axis or the price at which quantity demanded or supplied is zero. The coefficients of quantity demanded (−0.5) and quantity supplied (0.5) are the slopes of the demand and supply curves, respectively. The slope measures the change in price associated with a one-unit change in quantity demanded or supplied. The demand and supply functions are illustrated in Figure 2.18.

In market equilibrium, price is such that quantity demanded equals quantity supplied. The equilibrium quantity is determined by setting the demand price (p_d) and supply price (p_s) equal to each other and solving for quantity. The demand price is the price on the demand curve and the supply price is the price on the supply curve. Equilibrium price is determined by substituting the equilibrium quantity into either the demand or supply function and solving for price. Equating p_d and p_s gives:

$50 - 0.5Q_d = 5 + 0.5Q_s$.

Setting $Q_d = Q_s$, which is required for equilibrium, and solving for Q gives an equilibrium quantity of $Q^*_Y = 45$. Substituting Q^*_Y into either the demand function or supply function yields an equilibrium price of $p^*_Y = 27.5$.

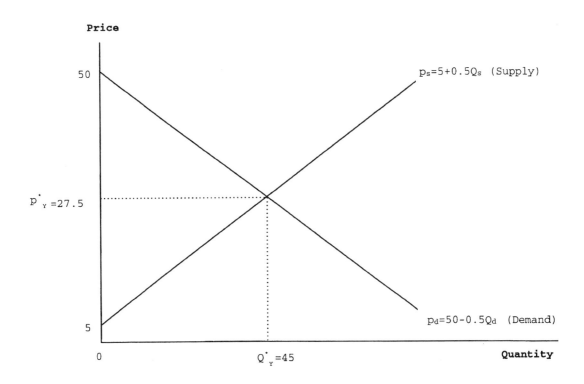

Figure 2.18. Market demand and supply functions for Y. p_Y = price of Y.

Present Value

Present value is the present worth of monetary values that occur over several time periods. Present values are used in determining the efficient allocation of natural and environmental resources over time (Chapters 7 and 8); efficient levels of pollution or pollution abatement (Chapter 9); and the economic feasibility of resource investments that generate benefits and costs which occur over several years (Chapter 11).

Present value is the opposite of *compounding*. Compounding is what happens to a fixed amount of money when it is placed in an interest-bearing savings or investment account. Suppose a present amount of money (A) is deposited in a savings account that pays interest annually at the rate of r percent. In this case, the savings account grows in value in discrete time intervals, namely, years. The future value (FV) of the account is $A(1 + r)$ at the end of the first year; $A(1 + r)(1 + r) = A(1 + r)^2$ at the end of the second year, and so forth. In general, the future value of A at the end of period t is:

$$FV_t = A(1 + r)^t.$$

The term $(1 + r)^t$ is the *discrete compound interest factor.* For A = $100, r = 8% and t = 5:

$FV_5 = \$100(1 + 0.08)^5 = \$146.93.$

The interest rate r must be expressed in decimal form, not in percentage form (0.08 instead of 8%).

What is the present value (PV) of a future amount of money? Since a present value is the same as a present amount, A can be replaced with PV in the FV_t formula to obtain:

$FV_t = PV(1 + r)^t.$

Solving for PV gives:

$PV = FV_t(1 + r)^{-t}.$

When r appears in a present value formula, it is called the *discount rate*. The term $(1 + r)^{-t}$ is called the *discrete discount factor*. Hence, PV equals FV_t times the discount factor.

Consider the present value of the benefits provided by a park. Benefits are $500,000 in year 1, $800,000 in year 2 and $1,000,000 in year 3. If the benefits are received at the end of each year, then the present value of benefits for the park is:

$PV = \$500,000(1.08)^{-1} + \$800,000(1.08)^{-2} + \$1,000,000(1.08)^{-3}$
$= \$462,963 + \$685,871 + \$793,832$
$= \$1,942,666.$

Each term on the right-hand side of the equality is the benefit times the discount factor for a particular year, which is called the *discounted value* of the benefit. In discrete time, present value is a sum of discounted values. As the discount rate increases (decreases), present value decreases (increases). The present value of park benefits is $1,867,017 for a 10 percent discount rate and $2,073,692 for a 6 percent discount rate.

Present values are calculated differently when the interest rate or discount rate changes continuously rather than discretely over time. Future value of a present amount of money with continuous compounding is:

$FV(t) = Ae^{rt}$

where e is the base of the natural logarithm, namely, 2.718, and e^{rt} is the *continuous compound interest factor*. For example, if the park provides an initial benefit of $500,000 and the benefit grows continuously over a three-year period, then the future value of the park at the end of year 3 is:

$FV(3) = \$500,000e^{(0.08)(3)} = \$635,500.$

The present value of the park with continuous discounting is:

$PV = FV(t)e^{-rt},$

where e^{-rt} is the *continuous discount factor*. Therefore, the present value of the park is:

PV = $635,500$e^{(-0.08)(3)}$ = $500,000.

Summary

Determining the economically efficient use of natural or environmental resources involves the application of economic and financial concepts, including consumption and demand theory, production and supply theory, and present value. Consumption and demand theory explains household purchases of commodities in terms of the household's utility function and budget set. A utility function ranks commodity bundles according to household preferences. It is an ordinal, single-valued function that exhibits diminishing marginal utility for individual commodities. The utility function is graphically represented by an indifference map that consists of a set of indifference curves. Each indifference curve is convex, nonintersecting and monotonically increasing. The budget set includes all commodity bundles that the household can purchase with a given income. A household is assumed to purchase the commodity bundle that maximizes utility and utilizes all of the household's income. This bundle is determined by the tangency between the budget line and the indifference curve.

Household demand expresses quantity demanded of a commodity as a function of its own price while holding constant the prices of other commodities and household income. Because the demand curve is downward sloping, quantity demanded is inversely related to price. Responsiveness of quantity demanded to changes in the price of a commodity is measured by the elasticity of demand for a commodity. For a normal good, increases in household income and/or the price of substitutes shift the demand curve to the right. Conversely, decreases in household income and/or increases in the price of complements shift the demand curve to the left.

Production and supply theory explains a firm's selection of inputs and outputs in terms of the production function and cost set. The production function indicates the maximum output resulting from different input levels. It embodies the most efficient technology available to the firm. A production function is a cardinal, single-valued function that exhibits diminishing marginal returns in the short run, and increasing, decreasing or constant returns to scale in the long run. In the short run, at least one input is fixed. All inputs are variable in the long run. The firm's cost set contains input bundles which can be purchased for the same cost.

An isoquant map is a graphical representation of the firm's production function, which consists of a set of isoquants. Each isoquant is convex, nonintersecting and monotonically increasing. The firm is assumed to maximize profit within the limits imposed by the production function and the cost set. In the short run, the most efficient input bundles for producing different outputs are given by the points of tangency between the isocost lines and the isoquants up to the level of the fixed input. These input bundles are used to derive the short-run marginal cost curve. The short-run supply curve for the firm is the short-run marginal cost curve above minimum average variable cost. In the short run, the firm maximizes profit by selecting the output where price intersects the firm's short-run supply curve.

The expansion path defines the efficient input bundles for producing different outputs when all inputs are varied simultaneously. It is derived by connecting the points of tangency between the isoquants and the isocost lines. Input bundles on the expansion path are used to derive the long-run marginal cost curve. The latter is downward sloping with respect to output for increasing returns to scale, constant for constant returns to scale, and upward sloping for decreasing returns to scale. The firm's long-run supply curve equals the long-run marginal cost curve when there are decreasing returns to scale. The long-run supply curve is not defined for increasing returns to scale or constant returns to scale. The firm maximizes profit in the long run by selecting the output at which commodity price intersects the long-run supply curve.

The firm's input demand curve indicates the relationship between quantity demanded and price of an input when prices of all other inputs and the production function are constant. The elasticity of supply measures the responsiveness of quantity supplied of a commodity to its own price when all other input prices and the production function are held constant.

The market demand curve for a commodity is derived by summing the demand curves for all households that are likely to purchase that commodity. Likewise, the market supply curve is found by summing the supply curves for all firms that are likely to produce that commodity. In market equilibrium, quantity demanded equals quantity supplied. Equilibrium price occurs at the intersection of the market demand curve and market supply curve. Above the equilibrium price, there is a surplus of the commodity because quantity supplied exceeds quantity demanded. Below the equilibrium price, there is a deficit of the commodity because quantity demanded exceeds quantity supplied.

Present value is the present worth of monetary values that occur over several time periods. It is the opposite of compounding. The present value of a series of monetary values is the sum of the discounted monetary values. A discounted value is the original value multiplied by the discount factor that is $(1+r)^{-t}$ in discrete time and e^{-rt} in continuous time, where r is the discount rate and t is the designation for time period. Present value increases (decreases) as the discount rate decreases (increases).

Questions for Discussion

1. Discuss the relationship between an indifference map and a utility function and between an isoquant map and a production function.

2. Explain a) why the tangency between the budget line and indifference curve is an equilibrium for a household and b) why the tangency between the isocost line and isoquant is an equilibrium for a firm.

3. Use an indifference map and budget line for commodities X and Y to illustrate what happens to household equilibrium when the price of Y decreases while holding the price of X and household income constant.

4. How is production technology accounted for in determining the efficient use of inputs by a firm?

5. Identify the kinds of inputs that are likely to limit the generation of electricity from solar energy in the short run.

6. Explain how the demand curve for an input is derived.

7. Describe the relationship between the expansion path and the long-run marginal cost curve.

8. Suppose the demand function for an input is:

$p_d = 24 - 2Q_d$

where p_d is the price and Q_d is the quantity demanded of the input. Graph the input demand function and determine equilibrium input use when the market price of the input is $6.

9. Calculate future value at the end of the fifth year of placing $100 in a savings account that pays 4 percent annual interest.

10. Suppose the costs of protecting wildlife habitat in a national refuge is $10,000 in year 1, $15,000 in year 2 and $25,000 in year 3. Calculate the present value of costs using an annual discount rate of 10 percent.

Further Readings

Baumol, William J. 1977. *Economic Theory and Operations Analysis,* 4th ed. Englewood Cliffs, New Jersey: Prentice-Hall, Inc.

Henderson, James M., and Richard E. Quandt. 1971. *Microeconomic Theory: A Mathematical Approach*, 2nd ed. New York: McGraw-Hill Book Company.

Leftwich, Richard H. 1976. *The Price System and Resource Allocation*, 6th ed. Hinsdale, Illinois: The Dryden Press.

Nicholson, Walter. 1978. *Microeconomic Theory: Basic Principles and Extensions.* Hinsdale, Illinois: The Dryden Press.

Solberg, Eric J. 1982. *Intermediate Microeconomics.* Plano, Texas: Business Publications, Inc.

Stigler, George J. 1987. *The Theory of Price,* 4th ed. New York: Macmillan Publishing Co.

Varian, Hal R. 1984. *Microeconomic Analysis*, 2nd ed. New York: W. W. Norton & Co.

CHAPTER 3

Historical Views of Natural and Environmental Resource Capacity

A great change in our stewardship of the Earth and the life on it is required if vast human misery is to be avoided and our global home on this planet is not to be irretrievably mutilated.

—Union of Concerned Scientists, 1992

The roots of natural resource and environmental economics can be traced to concerns about the capacity of natural and environmental resources to sustain economic growth and to the value of conserving natural and environmental resources. Natural resource limitations on economic growth were addressed by classical economists. Debate on limits to growth continues to this day. The capacity of environmental resources to assimilate residuals from economic activities, and the desirability of protecting nonuse values of the environment, such as solitude, beauty and biological diversity, have become just as limiting to economic activity as has depletion of natural resources. This chapter discusses the relationships between natural or environmental resource capacity and economic activity in terms of *classical economics, neoclassical economics* and *conservationism*; it also evaluates *contemporary views* and *indicators* of natural or environmental resource capacity.

Economic Views

Natural resource capacity was first discussed in terms of resource scarcity. Economic concern about resource scarcity and its effects on human welfare was expressed by classical economists. Neoclassical economists generally downplayed the idea that natural resources would limit economic growth. Conservationists were generally concerned about managing natural resources for the benefit of humankind and about protecting the resource base.

CLASSICAL ECONOMICS. Classical economics was concerned primarily with the effects of population growth and land use on resource scarcity, economic growth and the quality of life. Of particular interest was the "means of raising the physical productivity of labor and expanding the total volume of economic activity."[1] Economists making important contributions to this subject include Adam Smith, Thomas Malthus, David Ricardo, John Stuart Mill and William Stanley Jevons.

Adam Smith stated that the natural price of a commodity was determined by the amount of labor commanded in the market. This natural price was a more relevant indicator of wealth than were market prices. Smith argued that supply-side factors in the agricultural sector do not contribute to economic stagnation because capital accumulation in agriculture raises labor productivity. Agricultural products become relatively scarce because demand eventually exceeds supply, resulting in a whole host of social problems that lead to economic stagnation.[2]

In *An Essay on the Principle of Population*, Thomas Malthus[3] argued that population grows at a geometric rate and human subsistence (food supply) at an arithmetic rate, causing the human condition to decline over time. This is referred to as the Malthusian trap. In *Principles of Political Economy*,[4] Malthus solidifies the concept of *absolute scarcity* of land by arguing that geometric growth in population in combination with a fixed supply of land increases the intensity of land cultivation over time, resulting in diminishing per capita returns to land and higher food production and labor costs. The relationship between population and food-production capacity of land is depicted in Figure 3.1.

As long as the level of population remains below the maximum food-produc-

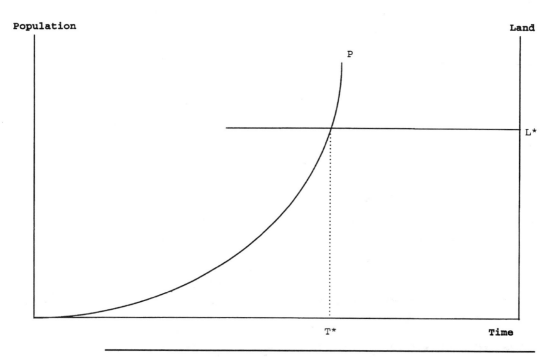

Figure 3.1. Malthusian model of absolute scarcity showing relationship between population growth (P) and food-production capacity of the land (L*).

tion capacity of the land (P < L*), there is sufficient food to maintain per capita food consumption. As soon as the level of population exceeds the maximum food-production capacity of land (P > L*), however, additional growth in population then causes per capita food consumption to decrease. If reductions in per capita food consumption slow down the rate of population growth, then population growth is likely to follow the logistic growth curve after T*, as shown in Figure 3.2. In this figure, population increases at an increasing rate up to T*. After T*, maximum food-production capacity of the land is reached and population increases at a decreasing rate. In summary, Malthus's absolute scarcity results when geometric growth in population eventually overtakes the capacity of the land to supply food.

David Ricardo[5] agreed with Malthus regarding rapid population growth, rising production costs and declining profits. Ricardo differed with Malthus regarding the timing of both diminishing returns to land and rising production costs. Malthus argued that returns to land would not diminish until the absolute limit on land availability is reached. Ricardo believed that average and marginal productivity of land decreases almost immediately because expansion in food production takes place on land of decreasing fertility. This concept is known as *relative scarcity*. One consequence of relative scarcity is that proportionately more labor is required per unit of land, resulting in higher wages and economic rent. Economic rent, which is central to Ricardo's macroeconomic analysis, equals the difference between price and average cost of production.

The implications of Malthusian absolute scarcity and Ricardian relative

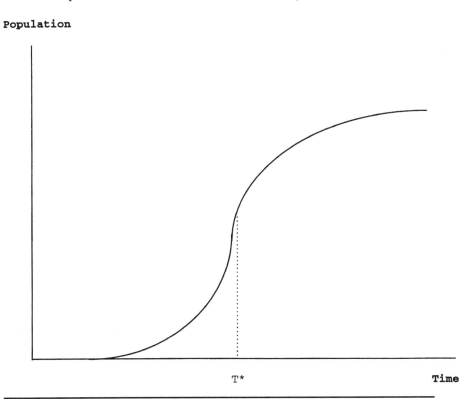

Figure 3.2. Logistic growth in population.

scarcity for marginal costs of food production are illustrated in Figure 3.3. With absolute scarcity, marginal cost is constant below an output of Q* but increases beyond Q*. Relative scarcity implies that marginal cost increases with the first and successive units of output as less fertile land is brought under cultivation. Ricardo subscribed to the labor theory of value (labor is the measure of all economic value) and considered technological change to be cost reducing. Malthus rejected the labor theory of value and believed that technological change increased output.

Economic stagnation from absolute or relative scarcity is delayed through the substitution of capital for labor, technological advancements and slower population growth. Substitution and technological progress increase the marginal productivity of labor and decrease marginal production cost. Slower population growth reduces total demand for food, provided there is not a compensating increase in food demand caused by income growth.

John Stuart Mill believed that economic growth was limited by the amount of land, diminishing marginal productivity of land, and increases in population spurred by growth in industrial production. Mill postulated that technological progress is sufficient to offset diminishing returns to land. He believed that diminishing returns in the mining of exhaustible resources, such as coal and metals, increase the cost of producing these resources, regardless of technological progress.

Mill appears to be the earliest economist to address other values associated with the environment, such as solitude, natural beauty and the provision of habitat for flora and fauna. Mill's philosophy regarding amenity values of the environment is similar to the philosophy espoused by conservationists such as Henry David Thoreau and John Muir. Regarding solitude, Mill observed, "Solitude in the pres-

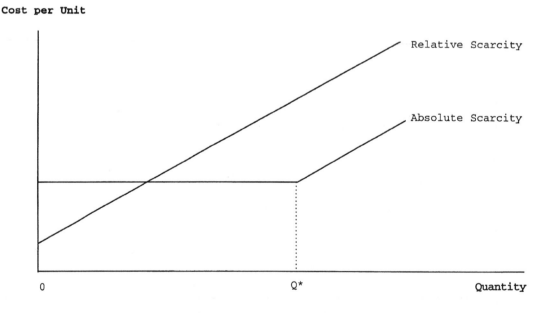

Figure 3.3. Marginal production cost with Malthusian absolute scarcity and Ricardian relative scarcity.

ence of natural beauty and grandeur is the cradle of thoughts of aspirations which are not only good for the individual, but which society could ill do without."[6]

NEOCLASSICAL ECONOMICS. Following classical economics, development of resource economics proceeded along two lines, neoclassical economics and institutional or land economics. The transition from classical to neoclassical economics occurred in the late nineteenth and early twentieth centuries. Neoclassical economic analysis is based on marginal analysis, which assumes that economic agents (households and firms) make individual decisions by comparing incremental changes in satisfaction or revenue to incremental cost. In neoclassical economics, the price of a product or service is determined by market equilibrium between demand and supply. Market price reflects a commodity's scarcity value.

William Stanley Jevons[7] argued that dependence on nonrenewable energy resources, notably coal, posed a serious threat to economic growth in industrialized countries, such as Britain. He argued that unlimited growth in coal consumption results in the exhaustion of coal reserves. As coal reserves are depleted, lower grade deposits are mined, causing mining costs and coal prices to increase. Prices would eventually increase to a point where British coal would no longer be competitive in world markets. Jevons feared that Britain's coal export market and economy might be devastated by higher domestic coal prices.

Jevons systematically rejected the possibility that this dilemma would be resolved by substitution of petroleum for coal, technology-driven decreases in coal mining and transportation costs, and the importing of coal from other countries. Jevons failed to understand that higher coal prices would stimulate development and use of petroleum products, and that technological advances reduce coal mining and transportation costs.

The cornerstone of neoclassical microeconomics was laid by Alfred Marshall. Other neoclassical economists making significant contributions include John Maynard Keynes in macroeconomics and Vilfredo Pareto, A.C. Pigou and Ronald Coase in welfare economics. Despite his seminal contributions to neoclassical economics, Marshall did not make a major contribution to natural resource economics. He agreed with classical economists that land is subject to diminishing marginal returns and believed that land prices rise in response to increasing land scarcity. He argued, however, that land scarcity is not a limit to economic growth because the higher prices for land would cause farmers to seek new knowledge and better management techniques, which increases agricultural productivity. Marshall's optimism formed the basis for the cornucopian view that technological progress and the substitution of manufactured capital for natural capital prevent the latter from constraining economic activity. Natural capital refers to the quantity and quality of natural resources.

Marshall recognized that the environment provides not only resources for production, but also "free gifts of nature," such as natural beauty, scenery, fresh air, light and fresh water. Marshall observed that the "services which land renders to man, in giving him space and light and air in which to live and work" are typically undervalued or not valued at all in the marketplace.[8]

While neither a classical or neoclassical economist, Karl Marx lines up with the neoclassical philosophy that natural resources are not a constraint to economic activity.[9] Marx's labor theory of value explains the exploitation of the labor class by capitalists. This theory is quite different from classical and neoclassical theories of value. Based on the labor theory of value, Marx concluded that "things supplied by nature 'gratis' and the services of capital proper have no value."[10] In a Marxian view of the world, value is created only when a natural resource is combined with labor.

Marx recognized that nature as well as labor contributed to wealth. He also provided what turned out to be an astute rationale for why privately optimal rates exceed socially optimal rates of soil erosion. Specifically, economic incentives for making long-term capital investments in soil conservation are lacking. This lack of incentives is perpetuated by the price system.

Institutional economics was pioneered by R. T. Ely, in collaboration with President Van Hise and J. R. Commons of the University of Wisconsin. Ely, along with Henry C. Taylor, Benjamin H. Hibbard, George C. Wehrwein and Lewis C. Gray, formed the core of the land economics tradition that evolved at the University of Wisconsin in the 1920s. Land economics focuses on the role of institutional factors, such as private property, public policy and public action, in the management of renewable natural resources. Early land economists studied land tenure (property rights) in agriculture, grazing and forestry. The institutionalist tradition at Wisconsin continues in the present-day writings of Daniel Bromley. Institutionalists generally associate environmental problems with economic growth and argue that social costs of environmental degradation require appropriate public action.

Conservationism

Concern about the depletion and degradation of America's natural resources has been voiced by many conservationists. The abundance of flora and fauna that existed at the time of the Lewis and Clark expedition was substantially reduced with the settlement of America. Native Americans criticized the profligate use of natural resources by white people. A classic example of this problem is the senseless slaughter of bison on the Great Plains for sport, which accompanied the westward expansion of the railroad. For decades, hunting and trapping took place without restrictions, resulting in the decimation of beaver in the Rocky Mountains. While some restrictions were eventually placed on resource use, extraction was accelerated by widespread mining of energy and precious minerals, livestock grazing and the damming of rivers to supply water for irrigated agriculture and commercial development. While taking a serious toll on natural resources and the environment, resource exploitation made possible the settlement and development of frontier regions in the United States.

It is not practical to trace all the contributions made to the philosophy and practice of conservationism in the United States. Fortunately, the thrust of the early conservation movement is captured in the lifestyles and achievements of a few noteworthy individuals. According to G. Tyler Miller, early actors in the conservation movement can be distinguished as being either scientific conservationists or preservationists. He asserts that "scientific conservationists see public lands as resources to be used now to enhance economic growth and national strength. They must be protected from degradation by being managed efficiently and scientifically for sustained yield and multiple use. Preservationists emphasize protecting large areas of public lands from mining, timbering, and other forms of development so they can be enjoyed by present generations and passed on unspoiled to future generations."[11]

George P. Marsh (1801–1882) was a scientist and Vermont congressman who laid much of the foundation for modern conservationism. In his influential book *Man and Nature*, Marsh expressed the diverse and complex interactions between humans and nature (environment) in terms of a "nature–man continuum." Along the continuum, humans affect nature and nature affects humans. In other words, there are feedback loops. Marsh contended that humans' complex dependence on nature extended far beyond the classical economics concept of diminishing marginal returns to agricultural land and mineral development. Nature can support human activities indefinitely, provided the ecological impacts of those activities are not excessive. Marsh argued that humans and the environment needed to be in balance.[12] His viewpoints represent an early manifestation of the ecological concept of environmental carrying capacity, which is generally ignored and even ridiculed by some contemporary economists. Marsh's concern about the need to achieve harmony between humans and the environment is echoed in the writings of contemporary conservationists such as Wendell Berry and Wes Jackson.

John Muir (1838–1914) was one of the premier preservationists in the United States. Muir received his motivation and strength through a close relationship with nature. While a successful fruit farmer in California's Alhambra Valley, Muir's first love was the study and preservation of the wilderness. His writings extol the "glories of nature" experienced during his many modest (stale bread and tea) but ambitious treks through the Sierra Nevada mountains, Alaska and many other parts of the United States. Muir devoted his wilderness odysseys to collecting field data for his many scientific investigations and enjoying nature. He is credited with discovering new species of plants and animals and the importance of glaciers in forming Yosemite Valley in central California.

Muir was instrumental in the 17-year struggle to establish a national park in Yosemite Valley and a national park system in the United States. Both efforts were successful. He received major credit for preserving the Grand Canyon. John Muir was the founder and first president of the Sierra Club, an activist grass roots organization started in 1892. His biggest defeat was the unsuccessful campaign to save Hetch Hetchy valley in Yosemite National Park from being flooded to form a water supply reservoir for San Francisco.[13]

President Theodore Roosevelt (1858–1919) was the first active conservationist in the White House. During his administration (1901–1909), Roosevelt initiated numerous programs to conserve America's natural resources. In 1903, he established the nation's first federal wildlife refuge to protect Florida's brown pelican. During his administration, the size of the national forest reserve tripled. In 1905, Congress created the United States Forest Service and President Roosevelt appointed Gifford Pinchot (1865-1946) as its first chief. Pinchot ushered in the era of scientific resource management based on multiple use-sustained yield management of national forest reserves (latter re-named national forests). Pinchot espoused the philosophy of managing natural resources for the greatest good of the greatest number of people for the longest time. In contrast to John Muir, Pinchot supported the flooding of the Hetch Hetchy valley.[14]

Aldo Leopold (1886–1948) operated in the same tradition as John Muir. Leopold received a college education in resource management (forestry and game management) and worked for Gifford Pinchot. Leopold is best known for his land ethic, which views land as a community of soils, water, plants and animals worth preserving and humans as responsible for land stewardship as an "ecological necessity." The land ethic does not require that land resources remain idle. Use is allowed as long as provision is made for the continued existence of land, "and, at least in spots, continued existence in a natural state." This latter concept spurred Leopold and Bob Marshall to establish the Wilderness Society in 1935.

Leopold considered it inappropriate to base land conservation decisions solely on economic criteria. He argued that many services provided by the land are essential for its proper functioning, even though such services have no commercial value. He wanted land managers to have an "ecological conscience" that causes them to take personal responsibility for maintaining the health of the land. Leopold reasoned that land managers needed to put aside the narrow mindset that "economics determines all land-use" and embrace what is ethically and aesthetically right for the land. This required preserving the land's integrity, stability and beauty.[15]

Contemporary Views

Prior to the late 1960s, the position that resource depletion and environmental degradation might limit long-term economic growth received little support for three reasons. First, Malthus's dire prediction that population growth would eventually cause economic stagnation did not materialize. Second, neoclassical economics provided a convincing argument that a well-functioning price system triggers higher prices for depleted resources, technological innovation and resource substitution, all of which sustain economic growth. Third, benefits of economic growth seemed to outweigh the resource and environmental costs.

Spurred by the birth of the environmental movement in the late 1960s and the concept of sustainable development that gained popularity in the late 1980s, scientists, policy-makers, government officials and the general public became concerned with the human and environmental costs of unbridled economic growth. This concern was heightened by the potential environmental impacts of adding another billion people to world population in 10 years (see Table 1.1) and the slowing of technological progress in many developed countries, such as the United States. The remainder of this section reviews some of the more important contemporary studies about the relationship between natural and environmental resource capacity and economic activity.

LIMITS TO GROWTH. In *The Limits to Growth*, Meadows et al.[16] used a dynamic simulation model to examine the potential environmental consequences of population and economic growth. Their simulation model indicated 1) that if present world trends in population, industrialization, pollution, food production and resource depletion continue, limits to growth would be reached in 100 years (between 2020 and 2100); 2) that present trends can be altered to achieve a sustainable balance between economic and ecological systems that is adequate and equitable for all persons; and 3) that the sooner society begins to work toward sustainability, the greater the likelihood of achieving it. The World3 model used in *The Limits to Growth* study assumes exponential growth in population, industrial production, pollution and consumption, and fixed quantities of land and exhaustible resources.

In the sequel to *The Limits to Growth*, entitled *Beyond the Limits*, Meadows et al.[17] observe that 1) many resource and pollution flows now exceed the limits of sustainability and, if unchecked, will eventually reduce food and industrial production and energy use; 2) economic decline can be averted by adopting policies to reduce growth in population and consumption and by increasing the efficiency of resource use; and 3) in addition to increased productivity and better technology, achieving sustainability requires maturity, compassion and wisdom.

The pessimistic conclusions of the limits-to-growth study have been criticized by Solow and others[18] on several grounds. First, technological innovation causes more or less exponential growth in the productivity of natural resources. Second, greater scarcity of exhaustible resources increases their prices as well as the prices of commodities that utilize these resources. Price increases reduce resource use and stimulate development of substitutes. Third, as countries become more affluent, population growth decreases; hence, population growth is not exponential. Fourth, industrial pollution does not have to increase in direct proportion to industrial production. Pollution can be reduced by such things as emission controls, effluent charges and taxes. Of course, regulations and financial disincentives for reducing pollution divert resources from production of commodities. Hence, pollution reduction involves a real social cost.

In support of Solow's viewpoints, Kransberg[19] contends that technological innovation helps to sustain economic growth. He argues that resource supplies are augmented and food and energy production are increased by technological change, both of which tend to moderate the effects of resource scarcity.

Several other studies have addressed natural and environmental resource limits to economic growth. Among the works supporting natural limits to growth are the *Global 2000 Report* by Barney, *Healing the Planet* by Ehrlich and Ehrlich, *Beyond Oil* by Gever et al., and *Steady-State Economics* by Daly.[20] Studies that reject the notion of resource-based limits to growth and that embrace the cornucopian view include *The Resourceful Earth* by Simon and Kahn, *The Global Possible* by Repetto and *The Next 200 Years* by Kahn.[21]

Other Factors

Resource development directly affects natural and environmental resource capacity. A comprehensive economic approach to resource development requires that a proposed activity not be undertaken unless it is economically feasible (benefits exceed costs, including resource and environmental costs). In an economic-based approach to resource development, natural and environmental resources are used in a socially efficient manner when resource prices reflect the full social cost of production. Resource development is, however, affected by factors other than economics. Legal, political, ethical, religious and sociological factors are especially important.

Consider the role of law and politics in government-funded projects such as construction of hydroelectric power plants. Project selection is based either directly or indirectly on authorizing legislation, which evolves from interaction and lobbying among executive and legislative branches of government, special interest groups and the general public. The resulting legislation is a political compromise in which economics often plays a minor role. For example, the Endangered Species Act of 1973 prohibits consideration of economic criteria in developing and implementing programs to protect threatened and endangered species. In the early 1990s, this stipulation resulted in a major clash between timber companies who wanted to harvest old-growth forests and environmental groups and federal agencies who wanted to protect the habitat of the northern spotted owl in the northwestern United States.

Even when resource development activities must pass the test of economic feasibility, politics influences the economic analysis of benefits and costs. When Jimmy Carter became President of the United States in 1977, he proposed increasing the discount rate used to evaluate the benefits and costs of water resource developments in the western United States. At the time, a low discount rate was being used. Many resource economists supported Carter's proposal because they believed the discount rate was too low, causing Congress to approve water resource development projects with short-term economic benefits but long-term environmental costs. A good example of this is the development of hydroelectric power facilities (dams and reservoirs) on the Columbia River. The higher discount rate proposed by President Carter would have disqualified many water resource development projects (costs would exceed benefits). Politically powerful interest groups persuaded the Carter administration to drop this proposal.

Another way that resources are misallocated is as a result of politically based decisions to subsidize certain activities. For example, the World Resources Institute claims that motor vehicle transportation in the United States is subsidized to the tune of more than $1,000 per person above and beyond what each person directly pays for motor transportation.[22] Without these subsidies, the use of motor vehicles would likely be much lower, as evidenced by the more energy-efficient transportation systems that have evolved in western Europe.

Judicial and administrative laws are also important in resolving legal conflicts between opposing parties and in establishing and protecting property rights to natural resources. Property rights specify who has the right to use a particular resource. For example, under the prior appropriation doctrine, western water users with earlier rights are given priority in water allocation.

Changes in political parties can have positive and/or negative effects on the environment. In the United States, the Democratic Party is more sympathetic to environmental protection than is the Republican Party. Green parties and green politics in many developed countries emphasize grassroots democracy, social responsibility and nonviolence, positions that align closely with the deep ecology movement.[23]

Equally significant is the growth in green businesses and the greening of businesses. In 1990, the world market for environmental goods and services amounted to $200 billion. By the year 2000, this market is projected to grow by 50 percent. In October 1992, the chairman and chief executive officer of Dow Chemical Company announced his belief that chemical prices should reflect environmental costs.[24] These and many other examples portend a "new" green revolution that aspires to achieve a balance between economic development and environmental quality. Other examples of events with negative environmental consequences include the defoliation of vast portions of the landscape from extensive use of napalm during the Vietnam War and air, water, and land pollution resulting from the conflagration of oil wells in Kuwait during the 1991 Persian Gulf War.

Indicators of Resource Capacity

The capacity of natural and environmental resources to sustain economic activity has been examined in several studies. Capacity indicators generally fall into two categories: physical measures and economic measures. Typical physical measures include projected production, amount of the resource discovered per unit of effort (barrels of oil discovered per foot [meter] of well drilled), the amount of labor and capital required to extract one unit of the resource and the energy profit ratio. Common economic measures include unit extraction costs and prices. Each measure of resource capacity has strengths and weaknesses.

PHYSICAL MEASURES. Most physical measures of resource capacity have been applied to exhaustible resources such as energy and minerals. The most famous theoretical projections of oil and natural gas production are based on Hubbert's bell-shaped curve developed in 1956.[25] Gever et al. updated Hubbert's curve using more advanced statistical techniques and more recent data, as shown in Figure 3.4. Both the original and revised curves indicate that oil and gas production grow slowly at first, increase rapidly, reach a peak and then decrease rapidly. The decline phase occurs because the quality of petroleum discoveries decreases over time. Predictions by Hubbert and Gever et al. proved to be very accurate in projecting the 1970 peak in oil production and the 1973 peak in natural gas production that occurred in the United States. Hubbert's projections indicate that United States capacity to produce oil and natural gas will dissipate by the year 2020.

Fossil fuel–production capacity, especially for oil and coal, has decreased on a worldwide basis, as shown in Table 3.1. Oil production increased from 1950 to 1979 but decreased from 1979 to 1992. Coal production increased from 1950 to 1989, but decreased from 1989 to 1992. Natural gas production increased throughout the 1950–1992 period.

Other physical measures of oil and gas production capacity are consistent with Hubbert's curve. Discoveries of crude oil per foot of well drilled followed a downward trend from 1930 to the mid-1980s. Additions to proven reserves per foot (or meter) of well drilled showed a marked downward trend from 1945 to 1956, increased slightly from 1957 to 1974, and fell sharply from 1975 to 1985. Using data

Table 3.1. World growth in production of fossil fuel, 1950–1992

Fuel	Growth Period		Decline Period	
	Years	Percent Annual Change	Years	Percent Annual Change
Oil	1950-1979	+6.4	1979-1992	−0.5
Coal	1950-1989	+2.2	1989-1992	−0.6
Natural gas	1950-1992	+6.2		

Source: Lester R. Brown, A Decade of Discontinuity, *World Watch* (Washington, DC: Worldwatch Institute, July/Aug 1993), pp. 19–26.

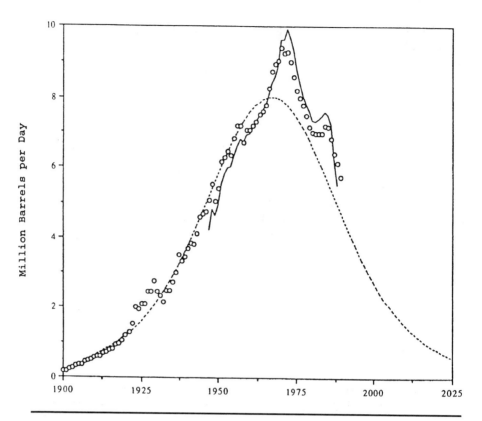

Figure 3.4. Actual and projected oil production in lower 48 states by Hubbert curve (*dashed line*), Kaufmann (1991) (*solid line*) and actual values (*open circles*).

SOURCE: John Gever, et al., *Beyond Oil: The Thrust to Food and Fuel in the Coming Decades* (Niwot, Colorado: University of Colorado, 1991), p. xxxi.

for the 1870–1957 period, Barnett and Morse[26] determined that there was a fourfold decrease in the quantity of labor and capital required per unit of resource extracted in the agricultural, mineral or fuels, forestry and fisheries sectors.

Physical efficiency of energy use has been measured using the net energy ratio, net energy return on investment and energy profit ratio.[27] While these measures can be applied to several types of resources, they are most often applied to energy resources such as coal, oil, natural gas, hydroelectric power and solar energy. Net energy return on investment is the energy content of a product minus all the energy required to produce that product, including the energy embodied in the capital and the raw materials used to produce that product. For example, if 25,000 barrels of oil equivalent (boe) are required to produce 75,000 boe from an oil shale deposit, then the net energy return on investment for that deposit is 50,000 boe. Production of the resource is energy efficient (inefficient) when the net energy return is positive (negative).

Net energy ratio or energy profit ratio is the ratio of the energy contained in a product to the energy required to produce it. A ratio greater (less) than 1 implies energy efficiency (inefficiency). The net energy ratio for the oil shale example is 3.

Gever et al.[28] determined that the energy profit ratio for crude oil decreased slowly from 27 in 1919 to 18 in 1955, increased to 23 by 1967, and decreased to about 18 in 1982. The long-term downward trend in the energy profit ratio is consistent with Hubbert's curve and the downward trend in both discoveries of crude oil and additions to proven reserves per foot (meter) of well drilled.

ECONOMIC MEASURES. Nordhaus examined the changes in mineral prices relative to wages in the manufacturing sector. Nordhaus concluded that minerals are becoming less scarce relative to labor based on declining relative mineral prices from 1900 to 1970.[29] In addition to their physical measure of resource capacity, Barnett and Morse[30] evaluated the prices of extractive resources (agricultural products, minerals, fuels, forest products and fish) relative to prices for nonextractive resources during the 1870–1957 period. Relative prices of agricultural and mineral resources showed a downward trend during this period, indicating reduced scarcity, whereas relative prices for forest products and fish rose in the early 1930s, indicating increased scarcity.

Johnson et al.[31] updated the original Barnett and Morse study by extending the database through 1970. Once again, prices of agricultural products, minerals, and fuels declined, indicating reduced scarcity. Compared with the increasing prices and greater scarcity found in the original 1870–1957 period, prices of forest products decreased from 1962 to 1970, implying less scarcity. Fish prices continued the upward trend that had begun in the early 1930s. Smith[32] examined resource prices using more advanced statistical methods and more recent data than Barnett and Morse. He determined that trends in resource prices depend on the time period selected and that Barnett and Morse's conclusion that resources are not becoming more scarce cannot be supported.

Cleveland[33] examined three economic indicators of petroleum scarcity: real well head prices of crude oil and natural gas, capital and labor (unit) cost per unit of extraction, and average total cost of extraction. Real crude oil prices followed a U-shaped time path. Prices fell from 1860 to 1890, remained relatively constant from 1890 to 1940, rose slightly from 1940 to the mid-1970s, and increased sharply from the mid-1970s to the early 1980s. Unit cost of petroleum fell through the early 1970s and increased through 1987. Average total cost decreased from the mid-1950s to the early 1970s and increased very rapidly from the mid-1970s through 1987.

Norgaard[34] argues that resource scarcity cannot be determined by studying trends in resource prices because of a logical fallacy. He points out that the theoretical models developed by Ricardo and Hotelling start with the major premise "if resources are scarce," and the minor premise "if resource allocators are informed of resource scarcity," and conclude that "then economic indicators will reflect this scarcity." The logical fallacy is that empirical studies have reasoned in an opposite manner. Norgaard contends that the model uses resource scarcity indicators (the conclusion) to determine whether resources are scarce (the major premise) and ignore whether resource allocators are informed (the minor premise). Rather than refuting Ricardo's predictions of increasing resource scarcity, empirical studies show that the assumption of informed resource allocators is false. If the theory is accepted, then resource allocators are informed about scarcity and resource scarcity can be determined by simply asking them whether resources are scarce.

Physical and economic measures of resource capacity have certain drawbacks. Although the effects of diminishing supplies are counteracted by increases in petroleum prices, judging the adequacy of petroleum reserves based on the number of years until exhaustion ignores this fact. On the demand side, higher prices for petroleum and petroleum-based products would increase conservation. Events of the mid-1970s and early 1980s bear this out. On the supply side, higher petroleum prices would stimulate petroleum exploration; use of technologies that increase the efficiency of petroleum exploration, development and processing; and replacement of petroleum with substitutes such as solar energy. Inferring resource capacity limits from economic indicators might be misleading because of price and cost distortions caused by federal price controls, depletion allowances, import restrictions, international cartels such as the Organization of Petroleum Exporting Countries, wars and other factors.

ENVIRONMENTAL RESOURCES. Many resource specialists believe that the capacity of the environment to assimilate residuals from production and consumption is a greater constraint to economic activity than are natural resource depletion and exploitation.

Unfortunately, physical and economic measures of natural resource capacity do not adequately reflect the impacts of resource extraction, processing, transportation and use on environmental quality. Burning of coal has created air quality problems, such as global warming and acid rain. In order to reduce damages from acid rain, the 1990 amendments to the Clean Air Act limit emissions of sulfur dioxide from power plants. Reducing carbon dioxide emissions, which are the major culprit in global warming, is more difficult due to the global nature of the problem. The environmental damage wrought by the *Exxon Valdez* oil spill in 1989 was not reflected in oil prices due to the competitive nature of the oil market. This incident did, however, bring about federally mandated safety standards for oil tanker construction and transportation.

Extraction and use of natural resources deplete and degrade other natural resources. By the late 1970s, 2,350 mi^2 (6,087 km^2) of various ecosystems in the United States were converted to drilling platforms, pipelines, refineries and other facilities that support petroleum use. Petroleum extraction and other industrial activities have claimed more than 100 mi^2 (259 km^2) of wetlands in Louisiana. Prior to the push for environmental protection legislation in the late 1960s, environmental costs were generally external, meaning they were not reflected in costs of production and the prices of goods and services. Legislation, such as the Clean Air Act, Clean Water Act and Endangered Species Act, has helped to internalize many environmental costs.

Summary

The capacity of natural and environmental resources to support economic activities is a major concern of resource economists, conservationists and others. Economic concern over resource capacity and its effects on economic activity was first expressed by classical economists. Thomas Malthus and David Ricardo argued that growth in population combined with a limited land base would eventually result in economic stagnation in the form of declining per capita food production. Malthus claimed that the marginal productivity of land declines when the absolute land limit is reached (absolute scarcity). In contrast, Ricardo believed that the marginal productivity of land declines almost immediately (relative scarcity). Both Malthus and Ricardo contended that economic stagnation could be delayed through the substitution of capital for labor, technological advancements and slower population growth. John Stuart Mill recognized that the environment provided natural services such as solitude and beauty.

Neoclassical economists generally sounded an optimistic note regarding resource scarcity. William Stanley Jevons claimed that dependence on exhaustible energy resources leads to economic stagnation. Alfred Marshall, the father of neoclassical economics, argued that increasing land scarcity causes land prices to increase, which stimulates farmers to seek new knowledge and better management techniques. As a result, agricultural productivity increases. Marshall believed that technological progress and the substitution of manufactured capital for natural resources eliminate natural resource constraints to economic activity. Institutional economists focused on the role of institutional factors, such as private property, public policy and public action in the management of rural, renewable natural resources.

Conservationists like George P. Marsh posited that people's dependence on nature extends far beyond diminishing marginal returns to agricultural land and mineral development. He argued that nature can support human activities indefinitely provided the ecological impacts of these activities are not excessive. Other conservationists such as John Muir, Theodore Roosevelt, Gifford Pinchot and Aldo Leopold contributed to the theory and practice of natural resource management and protection. Aldo Leopold developed a land conservation ethic and emphasized the ecological necessity of land stewardship and the protection of natural areas.

In the past 20 years, polar views have developed regarding the relationship between natural and environmental resource capacity and economic growth. At one extreme is the neo-Malthusian view that resource capacity limits economic growth. At the other extreme is the cornucopian view that technological progress and operation of the price system are sufficient to remove natural and environmental resource constraints to economic growth. Falling between these two extremes is the material balances and ecological economics approaches (discussed in Chapter 4), which provide an analytical framework for understanding the relationships between the economy and the ecosystem. Development of institutional mechanisms for resolving resource conflicts involve legal, political, ethical, religious and sociological considerations.

Various physical and economic indicators of resource capacity have been employed, including projections of production (Hubbert's curve), amount of the resource discovered per unit of effort, amount of labor and capital required to extract one unit of the resource, net energy ratio and energy profit ratio. Economic indicators include unit extraction costs and prices of resources. Physical and economic indicators of resource capacity are imperfect and often give contradictory evidence regarding the extent to which natural and environmental resource capacity constrains economic growth. Support is growing for the viewpoint that environmental resource capacity has become a greater potential constraint on economic activity than have natural resource depletion and exploitation.

Questions for Discussion

1. Suppose that instead of growing at an increasing rate, as depicted in Figure 3.1, population grows at a decreasing rate after T', where $T' < T^*$. What are the implications of this change in population growth for the Malthusian hypothesis that, over time, land becomes increasingly scarce and per capita food production decreases?

2. Marshall conceptualized that technological advancements increase the marginal productivity of labor, which offsets the effects of increasing land scarcity. Resource economists have pointed out that agricultural technologies have increased the intensity of land use, thereby accelerating soil erosion and dependence on fossil fuels. In particular, substitution of machinery power for animal power increased soil erosion on cultivated land and boosted use of fossil fuel per unit of agricultural production. Do technological improvements delay the effects of land scarcity at the expense of increased land degradation and higher energy use?

3. Aldo Leopold maintained that land managers need to develop a land ethic to preserve certain noneconomic benefits of the land such as keeping a wetland in its natural state rather than converting it to cropland. Some resource economists have estimated nonmarket values of different land uses such as the recreational and water quality benefits of a wetland. Is the determination of nonmarket values of land uses inconsistent with Leopold's land ethic?

4. If, as some economists argue, the assumptions made by proponents of limits to growth are faulty, then do their conclusions have any relevance for natural and environmental resource management? Explain.

5. How important are noneconomic factors in determining resource development and use?

6. What are some of the advantages and disadvantages of using physical versus economic measures of natural resource capacity?

7. Is environmental resource capacity or natural resource capacity likely to be a greater concern in developed or in developing countries? Explain.

Further Readings

Castle, Emery N., et al. 1977. "Natural Resource Economics, 1946–75." *A Survey of Agricultural Economics Literature, Vol. 3, Economics of Welfare, Rural Development, and Natural Resources in Agriculture, 1940s to 1970s*. Lee R. Martin, ed. Minneapolis, Minnesota: University of Minnesota Press.

Hall, Darwin C. and Jane V. Hall. 1984. "Concepts and Measures of Natural Resource Scarcity with a Summary of Recent Trends." *Journal of Environmental Economics and Management* 2:363–379.

Kneese, Allen K., Robert U. Ayres and Ralph C. d'Arge. 1970. *Economics and the Environment: A Material Balances Approach*. Baltimore, Maryland: The Johns Hopkins University Press.

Myers, Norman and Julian L. Simon. 1994. *Scarcity or Abundance? A Debate on the Environment*. New York: W. W. Norton & Co.

Pearce, David, Anil Markandya and Edward B. Barbier. 1989. *Blueprint for a Green Economy*. London: Earthscan Publications Ltd.

Tanner, Thomas, ed. 1987. *Aldo Leopold: The Man and His Legacy.* Ankeny, Iowa: Soil Conservation Society of America.

Notes

1. Hla Myint, "The Classical View of the Economic Problem," in *Essays in Economic Thought: Aristotle to Marshall,* Joseph J. Spengler and W. R. Allen, eds. (Chicago: Rand-MacNally, 1960), pp. 451–452.

2. Adam Smith, *An Inquiry into the Significance and Causes of the Wealth of Nations* (London: J. M. Dent & Co., 1910).

3. Thomas R. Malthus, *An Essay on the Principle of Population,* 1st ed. (London: J. Johnston, 1798), and Thomas R. Malthus, *An Essay on the Principle of Population,* 2nd ed. (London: J. M. Dent & Sons, 1914).

4. Thomas R. Malthus, *Principles of Political Economy: Considered with a View to Their Practical Application* (London: John Murray, 1820).

5. David Ricardo, *The Principles of Political Economy* (London: J. M. Dent & Sons, 1973).

6. John Stuart Mill, *Principles of Political Economy with Some of Their Application to Social Philosophy,* 5th ed., vol. II, book IV (London: Parker, Son and Bourn, 1862), p. 325.

7. William Stanley Jevons, *The Coal Question: An Inquiry Concerning the Progress of the Nation and the Probable Exhaustion of Our Coal Mines* (London: Macmillan, 1909).

8. Alfred Marshall, *Principles of Economics: An Introductory Volume,* 8th ed. (London: Macmillan, 1945), pp. 138–139.

9. Karl Marx, *Capital: A Critique of Political Economy,* vol. I (Harmondsworth, England: Penguin Books, 1976).

10. Nicholas Georgescu-Roegen, *The Entropy Law and the Economic Process* (Cambridge, Massachusetts: Harvard University Press, 1971), pp. 288–289.

11. G. Tyler Miller, Jr., *Resource Conservation and Management* (Belmont, California: Wadsworth Publishing Co., 1990), p. 38.

12. George P. Marsh, *Man and Nature* (New York: Charles Scribner, 1865).

13. Edwin Way Teale, *The Wilderness World of John Muir* (Boston: Houghton Mifflin Co., 1954).

14. Frank Graham, *Man's Dominion: The Story of Conservation in America* (New York: M. Evans, 1971).

15. Aldo Leopold, *A Sand County Almanac with Essays on Conservation from Round River* (New York: Ballantine Books, 1970).

16. D. H. Meadows et al., *The Limits to Growth* (New York: Universe Books, 1972).

17. Donella Meadows, Dennis L. Meadows and Jorgen Randers, *Beyond the Limits: Confronting Global Collapse, Envisioning a Sustainable Future* (Post Mills, Vermont: Chelsia Green Publishing Co., 1992), pp. xiii–xvi.

18. Robert M. Solow, "Is the End of the World at Hand?" in *The Economic Growth Controversy,* Andrew Weintraub et al., eds. (New York: International Arts and Science Press, 1973), pp. 39–61.

19. M. Kranzberg, "Can Technological Progress Continue to Provide for the Future?" *The Economic Growth Controversy,* Andrew Weintraub et al., eds. (New York: International Arts and Science Press, 1973), pp. 62–81.

20. G. O. Barney, *The Global 2000 Report to the President of the United States,* 2 vols. (Oxford, England: Permagon Press, 1980); Paul R. Ehrlich and Anne H. Ehrlich, *Healing the Planet* (Reading, Massachusetts: Addison-Wesley, 1991); John Gever et al., *Beyond Oil: The*

Thread to Food and Fuel in the Coming Decades (Niwot, Colorado: University Press of Colorado, 1991); Herman Daly, *Steady-State Economics,* 2nd ed. (Washington, D.C.: Island Press, 1991).

21. Julian Simon and Herman Kahn, *The Resourceful Earth: A Response to Global 2000* (Oxford, England: Basil Blackwell, 1984); Robert Repetto, ed., *The Global Possible: Resources, Development and the New Century* (New Haven, Connecticut: Yale University Press, 1986); Herman Kahn, *The Next 200 Years: A Scenario for America and the World* (New York: William Morrow, 1976).

22. Ed Ayers, "Breaking Away," *World Watch* (Washington, D.C.: Worldwatch Institute, January/February 1993), pp. 10–18.

23. Bill Devall, "Deep Ecology and Radical Environmentalism," in *American Environmentalism: The U.S. Environmental Movement, 1970–1990,* Riley E. Dunlap and Angela G. Mertig, eds. (Washington, D.C.: Taylor & Francis Inc., 1992), pp. 51–62.

24. Christopher Flavin and John E. Young, "Will Clinton Give Industry a Green Edge?" *World Watch* (Washington, D.C.: Worldwatch Institute, January/February 1993), pp. 26–33.

25. M. K. Hubbert, "Nuclear Energy and the Fossil Fuels," *Drilling and Production Practice* (Washington, D.C.: American Petroleum Institute, 1956); R. K. Kaufmann, "Oil Production in the Lower 48 States: Reconciling Curve Fitting and Econometric Methods," *Resources and Energy* 13(1991):111–127.

26. Harold Barnett and Chandler Morse, *Scarcity and Growth: The Economics of Natural Resource Availability* (Baltimore, Maryland: Johns Hopkins University Press, 1963).

27. Howard T. Odum, *Environment, Power and Society* (New York: John Wiley & Sons, 1971); Charles Hall, Cutler Cleveland and Mitchell Berger, "Yield Per Effort as a Function of Time and Effort for United States Petroleum, Uranium and Coal," in *Energy and Ecological Modelling,* W. J. Mitsch, R. W. Bosserman and J. M. Klopatek, eds. (New York: Elsevier, 1981); and John Gever et al., *Beyond Oil: The Threat to Food and Fuel in the Coming Decades* (Niwot, Colorado: University Press of Colorado, 1991).

28. Gever et al. (1991).

29. W. D. Nordhaus, "The Allocation of Energy Resources," *Brookings Papers on Economic Activity* 3(1973):529–570.

30. Barnett and Morse (1963).

31. Manual H. Johnson, Frederick W. Bell and James T. Bennett, "Natural Resource Scarcity: Empirical Evidence and Public Policy," *Journal of Environmental Economics and Management* 7(1980):256–271.

32. V. Kerry Smith, "The Evaluation of Natural Resource Adequacy: Elusive Quest or Frontier of Economic Analysis?" *Land Economics* 56(1980):257–298; V. Kerry Smith, "Measuring Natural Resource Scarcity: Theory and Practice," *Journal of Environmental Economics and Management* 5(1978):150–171; and V. Kerry Smith, "Natural Resource Scarcity: A Statistical Analysis," *Review of Economics and Statistics* 61(1979):423–427.

33. Cutler J. Cleveland, "An Exploration of Alternative Measures of Natural Resource Scarcity: The Case of Petroleum Resources," *Ecological Economics* 7(1993):123–152.

34. Richard B. Norgaard, "Economic Indicators of Resource Scarcity: A Critical Essay," *Journal of Environmental Economics and Management* 19(1990):19–25.

CHAPTER 4

Economy and Environment

The capacity of the environment to rearrange matter (free energy), as well as the stock of matter (resources) . . . are both irrevocably used up in the economic process. Replenishment of the physical basis of life is not a circular affair.

—HERMAN DALY, 1991

Our understanding of the relationship between the economy and the environment has evolved from the classical economic view that natural resource capacity is a constraint on economic activity to the neoclassical position that a well-functioning market and technological innovation eliminate natural and environmental resource constraints on economic growth. Critics argue that the neoclassical paradigm epitomizes the fallacy of misplaced concreteness. This fallacy occurs when abstractions are used to draw conclusions about reality without considering the limitations of those abstractions and the validity of the resulting conclusions.[1] One way to deal with this criticism is to employ a holistic model of the economy and environment that reflects the interplay between economic activities, resource depletion and environmental degradation. This chapter discusses the relationships between the economy and the environment in terms of the *circular flow of economic activity, material balances, ecological economics* and *sustainable development.*

Circular Flow Economy

Science utilizes models to describe and predict the behavior of physical and behavioral systems. A model is a simple abstract representation of a system. The language of models is words, diagrams and equations. The complexity of a model depends on the purpose for which it is developed and the consequences of making a wrong prediction based on the model. For example, a model to determine whether a nuclear reactor is safe to operate should be fairly complete to ensure that it accurately simulates the reactor's operation. Completeness is paramount because the potential human and environmental consequences of inaccurate model predictions of reactor safety are simply too great. In comparison, a model to predict soil erosion in a field of soybeans would not have to be as complete because the socioeconomic consequences of an inaccurate prediction are relatively minor.

Economists use models to explain and predict economic behavior of households and firms and the economy as a whole. Because economics deals with behavior, economic models tend to focus on behavioral relationships. The *circular flow economy*, illustrated in Figure 4.1, represents the traditional explanation of how firms and households interact in a market economy.

The model consists of a physical flow in the inner loop and a monetary flow in the outer loop. In the physical loop, firms employ resources supplied by households (A) to produce commodities (B) that are sold to households. In the monetary loop, households make expenditures on commodities (C) using income (D) derived from selling resources to firms. Income is in the form of wages, salaries and rents. Household expenditures equal household income (C = D) and expenditures made by firms equal revenues received by firms in a closed economy. While the completeness of the model can be enhanced by adding government and import–export sectors, such additions do not change the basic structure of the model.

The circular flow model captures the salient behavioral features of production and consumption in a market economy. Unfortunately, it ignores several important relationships. First, it does not take account of the physical realities of production and consumption. The model implies a circular flow of resources and commodities between households and firms. Physical laws dictate that energy and materials are transformed and degraded (entropy), which suggests a unidirectional flow rather than a circular flow of resources and commodities. In addition, production and consumption result in by-products and residuals. Second, the model does not distinguish between natural and human resources and between different types of natural

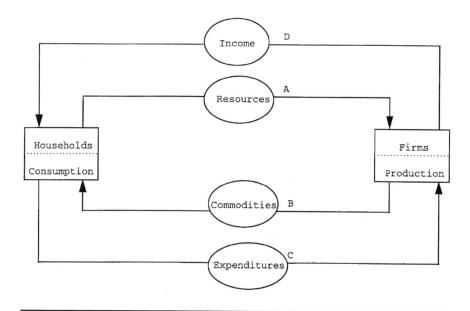

Figure 4.1. Circular flow model of economy.

resources. Lumping natural resources together glosses over the important distinction between the depletion of exhaustible resources and the overexploitation of renewable resources. In addition, the model ignores the satisfaction received by households and firms from consuming or preserving the option to consume amenity services provided by natural and environmental resources.

Third, depicting monetary flows in a circular manner suggests that the economy can grow indefinitely. In fact, neoclassical economics assumes that maximizing economic output is an appropriate goal of the economy. This goal is consistent with the human-centered orientation of neoclassical economics, but it contradicts the growing body of evidence that unbridled economic growth adversely affects natural and environmental resource capacity. The circular flow model abstracts from the physical and biological setting in which the economy operates and ignores the principles governing the use and transformation of material and energy. The material balances, ecological economics and sustainable development approaches discussed in the remainder of this chapter attempt to integrate economic, physical and biological processes.

Material Balances

The *material balances* (MB) model provides a more comprehensive description of the relationship between the economy and the environment by considering natural resource depletion and pollution of environmental resources. A simplified MB model that accounts for the flows of material and energy within the economy and between the economy and the environment is depicted in Figure 4.2. The left side of the diagram illustrates the interactions between firms and households and is identical to the circular flow model shown in Figure 4.1. The right side of the diagram depicts the environmental sector, which incorporates residuals, assimilative capacity of natural systems, recycling and environmental pollution.[2]

Firms use material and energy in the form of human and natural resources (A) to produce commodities that are sold to households (C). Production activities generate material and energy residuals that are released to the environment (B). Households satisfy their wants and needs by consuming commodities (C) produced by firms and environmental and amenity services (D). Environmental and amenity services include such things as purification of air and water by natural processes, protection from ultraviolet (UV) radiation by the ozone layer, and biodiversity. Amenity services include outdoor recreation, aesthetic beauty and an attractive place to live.

Consumption generates residuals that enter the environment (E). Residuals are stored, dispersed, recycled or assimilated by the natural systems. Recycling of residuals (F) extends the resource base. Natural resources can be depleted or overexploited. Use of exhaustible petroleum resources depletes this resource. Renewable resources such as soil are overexploited when rates of use exceed rates of regeneration. Pollution of natural resources such as air and water occurs when the flow of residuals into the environment exceeds the rate at which natural systems can assimilate those residuals (G > 0). Pollution decreases the capacity of natural systems to

76 Natural Resource and Environmental Economics

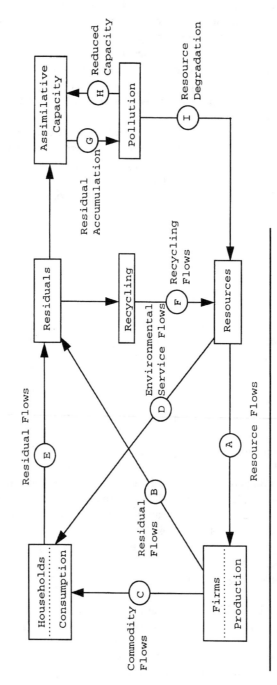

Figure 4.2. Simplified material balances model. *Arrows* indicate the direction of flow.

4. Economy and Environment

assimilate residuals (H) and results in resource degradation (I). For example, high nitrate–nitrogen concentrations in runoff from agricultural fields can impair reproduction and growth of plants and animals in rivers and lakes and can contaminate drinking water. A more complete description of the environmental component of the material balances model is contained in Chapter 9.

The *principle of material balances* states that in a closed economy (no imports or exports) and in the absence of accumulation of stocks of buildings, equipment, inventories and consumer durables, material and energy used in production and consumption equal material and energy in residuals, namely, A + D = B + E. The first two laws of thermodynamics, a branch of physics, form the theoretical basis for the material balances model. The first law of thermodynamics says matter and energy can neither be created nor destroyed (law of conservation of matter and energy).[3] In other words, the amount of energy and matter is constant.

Implications of the first law were illustrated by Boulding.[4] He used a word picture of the earth as a self-contained spaceship having an initial stock of resources and a limitless external supply of solar energy. If the spaceship's internal resources and residuals are not managed in a sustainable manner, then the spacemen eventually die when natural resources are depleted and environmental resources are degraded. If the spaceship's limited resources are used in a sustainable manner, however, then life on spaceship earth continues indefinitely. The message of Boulding's metaphor is that the earth is a closed system whose ability to support life depends on minimizing rather than maximizing the throughput of material and energy.

The second law of thermodynamics states that entropy increases over time. Entropy is the amount of unavailable energy and disorder in a system. Available energy is high (low entropy) in fossil fuel deposits. These sources of energy are of high quality because they can be used to do useful work. Unavailable energy increases as available sources of energy are used up. When coal is burned in a power plant, it generates heat that is used to produce steam. Residuals from coal combustion include heat, carbon dioxide, sulfur dioxide and particulate matter that go up the smokestack, and coal ash, which accumulates at the bottom of the boiler. The energy content of the useful heat is eventually dissipated, and the energy content of residuals is immediately dissipated (high entropy). Entropy increases as coal and other fossil fuels are burned. In this regard, the only way to reduce energy entropy is to reduce use of fossil fuels.

While the second law is formulated in terms of energy, it has been extended to material by Georgescu-Roegen. A more general economic interpretation of the second law implies that economic activities transform useful (low entropy) material and energy into useless (high entropy) material and energy. As Daly[5] points out, "The throughput flow [of material and energy] maintains or increases the order within the human economy, but at the cost of creating greater disorder in the rest of the natural world, as a result of depletion and pollution." People use material and energy from the natural world (environment) to produce goods and services that raise living standards. Unfortunately, goods wear out, services are reoccurring, and energy dissipates, placing further demands on natural systems. An important distinction exists between material entropy and energy entropy. While some residuals can be recycled (aluminum cans, glass bottles and newspapers), energy residuals, such as waste heat from a power plant, cannot be recycled. Recycling extends the life of the natural resources used in production.

Ecological Economics

Costanza et al.[6] have defined *ecological economics* as "a new transdisciplinary field of study that addresses the relationships between ecosystems and economic systems in the broadest sense." Ecological economics takes a more comprehensive view of this relationship than do neoclassical economics, resource economics, environmental economics or conventional ecology. Neoclassical economics views the satisfaction of human wants as the ultimate goal of economic activity. Natural resources are not considered to be a constraint on economic growth because of unlimited potential for technological innovation and resource substitution. Environmental economics deals primarily with the management of environmental residuals and the valuation of amenity services provided by the environment. Natural resource economics concentrates on efficient allocation of renewable and exhaustible resources over time.

Conventional ecology is that subfield of biology that deals with the interrelationships between organisms in a particular location. Ecologists focus on the productivity, stability and resilience of natural ecosystems. Humans are just one of many species; however, unlike other species, they have the unique ability to manage the environment for personal gain. As Odum[7] points out, the exploitation of the environment by humans is "increasingly disrupting, even destroying, the biotic components which are necessary for [their] physiological existence." Some ecologists study human-perturbed ecosystems such as agroecosystems.

In terms of the relationship between ecosystems and economic systems, which is at the core of ecological economics, Ehrlich and Ehrlich[8] identify several ways an ecosystem supports the economy, including the following: a) by maintaining atmospheric quality by regulating gas ratios and filtering dust and pollutants; b) by controlling and ameliorating climate through the carbon cycle and effects of vegetation in stimulating local and regional rainfall; c) by regulating freshwater supplies and controlling flooding; d) by generating and maintaining soils through the deposition of organic matter and the relationships between plant roots and mycorrhizal fungi; e) by disposing of residuals, including domestic sewage and by-products from industrial and agricultural production, and cycling of nutrients; f) by controlling pests and diseases; and g) by pollinating crops and useful wild plant species by insects, bats, hummingbirds and other pollinators.[9]

Ecological economics takes the holistic view that humans, culture and biological systems co-evolve. Co-evolution implies that the cultural objective of maximizing short-term economic output be made subservient to long-term biological constraints on economic activity. Constraints include the capacity of natural systems to assimilate residuals, ethical concerns for future generations (intergenerational equity), and the survival of nonhuman species (biodiversity). Ecological economics views the economy as a subsystem of a larger finite and nongrowing ecosystem. While neoclassical economics has a human-centered (anthropocentric) perspective of the economy, ecological economics takes a biologically based (biocentric) or ecologically based (ecocentric) view.

A distinguishing feature of ecological economics is the purposely conservative assumption that technological innovation does not eliminate resource constraints to economic growth. This assumption is consistent with a *no-regrets public policy* for dealing with specific environmental problems. Consider a no-regrets policy for global warming. The policy requires development of cost-effective technologies and

4. Economy and Environment

pricing policies to reduce global emissions of carbon dioxide. Global warming is attributed primarily to emissions of carbon dioxide. If the policy is carried out and the threat is real, then the severity of the threat is reduced. There is no regret because the policy reduced the likelihood of global warming. If the policy is not pursued and the threat is real, then action could have reduced the threat. There is a regret for not taking action. Of course, a no-regrets policy has to specify the types of technologies and pricing policies to be undertaken. The conservative technology assumption of ecological economics is quite different from the optimistic assumption of neoclassical economics that technological innovation is sufficient to ensure that natural resources are not a constraint on economic growth.

A major goal of ecological economics is to determine ways to maintain long-term sustainability of the integrated ecological–economic system. Achievement of this goal requires establishing the optimum scale of the economy relative to the ecosystem and implementing policies and institutions that maintain the optimum scale. Daly[10] defines optimal scale in terms of the physical size of the total capital stock, which includes people and their physical artifacts. The latter includes service-yielding capital assets such as automobiles, buildings and natural resources. Total capital stock is directly proportional to population growth and the accumulation of artifacts (inflow) and inversely proportional to death of people and depreciation of artifacts (outflow). Maintaining the optimal scale of the economy requires balancing inflow and outflow of matter–energy in the economy.

Choice of optimal scale is subjective and difficult because it entails social, ethical and moral considerations. Daly opts for a scale that achieves a sufficient level of wealth to be sustainable in the long run. Sustainability requires the rate of throughput flow (depletion of natural resources and pollution) to be within the assimilative and regenerative capacities of the ecosystem. Because throughput flow represents a cost, it should be minimized subject to the constraint that the optimal scale of the economy be maintained.

Ecological economic theory differs from neoclassical economic theory in two respects. First, neoclassical economic theory is not concerned with the optimum scale of the economy. The only place neoclassical theory addresses optimum scale is in terms of the optimal scale of the physical facilities chosen by a firm. Second, neoclassical macroeconomics adopts the overall goal of maximizing gross national product (GNP), which is consistent with the philosophy that more is best. Maximizing GNP implies maximizing the flow of matter and energy through the economy. This objective contradicts the goal of ecological economics, which is to minimize the cost of the throughput flow of matter and energy needed to sustain the optimum scale of the economy. Ecological economics embraces the philosophy that enough is best.

The economy does not automatically achieve an optimum scale. As Daly[11] points out "... just as there is nothing in the price system that can identify the best distribution of ownership according to criteria of justice, neither is there anything that allows the price system to determine the best scale of throughput according to ecological criteria of sustainability." Therefore, achieving an optimal scale requires that specific actions and policies be undertaken. Daly proposes three policies or institutions for this purpose: stabilizing the population and stock of physical artifacts; keeping throughput flow below ecological limits; and controlling the degree of inequality in the distribution of services among people.

Population can be stabilized by transferable birth quotas (as first suggested by Boulding), raising educational and income levels of women in developing countries (which reduces fertility rates), family planning, and providing financial disincentives for increasing family size. Keeping throughput of matter and energy below ecological limits can be achieved by imposing severance taxes on, and/or auctioning depletion quotas for, extraction of exhaustible resources; imposing taxes on renewable resources that take effect when the rate of use exceeds the rate of regeneration; setting harvest quotas or limitations on exploitation of renewable resources; limiting disposal of residuals and potential pollutants; taxing emissions of potential pollutants; protecting critical habitats of rare and endangered species; and others. Inequities in the distribution of services among people (distributive justice) can be alleviated by tax policies designed to redistribute income (already done in many countries) and international aid from developed to developing countries.

Ecological economics is similar to the material balances approach. Both approaches account for the two major linkages between the economy and the environment; namely, the environment supplies the economy with productive natural and environmental resources and amenities, and the economy generates residuals that enter the environment. Material balances and ecological economics approaches are compatible with the first and second laws of thermodynamics. There are two fundamental differences between the two approaches. First, ecological economics views the economy as a subsystem that is capable of exceeding natural and environmental resource capacities. Ecological economics supports actions and policies to keep the economic subsystem within the biophysical limits imposed by the parent ecosystem. Second, ecological economics is concerned about the optimal scale of the economy and development of natural and environmental resource accounts that trace the adverse effects of economic activity on natural and environmental resources. Neither of these issues is addressed in the material balances approach.

Sustainable Development

The *sustainable development* approach attempts to balance the cornucopian view that continued economic growth is essential and possible with the neo-Malthusian view of ecological limits to economic growth. The cornucopian element recognizes that economic growth is needed to increase per capita household incomes and living standards in the face of population growth. Economic growth appears to be essential in developing countries because of their generally low incomes, substandard living conditions and rapid population growth. The neo-Malthusian element of sustainable development acknowledges that economic growth typically results in the degradation of natural and environmental resources, which, if unchecked, reduces long-term prospects for economic development. Because developed countries account for a disproportionate share of natural resource use and environmental degradation, global sustainable development addresses sharing of benefits and costs of reducing global environmental impacts between developed and developing countries.

Sustainable development originated in 1968 with the Paris Biosphere Conference and the Conference on the Ecological Aspects of International Development held in Washington, D.C.[12] Both conferences pointed out that policy planning and development generally ignore environmental impacts because of lack of interest and/or limited financial and technical resources. Sustainable development is an integral part of the World Conservation Strategy, which promotes the "maintenance of essential ecological processes and life-support systems, the preservation of genetic diversity, and the sustainable utilization of species and ecosystems."[13] Economists have criticized the definitions and objectives of the World Conservation Strategy for being impractical and for not paying attention to tradeoffs between economic and conservation objectives.[14]

The concept of sustainable development achieved prominence during the United Nations Conference on the Human Environment held in Stockholm in 1972. It was firmly established as the focal point for international development by the United Nations Conference on Environment and Development held in Rio de Janeiro in 1992. The most often quoted definition of sustainable development appears in the Brundtland Commission report:[15] "Sustainable development is development that meets the needs of the present without compromising the ability of future generations to meet their own needs." The definition implies a balancing of resource use between current and future generations. The Brundtland report indicates that sustainable development requires keeping consumption levels within ecological bounds, ensuring increased production potential and equity, protecting life-support systems, using nonrenewable resources at rates below the limits of regeneration and regrowth, depleting exhaustible resources at rates that minimize foreclosure of future options, and minimizing adverse impacts on environmental quality. The Brundtland Commission's definition of sustainable development has been endorsed by the World Bank.

While the Brundtland Commission's definition of sustainable development has broad-based appeal, its implementation is no simple matter. For example, what is meant by *meeting the needs of the current generation*? How is it to be determined whether the ability of future generations to meet their own needs is being compromised? These are difficult questions.

Pearce et al.[16] believe that sustainable development should bolster indicators of economic development over time. Typical economic development indicators include rising levels of per capita income, higher levels of educational attainment, improved health care and nutrition, a more equitable distribution of income, and growth in personal freedoms. According to Pearce et al.,[17] a necessary condition for sustainable development is that the current level of natural capital stock (defined as the "stock of all environmental and natural resource assets") be maintained. Maintaining the current level is preferable to achieving the economically optimal level of natural capital stock—which is likely to be lower—because of the multiple uses of natural resources, uncertainty and irreversibilities regarding the consequences of natural resource depletion and environmental degradation, and the nonefficiency benefits of natural capital, such as provision of life-support systems.

Clugston[18] states that "sustainable development does not endanger the atmosphere, water, soil, and ecosystems that support life on earth. It is a dynamic process in which resource use, economic policies, technological development, population growth, and institutional structures are in harmony, contributing to both the present and future potential for human progress." He also points out that sustainability implies a deep ethical and moral connection to all of creation.

Clark has identified three obstacles to sustainable development.[19] First, a tragedy of the commons occurs at the international level because individual countries are often unwilling to cooperate for the purpose of alleviating global environmental problems, such as climate change and ozone depletion. While there have been successes in international cooperation, frustrations have been more common. A successful international agreement is the 1984 Montreal Protocol, which established targets for reducing emissions of cholorofluorocarbons (CFCs), the primary chemical agent in ozone depletion. In contrast, the 1992 United Nations Conference on Environment and Development (Earth Summit), which focused on sustainable development, left much to be desired in the area of climate change and biodiversity. The Earth Summit's agreement on climate change did not include specific targets and timetables for reducing carbon dioxide emissions, and the treaty to protect biological diversity was weakened by lack of support from the United States.[20] These events underscore the difficulty of securing international agreement to alleviate global environmental problems.

Second, uncertainty regarding the environmental benefits of reducing pollution stymies the willingness of countries to take appropriate action. President Bush was unwilling to agree to specific reductions in carbon dioxide emissions at the 1992 Earth Summit because the cost to the United States economy was thought to be too high and the benefits of reducing global warming, too uncertain. William Nordhaus estimates the worldwide cost of controlling global warming to be $250 billion per year.[21] According to Flavin and Lenssen,[22] implementing technologies to reduce carbon dioxide emissions in the United States could cost between several hundred billion dollars and $3.6 trillion.

Third, calculating the present value of a resource investment using a high discount rate (as opposed to a low discount rate) could lead to a shift in resource use from the future to the present, provided that it does not render the investment economically infeasible. If this shift occurs, then the welfare of the present generation increases and the welfare of future generations decreases because the resource is used up more quickly. d'Arge[23] argues for "exceedingly low, if not zero," discount rates in cases involving intergenerational choices. Relative to a high discount rate, a low discount rate generally lengthens the time period over which natural resources are used, which benefits future generations. A low discount rate can, however, also increase the economic feasibility and selection of resource investments that have a high initial capital cost and economic benefits spread out over several years, such as development of water resources for hydroelectric power generation. Should this occur, the present generation benefits at the expense of future generations. It is, therefore, difficult to generalize about how different discount rates influence resource use and intergenerational equity.

Howarth and Norgaard[24] suggest that the dilemma regarding intergenerational consequences of different discount rates can be avoided by assigning future generations specific rights to natural and environmental resources. In their scheme, future generations are given a right to specific quantities and qualities of fossil energy, minerals and water resources, as well as environmental resources. This approach is quite similar to the way by which individuals are protected from criminal acts. All individuals, regardless of generation, have been granted rights of protection from thieves and murderers. Levels of protection are not based on the present value of future benefits and costs, but rather on the decision that such protection is a right.

4. Economy and Environment

SUSTAINABLE RESOURCE USE. While there is broad support for sustainable development, there have been few attempts outside of natural resource and environmental accounting (see Chapter 10) to develop an economic framework that incorporates sustainable resource use. This section discusses such a framework and its practical implementation. Sustainability implies a balancing of economic and environmental objectives. The economic objective can include several components, such as income, employment and inflation. The environmental objective reflects the quality of air, soil, water and wildlife resources. Therefore, a multiple objective decision-making framework is well suited for evaluating sustainable resource use.

In a multiple objective framework, resources are devoted to the achievement of economic and/or environmental objectives. Resources are denoted by the vector **x**. The elements of **x** include natural resources, labor and manufactured capital. Because resources are limited, **x** is finite which constrains the attainment of economic and environmental objectives. For a given technology and finite resources, allocating more resources to attainment of the economic (environmental) objective eventually curtails attainment of the environmental (economic) objective.

The relationships between attainment of economic (EC) and environmental (EV) objectives and resource use are as follows:

$$EC = EC(x_{EC}) \text{ and } EV = EV(x_{EV}),$$

where x_{EC} and x_{EV} are the resources allocated to achievement of economic and environmental objectives, respectively, and $x_{EC} + x_{EV} = x$. EC (EV) increases with respect to x_{EC} (x_{EV}).

Given these production relationships, mathematical optimization techniques can be used to determine the efficient allocation of resources to the economic and environmental objectives. Efficient combinations of objectives lie on the efficiency frontier illustrated in Figure 4.3.[25] The frontier is concave (bowed outward) because allocating more resources to achievement of the economic (environmental) objective decreases achievement of the environmental (economic) objective.

The combination of economic (EC) and environmental (EV) objectives chosen by an individual depends on that individual's preferences for the two objectives. The term *individual* is used here in a generic sense. It encompasses a wide range of economic agents including businesses, consumers, government and public interest groups. The relevant point here is not who the individual is, but rather that individuals have different preferences for objectives. Suppose the satisfaction that individual i and individual j receive from the economic and environmental objectives can be represented by the following utility functions:

$$U_i = U_i[EC(x), EV(x)]$$

$$U_j = U_j[EC(x), EV(x)].$$

Corresponding to each utility function is an indifference map. Superimposing the indifference maps for individual's i and j on the efficiency frontier, and locating the tangency between the efficiency frontier and the highest indifference curve for each individual, give the equilibria shown in Figure 4.4. Individual i selects EC_i and EV_i and individual j selects EC_j and EV_j of the economic and environmental objectives, respectively. In this example, individual i's preferences favor the economic

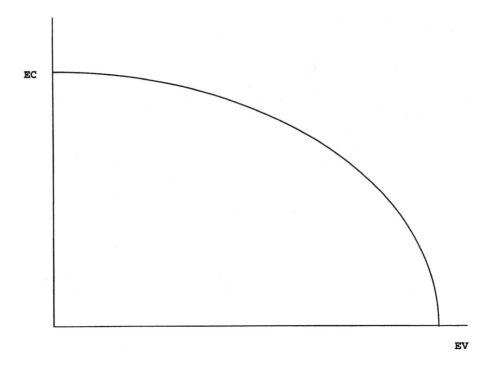

Figure 4.3. Efficiency frontier for economic (EC) and environmental (EV) objectives.

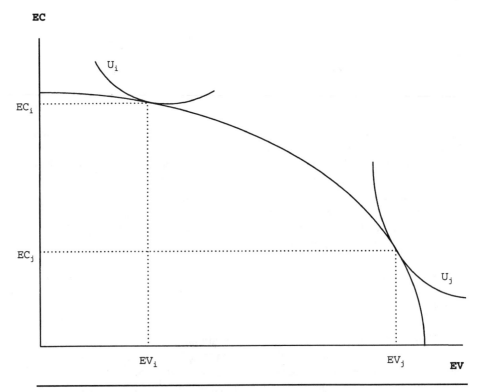

Figure 4.4. Equilibrium choice of objectives for individuals i and j. EC = economic objective; EV = environmental objective; and U = utility.

objective and individual j's preferences favor the environmental objective. Therefore, individual i selects relatively more of the economic objective and individual j selects relatively more of the environmental objective. For each combination of objectives on the efficiency frontier, there is a unique and efficient allocation of resources that achieves that combination.

Sustainable resource use can be considered in the context of the example just described by posing the following question. Are the combinations of objectives and underlying resource allocation that maximize utility for individuals i and j sustainable? The answer requires specifying those combinations of objectives that imply unsustainable use of resources. For example, combinations of economic and environmental objectives that provide high (low) achievement of the economic objective and low (high) achievement of the environmental objective might be considered unsustainable. This would eliminate, for example, combinations of objectives above the upper sustainability contour (C_U) and below the lower sustainability contour (C_L) in Figure 4.5. Determination of the sustainability contours is not a simple matter. Ideally, it should be based on feedback from scientists, policy-makers, government, business and the general public.

For the placement of the sustainability contours shown in Figure 4.5, neither individual i's nor individual j's equilibrium combination of objectives is sustainable. The segment of the efficiency frontier between the sustainability contours (AB) contains the efficient and sustainable combinations of objectives and corresponding

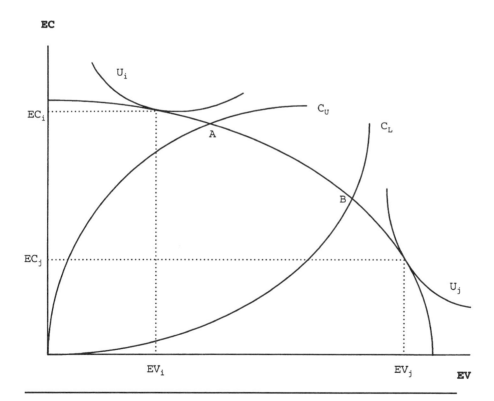

Figure 4.5. Sustainable and efficient combinations of economic (EC) and environmental (EV) objectives.

resource allocations. Therefore, the extent to which sustainable resource use constrains the choice of objectives depends on the nature of preferences and the availability of resources.

Except when segment AB is very short, there is still a need to resolve resource use conflicts between individuals. For the sustainability contours given in Figure 4.5, individual i would prefer to be at A and individual j would prefer to be at B when choices are constrained to the sustainable segment of the efficiency frontier. Because there is still a relatively large discrepancy between combinations A and B, conflict resolution would be needed to determine a compromise combination of objectives.

This conceptual framework provides a basis for implementing sustainable resource use. The three major elements of the framework are the efficiency frontier, the sustainability contours and individuals' preferences. The efficiency frontier can be determined for given technologies and resource endowments. For example, Xu et al.[26] estimated an efficiency frontier for farming systems that achieve three objectives: maximum profit, minimum soil erosion and minimum nitrate leaching through the soil. Determination of the sustainability contours is not a simple matter because it involves consideration of the long-term economic and environmental impacts of different patterns of resource use. Concepts being applied in natural resource and environmental accounting, such as adjustments to GNP for depletion of exhaustible resources (see Chapter 10), would be useful in evaluating such consequences. Finally, preferences for the multiple objectives would ultimately determine which of the sustainable and efficient combinations of objectives and underlying resource allocations would be preferred by different individuals.

Summary

The traditional circular flow model of the economy does not account for biophysical limits to economic activity. Extensions of the circular flow model based on material balances and ecological economics offer a more integrated view of the economy and the environment. The material balances model focuses on the depletion of exhaustible resources and exploitation of renewable resources in economic production and consumption, as well as residuals and pollution generated by natural resource use, production and consumption. Ecological economics views the economy as a subsystem of a finite and nongrowing ecosystem, which implies that economic growth is constrained by the capacity of natural resources to support economic activity and of environmental resources to assimilate residuals. Ecological economics is concerned about the optimal scale of the economy and accounting for resource depletion and degradation in measuring economic progress.

Sustainable development attempts to balance the economic benefits with the resource and environmental costs of growth. It is achieved when the needs of the current generation are met without compromising the ability of future generations to meet their own needs. Sustainable development requires keeping consumption levels within ecological limits, ensuring increased production potential and equity, protecting life-support systems, using exhaustible resources at rates below the limits of regeneration and regrowth, depleting nonrenewable resources at rates that minimize foreclosure of future options, and minimizing adverse impacts on environmental quality. Achievement of sustainable development is often thwarted by discounting, a procedure that reduces the value of future benefits and costs from the development and use of natural and environmental resources. An alternative to discounting is to give future generations rights to a minimum quantity and quality of natural and environmental resources.

Questions for Discussion

1. The circular flow model of an economy is suited to conditions in a frontier economy, which has a relatively small population, primitive technology and abundant supplies of natural resources. The model is less appropriate for an industrial economy, where population and resource use are high relative to ecosystems, and technologies have a greater capacity to degrade natural and environmental resources. Are you in agreement or disagreement with this statement? Explain.

2. Would someone who embraces the concept of ecological economics be content with the material balances approach? Explain.

3. Are the following activities sustainable: a) imposing a tax on the sale of petroleum products and using the tax revenue to develop a substitute source of energy and b) harvesting timber at a rate that exceeds new tree planting and natural regrowth? Explain.

Further Readings

Costanza, R. and H. E. Daly. 1987. "Toward an Ecological Economics." *Ecological Modeling* 38:1–7.

Dorfman, Robert and Nancy S. Dorfman, eds. 1977. *Economics of the Environment*, 2nd ed. New York: W. W. Norton & Co.

Kneese, Allen V., Robert U. Ayers and Ralph C. d'Arge. 1970. *Economics and the Environment: A Materials Balance Approach*. Washington, D.C.: Resources for the Future.

Kneese, Allen V. 1977. *Economics and the Environment*. New York: Penguin Books.

Olson, Richard K., ed. 1992. *Integrating Sustainable Agriculture, Ecology, and Environmental Policy*. Binghamton, New York: Food Products Press.

Pearce, David W. and R. Kerry Turner. 1990. "The Circular Economy." Chapter 2 in *Economics of Natural Resources and the Environment*. Baltimore, Maryland: The Johns Hopkins University Press, pp. 29–42.

Notes

1. A. N. Whitehead. *Process and Reality* (New York: Harper, 1929).

2. A. Myrick Freeman III, Robert H. Havemen and Allen V. Kneese, *The Economics of Environmental Policy* (New York: John Wiley & Sons, 1973), pp. 11–16.

3. Nicholas Georgescu-Roegen, *The Entropy Law and the Economic Process* (Cambridge, Massachusetts: Harvard University Press, 1970).

4. Kenneth Boulding, "The Economics of Coming Spaceship Earth," in *Environmental Quality in a Growing Economy,* Henry Jarrett, ed. (Baltimore, Maryland: The Johns Hopkins University Press, 1966).

5. Herman E. Daly, "Entropy, Growth and the Political Economy of Scarcity," in *Scarcity and Growth Reconsidered,* Kerry Smith, ed. (Baltimore, Maryland: The Johns Hopkins University Press, 1979), p. 76.

6. Robert Costanza, Herman E. Daly and Joy A. Bartholomew, "Goals, Agenda, and Policy Recommendations for Ecological Economics," in *Ecological Economics: The Science and Management of Sustainability,* Robert Costanza, ed. (New York: Columbia University Press, 1991), p. 3.

7. Eugene P. Odum, *Fundamentals of Ecology,* 3rd ed. (Philadelphia, Pennsylvania: W. B. Sanders, 1971), p. 251.

8. P. R. Ehrlich and A. H. Ehrlich, *Extinction: The Causes and Consequences of the Disappearance of Species* (New York: Random House, 1981).

9. Reed F. Noos and Allen Y. Cooperrider, *Saving Nature's Legacy: Protecting and Restoring Biodiversity* (Washington, D.C.: Island Press, 1994), p. 21.

10. Herman Daly, *Steady-State Economics,* 2d ed. (Washington, D.C.: Island Press, 1991).

11. *Ibid.,* p. 190.

12. Lynton K. Caldwell, "Political Aspects of Ecologically Sustainable Development," *Environmental Conservation* 2(1984):299–308.

13. International Union for the Conservation of Nature (IUCN), *World Conservation Strategy: Living Resource Conservation for Sustainable Development* (Gland, Switzerland: IUCN-NEP-WWF, 1980).

14. Edward Barbier, *Economics, Natural-Resource Scarcity and Development: Conventional and Alternative Views* (London: Earthscan Publications, Ltd., 1989).

15. World Commission on Environment and Development, *Our Common Future* (New York: Oxford University Press, 1987), p. 43.

16. David Pearce, Edward Barbier and Anil Markandya, *Sustainable Development: Economics and Environment in the Third World* (London: Earthscan Publications, 1990), pp. 2–3.

17. *Ibid.,* p. 1.

18. Richard M. Clugston, "A Turning Point," *Earth Ethics* (Washington, D.C.: Center for Respect of Life and Environment, Winter 1992), p. 5.

19. Colin Clark, "Economic Biases Against Sustainable Development," *Ecological Economics: The Science and Management of Sustainability,* Robert Costanza, ed. (New York: Columbia University Press, 1991), pp. 319–330.

20. Robert W. Fri, "Environment and Development: The Next Step," *Resources* (Washington, D.C.: Resources for the Future, Winter 1993), pp. 16–18.

21. As quoted in Christopher Flavin and Nicholas Lenssen, "Saving the Climate Saves Money," *World Watch* (Washington, D.C.: Worldwatch Institute, November/December 1990), pp. 26–33.

22. Flaven and Lenssen (1990).

23. Ralph d'Arge, "Marking Time with CERCLA: Assessing the Effect of Time on Damages from Hazardous Wastes," chapter 12 in *Valuing Natural Assets: The Economics of Natural Resource Damage Assessment,* Raymond J. Kopp and V. Kerry Smith, eds. (Washington, D.C.: Resources for the Future, 1993), p. 248–263.

24. Richard B. Howarth and Richard B. Norgaard, "Intergenerational Resource Rights, Efficiency, and Social Optimality," *Land Economics* 66(1990):1–11.

25. V. Changkong and Y. Y. Haimes, "The Surrogate Worth Trade-off Method and Its Extensions," chapter 8 in *Multiobjective Decision Making: Theory and Methodology* (New

York: Elsevier-North Holland, 1983), pp. 352–382; J. L. Cohon, *Multiobjective Programming and Planning* (New York: Academic Press, 1978); J. L. Cohon and D. H. Marks, "Multiobjective Screening Models and Water Resources Investment," *Water Resources Research* 9:(1993):826–836; and Y. Y. Haimes and W. A. Hall, "Multiobjectives in Water Resource Systems Analysis: The Surrogate Worth Tradeoff Method," *Water Resources Bulletin* 10(1974):615–624.

26. Feng Xu, Tony Prato and Jian C. Ma, "A Farm-Level Case Study of Sustainable Agricultural Production," *Journal of Soil and Water Conservation* 50(1995):39–44.

CHAPTER 5

Property Rights and Externalities

A resource whose rights are unassigned is likely to be abused.

—RICHARD L. STROUP AND JOHN A. BADEN, 1983

A well-known conclusion of welfare economics is that a purely competitive market economy automatically achieves an efficient use of resources. Where markets do not exist for resources, the markets that do exist are incomplete, or externalities are present, and the use of natural and environmental resources is not socially efficient. These inefficiencies can take the form of excessive depletion of exhaustible resources, overexploitation of renewable resources, and environmental pollution, all of which decrease economic welfare. When this occurs, the economy experiences a loss in economic welfare. This chapter considers the distinction between a *centralized economy* and a *market economy* and explains the role and importance of *efficient property rights* and *transaction costs* in achieving an efficient use of resources. Causes and consequences of *market failures,* particularly *externalities,* are discussed. Finally, the pros and cons of *government intervention* to alleviate externalities are discussed.

Markets and Efficient Property Rights

MARKETS. While it is possible to achieve efficient resource use without reliance on markets, the informational and computational requirements are very high. An alternative to a market economy is a centrally planned economy. A *centrally planned economy* is a form of economic organization in which decisions regarding what to produce, how much to produce, where to produce, how to distribute production, and the level of resource and commodity prices are determined by government committees. Efficient use of resources in a centrally planned economy requires information on input requirements and output possibilities of alternative production technologies for all locations and points in time; consumers' willingness to pay for all commodities over time and space; and transportation, marketing and distribution requirements and costs. Based on this information, the government committees have to determine the production, consumption and prices for all resources and commodities for all regions and time periods. Therefore, a centrally planned economy requires the orchestration of, literally, thousands of economic decisions.

Purely competitive markets automatically achieve efficient resource use through the independent profit-maximizing behavior of firms and the independent utility-maximizing behavior of households, provided there exists an efficient set of property rights. *Property rights* are institutional rules that govern and facilitate the use and exchange of resources and commodities. Efficient property rights are a prerequisite for market transactions.

Market equilibrium prices and quantities are influenced by the specification of property rights. This is illustrated in Figure 5.1, which shows the market demand and supply curves for an agricultural crop under pure competition. Under *pure competition* a) there is a large number of producers and consumers; b) no one producer or consumer can influence the price received or paid for the product; and c) there is no uncertainty regarding production technologies, consumer preferences and prices. In Figure 5.1, D is the market demand curve. It has a negative slope because buyers are willing to purchase additional units of the crop as the price of the crop decreases, other things equal. Marginal cost (MC) is the market supply curve that is the sum of the marginal cost curves above average variable costs for all firms producing the crop. It is positively sloped because sellers are willing to sell more of the crop as the price increases. The profit maximizing level of crop production is Q_r where $p_r = MC$. Q_r is also known as the privately efficient level of production.

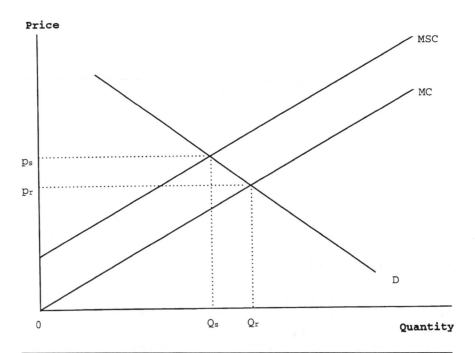

Figure 5.1. Privately efficient crop production (Q_r) and price (p_r), and socially efficient crop production (Q_s) and price (p_s), in a purely competitive market. D = demand; MC = marginal cost; and MSC = marginal social cost.

In selecting the profit maximizing output, firms typically do not consider environmental damages caused by production. For example, runoff from agricultural fields carries sediment, fertilizers and pesticides to nearby streams and lakes, which can result in water pollution. Marginal cost excludes damages caused by water pollution. Adding the marginal environmental damages from crop production to MC gives the marginal social cost (MSC) of crop production. The socially efficient level of crop production is Q_s where p_s = MSC.

When farmers have a right to pollute water, they ignore the environmental damages from crop production and select the privately efficient level of crop production, which is Q_r. Because MSC > MC, $Q_s < Q_r$ and $p_s > p_r$, production and pollution are higher and prices are lower when producers have the right to pollute water. When producers do not have the right to pollute water, society has the right to take action to reduce pollution damages. Suppose this action is in the form of a tax on production. Imposition of the tax causes the MC curve to shift upward by the amount of the tax. An increase in MC causes the profit maximizing production level to decrease. If the tax equals the difference between MSC and MC, then the market achieves equilibrium where p = MSC and the privately efficient level of production equals the socially efficient level of production.

EFFICIENT PROPERTY RIGHTS. Property rights are efficient if they satisfy four basic properties: *ownership, specificity, transferability* and *enforceability*. Ownership is the legal mechanism for conveying property rights to a resource. It determines who has the legal right to use a resource. In cases where resources are privately owned, ownership is secured by payment in an amount that is mutually agreeable to buyer and seller. Payment can be in the form of legal tender (currency) or other resources (barter). Several types of ownership are possible. At one extreme is exclusive ownership, which is the most restrictive form of ownership. Exclusive ownership of a resource requires that all benefits and costs associated with the ownership and use of the resource be borne by the owner. Suppose odor from a paper mill pollutes the air in a nearby community. If the cost of air pollution is not considered by the paper mill owner, then exclusivity is violated.

Exclusive ownership allows the resource owner to let another party use the resource. For example, an absentee landowner can lease the farming rights to his or her land to another party in return for a share of the crop or a flat fee. Leasing of land and other resources involves certain terms and conditions. Violation of these terms and conditions can result in remedial action, including termination of the lease. At the other extreme of the ownership spectrum is *res nullius,* which means that the property belongs to no one. While exclusive ownership allows the resource to be bought and sold, *res nullius* prevents it. Another form of ownership is *res communis,* which refers to common or communal property that is owned and/or managed by a group of kindred individuals, such as a tribe or clan.

Specificity of property rights refers to the bundle of rights that apply to a particular property. It determines what can and cannot be done with the resource. Consider what happens when the cattle owned by a rancher graze on a neighbor's land, causing damages to the land. The neighbor's property rights are being violated if the rancher does not have permission to graze cattle on the neighbor's land. Under this specification of property rights (lack of permission to graze), the rancher is legally required to keep the cattle off the neighbor's land. If the neighbor's land is designated as open range, however, then the rancher has a right to graze cattle on the neighbor's land. Under this specification of property rights, the only feasible way to keep the cattle off the neighbor's land is for the neighbor to install fencing, which is costly. If cattle grazing is not covered by a property right, then a conflict is likely to arise between the rancher and the neighbor. Resource conflicts can be resolved when the rights of the rancher and the neighbor are completely specified.

An efficient property right is *transferable,* which allows resources to be allocated to their highest valued use. For example, in an effort to reduce erosion rates on highly erodible land, the United States government created the Conservation Reserve Program (CRP) in 1985. The CRP allows the government to purchase the cropping rights on highly erodible land for a period of 10 years. Landowners who participate in the program relinquish cropping rights in return for an annual rental payment. The CRP was possible because landowners could transfer cropping rights for specific parcels of land to the government. If the cropping rights to land could not be transferred independently of other rights, then the only way the federal government could have achieved the benefits of the CRP would have been to buy the land outright or to prohibit agricultural production on highly erodible land. The former would be very expensive and cumbersome and the latter very unpopular. Transferability of one or more elements in a bundle of property rights makes possible the leasing of private land, creation of utility easements and the existence of restrictive covenants.

Ownership, specificity and transferability of property rights are of limited value in achieving efficient resource use without enforceability. *Enforceability* requires that the property right be enforceable and enforced when there is a violation. A right that is not enforceable is of little value. Consider eutrophication of a lake, which occurs when some of the phosphorus in livestock manure applied to the land surrounding the lake washes into the lake. Even if a phosphorus loading standard was established for the lake (phosphorus loads below the standard do not cause eutrophication), it would be very difficult to enforce the standard without detailed assessments of manure application rates and phosphorus transport to the lake.

In contrast, it is relatively easy to determine the sources of phosphorus loading to the lake and to enforce the standard when phosphorus enters the lake through pipelines connected to industrial facilities. In both cases, enforceability requires that water quality in the lake be monitored for violation of the standard and that sources contributing to the violation be identified. Enforcement requires that appropriate legal action be taken when a violation occurs. For example, offenders (sources violating the standard) could be required to reduce their phosphorus loading to the lake or pay a fine.

Even when enforceability is feasible, violations can occur. Two types of violations are possible: unintentional and intentional. An unintentional violation is one that occurs even though best management practices are being followed. Phosphorus pollution of the lake by nearby farms is likely to occur when there is a heavy rain on the day following application of manure. This type of pollution is unintentional. Pollution of the lake by industrial sources would be intentional when the industry knowingly releases an amount of phosphorus to the lake that exceeds the standard. Industrial sources might risk paying the fine when the expected penalty (probability of a being caught times the amount of the fine) is less than the expected benefit of exceeding the standard (reduction in phosphorus disposal costs).

5. Property Rights and Externalities

Transaction Costs

While establishment of efficient property rights is a necessary condition for efficient resource use under pure competition, it is not costless. The costs of achieving efficient property rights are called *transaction costs*. Transaction costs are usually small when only a few individuals are involved and the conflict is minor. However, transaction costs can be high when many parties are involved. Consider the tremendous social, economic and political cost incurred by countries in the former Soviet Union when socialism was renounced in favor of capitalism. The high initial cost of establishing a market-based, capitalistic system was essentially a front-end transaction cost of establishing efficient property rights. In addition, market-based economies incur a perennial cost of maintaining the property rights system once it is established. Perennial transaction costs include the cost of deciding which rights will be determined by market versus nonmarket (political) forces, resolving conflicts in resource use, and enforcing property rights.

The costs of deciding which rights are determined by market versus nonmarket forces or a mixture of the two is illustrated for health care in the United States. Historically, access to health care in the United States is determined by market forces. Persons who have the opportunity and financial means to enroll in private health care programs have greater access to health care than do those who cannot afford such programs. As with all market-based solutions, those who desire and can afford health care purchase it. Others either do without or make do with lower quality health-care services. An alternative to private health care is publicly supported health care, sometimes referred to as national health care or socialized medicine. Under national health care, the federal government has a much greater role in determining health care rights for citizens and in controlling access to health care providers. Property rights have been more difficult to establish in health care than in more traditional areas such as national defense, space exploration, social security, national parks and wilderness preservation.

Resolving resource conflicts constitutes a major perennial transaction cost. Examples of resource conflicts in the United States include competition between agricultural, energy and urban water uses; access to and/or use of public lands for grazing, mining, timber harvesting, water supplies, recreation and endangered species; encroachment of prime agricultural areas by urban development; pollution of air, soil and water by agricultural, urban and industrial activities; and others. On a global scale, resource conflicts are caused by global warming, deforestation, ozone depletion and loss of biological diversity. Finally, enforcement of property rights entails major costs. Conflicts exist at all political levels (local, state, regional, national and international) and are increasing due to development, population and environmental pressures.

Do the benefits of establishing efficient property rights outweigh the costs? Even in cases where market allocation of resources is desirable, it may not be practical. Returning to the eutrophication example, phosphorus loading to the lake can originate from any of the land that drains into the lakes. This makes it difficult to pinpoint each land parcel's phosphate loading to the lake. Under these conditions, it would be very time-consuming and costly for landowners to negotiate efficient property rights in clean water and to enforce those rights. Providing landowners with technical and financial assistance for best management practices for manure disposal is likely to be a more practical and cost effective than a market solution.

Market Failures

There are three conditions under which operation of free or incomplete markets fails to achieve efficient resource use:

1. The market is not purely competitive.
2. The resource is a common property or open access resource.
3. There are externalities.

Each of these conditions constitutes a *market failure*. The failure does not mean there is something morally or ethically wrong with the market. Rather, it implies that the prices generated in the market do not provide firms and households with the incentives needed to achieve socially efficient resource use. This section considers market failures due to imperfect competition and common property or open access resources (conditions 1 and 2). The next section considers market failures due to externalities.

A market failure occurs when the market for a resource is not purely competitive. Suppose there is only one buyer of a resource (monopsony) instead of many independent buyers (pure competition). The demand curve for the commodity produced using the resource (D), the marginal cost of production for a purely competitive industry (MC_c), and the marginal cost of production for a monopsonist (MC_m) are depicted in Figure 5.2. Under pure competition, individual firms cannot influence the price paid for the resource. Each firm can purchase as many units of the re-

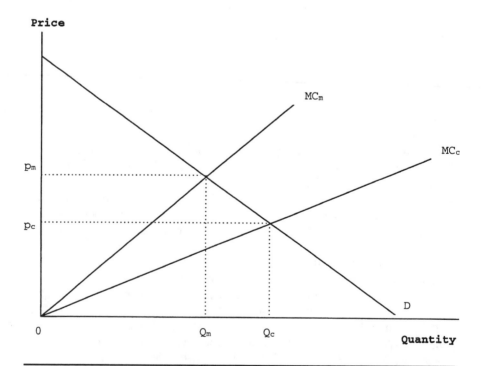

Figure 5.2. Market failure from monopsony. D = demand; and MC = marginal cost.

source as needed at the market price. Therefore, the marginal cost of production under pure competition (MC_c) is determined by adding up the marginal cost curves for all firms in the industry.

When there is only one buyer of the resource, namely, monopsony, additional units of the resource can only be purchased by paying a higher price. The higher price applies not only to the last unit purchased, but to all units purchased. Therefore, $MC_m > MC_c$, which causes the equilibrium price to be higher ($p_m > p_c$), and the equilibrium quantity to be lower ($Q_m < Q_c$), with monopsony than with pure competition. Resource use is not efficient with monopsony. Inefficiencies caused by monopsony and other forms of imperfect competition reduce economic welfare.

COMMON PROPERTY AND OPEN ACCESS RESOURCES. Natural resources can be managed as *common property* or *open access* resources. *Common property resources* are resources that are owned in common and managed for a common purpose. Owners have exclusive rights to the property but cannot exclude one another from using it. There may or may not be restrictions on how frequently owners may use the resource. If frequency of use is not restricted, as in a city park, then the resource tends to be overexploited, which results in Garrett Hardin's *tragedy of the commons*.[1] If frequency of use by owners is restricted, as with tribal grazing lands in African countries, then overexploitation can be avoided. *Open access resources* are not owned by anyone (*res nullius*). Thus, it is not practical to exclude others from using them, and there is generally no incentive for an individual to limit his or her use of the resource. An example of an open access resource is the ocean fisheries.

Overexploitation of a common property resource is illustrated for cattle grazing on rangeland. The efficient stocking rate (number of cattle per acre) for rangeland owned in common by a group of ranchers is demonstrated in Figure 5.3. Suppose the ranchers do not control the stocking rate selected by each rancher. This means each rancher has complete freedom to select his or her own stocking rate. Define the marginal net private benefit (MNPB) of cattle grazing as the difference between the price and the private marginal cost of production for cattle. Profit for each rancher is maximized by selecting a stocking rate of Q_r, where MNPB = 0. At Q_r, the price of cattle equals the marginal private cost of production.

Suppose that when a certain stocking rate is exceeded, namely, Q_h, increasing the number of cows grazed in the common area reduces the quantity and quality of forage for each cow, which in turn decreases weight gain per cow. In other words, overgrazing occurs. The marginal damage (MD) from overgrazing equals the loss in income from grazing an additional cow when the stocking rate exceeds Q_h. For simplicity, let MD be a linear function of the stocking rate when the rate exceeds Q_h. Because the profit maximizing stocking rate exceeds the threshold rate ($Q_r > Q_h$), there is no incentive for an individual rancher to select a stocking rate below Q_h.

When each rancher independently selects the profit maximizing stocking rate, namely, Q_r, damages to the rangeland are a maximum. This suggests there are potential benefits from joint management of the rangeland. Under joint management, the ranchers agree to a stocking rate that maximizes net social benefit, which equals total profit minus damages from overgrazing, which equals the area between the MNPB curve and MD curve. Consider a stocking rate of Q_s. At this rate, MNPB =

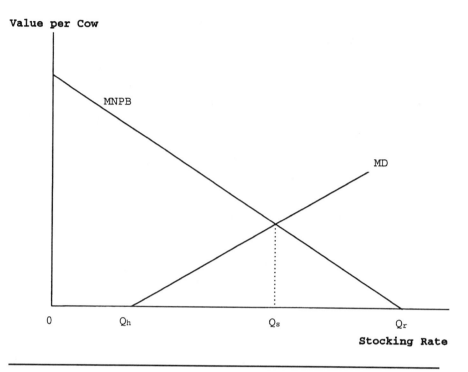

Figure 5.3. Privately (Q_r) and socially (Q_s) optimal stocking rate for a commonly owned grazing area. MD = marginal damage; and MNPB = marginal net private benefit.

MD and net social benefit is greater than it is for any other stocking rate. Because Q_s maximizes net social benefit, it is the socially efficient stocking rate. Notice that the socially efficient stocking rate (Q_s) is less than the privately efficient stocking rate (Q_r). There is no guarantee that the ranchers will agree to limit the stocking rate to Q_h.

Even in cases in which exclusive property rights to a common property resource are granted, overexploitation can occur. For example, ranchers in the western United States are able to secure exclusive grazing rights to specific parcels of public rangeland by paying a grazing fee based on the number of cattle grazed per month. There is considerable evidence that even this controlled grazing arrangement results in overgrazing and significant ecological degradation.

One way to restrict access to a common property or open access resource is for a designated resource manager to ration the resource. In the case of offshore oil, which is a common property resource, the federal or state government leases exploration and development rights to offshore areas. In the United States, competitive lease sales give qualified companies the opportunity to bid for specific tracts of offshore land. If the bid of a qualified company or consortium of companies is accepted, that company or consortium is assigned exclusive development rights to any oil and/or natural gas located on the tract for a specified period of time. Leases can be bought and sold.

This competitive leasing scheme not only establishes efficient property rights for offshore oil, but also limits the total area open for development through a congressionally approved leasing schedule. The schedule indicates when and where lease sales take place and the amount of land offered for lease. In addition to limiting the access to offshore tracts, the leases stipulate the safety and environmental precautions that must be observed in developing each lease tract. Lease terms and conditions are designed to reduce accidents and oil spills.

Fish and wildlife are common property resources whose use is controlled by state or provincial governmental agencies. Consider big game hunting. Management agencies convey hunting rights to a particular species through the issuance of fee licenses that stipulate the terms and conditions for different types of hunting. The license usually restricts the number and/or sex of animals harvested in a given period (day or season), the specific hunting area and the type of weapon. The number of hunters is controlled by limiting the number of licenses offered for sale or the number of permits issued in a hunting area. Some hunting does occur in private game preserves that charge rather high fees for a quality experience. Commercial and recreational fishing licenses operate in a similar manner to publicly provided big game hunting.

In the case of *open access resources,* no one owns the resource and overexploitation occurs. A good example of an overexploited open access resource is the harvesting of whales. Because whale products are very valuable and there are no restrictions on harvest rates, whale harvesting exceeds rates of regeneration. As a result, certain species of whale have been almost hunted to extinction, such as the blue whale in the southern oceans of Antarctica. In an effort to prevent extinction, the International Whaling Commission succeeded in getting a group of nations to agree to voluntary restrictions on whale harvesting. While several nations did not sign the agreement, it did reduce the harvesting of whales.

Another example of overexploitation of an open access resource is air pollution. As long as air emissions in an airshed remain below the capacity of the air to assimilate emissions, air pollution does not occur. Rapid growth in urbanization and industrial development in several cities in the United States, notably Los Angeles, California, and Denver, Colorado, has increased air pollution, which has contributed to health-related problems, ecological and property damages and reduced visibility. Environmental pollution is typically controlled by restricting total emissions and/or taxing emissions. The Clean Air Act of 1970 placed limits on air emissions in so-called nonattainment areas within the United States. These are areas in which concentrations of pollutants exceeded health-based standards established by the United States Environmental Protection Agency (U.S. EPA). The Clean Air Act significantly improved air quality in many nonattainment areas, although several of these areas are still not in compliance with the standards.

The Clean Air Act Amendments of 1990 restricted the amount of sulfur dioxide emissions in an airshed by issuing permits that restricted emissions to an acceptable level. By allowing firms to trade emission permits, the target level of emissions is achieved at least cost. This is the concept of tradable emissions permits. A more complete discussion of methods for controlling environmental pollution is given in Chapter 9. While schemes to improve the management of common property and open access resources improve the efficiency of resource use compared to what it would be in the absence of management, they do not necessarily achieve socially efficient resource use.

PUBLIC GOODS. A *public good* has two characteristics: property rights to the good are not exclusive, and use of the good by one person does not diminish the benefits that the good provides to other persons. The term *public good* is generic and encompasses both natural and environmental resources. A public good is different from a common property resource. Use of a common property resource by one person decreases its availability to other persons. Examples of public goods include national defense, interstate highways, public education and national parks. While these resources could be managed by the private sector, it would lead to a more limited supply of the resource and a higher price. Consequently, net social benefit would be lower than with public management. Society has decided that public goods should be managed for the benefit of the general public rather than for private gain.

Think about public goods, such as wildlife viewing in a public park. If additional viewers do not diminish the value of the experience to existing viewers (congestion does not occur), then the marginal cost of wildlife viewing is very low. Under these conditions, it is inefficient to charge anything other than a very low fee for wildlife viewing. A high fee is inappropriate because it excludes some users without decreasing the cost of providing wildlife viewing. Hence, the efficient price for wildlife viewing is very low. Because a low price would not generate sufficient revenue to cover the costs of providing the viewing, a profit-motivated firm is not likely to be interested in managing the park for wildlife viewing. The park is clearly a public good that needs to be managed by a public agency. Under public management, access to the park would require payment of a very low user fee. The public agency could recoup the bulk of the cost of operating the park from general tax revenue. Revenue generated by the access fee would cover some of the administrative cost of public management.

Externalities

Of the three sources of market failure just described, externalities have received the most attention in natural resource and environmental economics. An externality exists when the activities of an acting party influence the welfare of an affected party and the acting party does not consider how its activities affect the welfare of the affected party. The acting party is the party engaged in the activity responsible for the externality, and the affected party is the party whose welfare is influenced by the externality. The acting and affected party can be a household or a firm. Hence, externalities can take place between firms, between households, and between households and firms. If the acting party engages in an activity for the sole purpose of harming or benefiting the affected party, then the activity does not constitute a true externality.

When an externality is present, the welfare of the affected party is influenced not only by its own activities, but also by the activities of the acting party. Suppose the Cone family plays loud music late at night without concern for how the music affects its neighbors. If the loud music prevents the neighbors from sleeping, then there is an externality. In this case, the Cone family is the acting party and the neighbors are the affected party.

An externality is *relevant* when the affected party wants the acting party to change the activity that causes the externality. The externality is not relevant when the affected party does not care whether the acting party changes the activity causing the externality. If the neighbors do not care whether the Cone family plays loud music at night, then the externality is not relevant. An externality is said to be Pareto relevant when the activity causing a relevant externality can be changed so that the welfare of the affected party can be increased without decreasing the welfare of the acting party. When an externality is Pareto relevant, modifying the activity that causes the externality offers a potential improvement in economic welfare. If lowering the volume of the music does not reduce the welfare of the Cone family, then the neighbors would be better off and the Cone family would be no worse off. The externality is Pareto relevant.

TYPES OF EXTERNALITIES. Not all relevant externalities are of interest from an efficiency standpoint. Of the two types of externalities, *pecuniary externality* and *technological externality,* only the latter adversely affects economic efficiency. A pecuniary externality occurs when a company develops a computer software package that significantly reduces the time required for electronic communication. Other firms who do business with this company derive benefits from the time savings even though they did not bear the cost of developing the software. Because a pecuniary externality only changes the relative prices and financial conditions faced by affected parties, it does not result in an inefficient use of resources. In this regard, a pecuniary externality is no different than other market forces that influence resource prices.

A technological externality affects the level of production or satisfaction achieved by the affected party, which results in inefficient resource use. The playing of loud music by the Cone family is a technological externality. There are two forms of technological externality: *external economies* and *external diseconomies.* An external economy increases the welfare of the affected party, whereas an external diseconomy decreases the welfare of the affected party. The term external means the activity generating the externality is external to the affected party. Activities that do not generate external or third-party effects are not considered externalities. Because of their negative welfare effects, external diseconomies are of greater concern than external economies. Externalities can be classified according to whether there is a conflict between the acting and affected parties and whether they cause inefficient resource use as shown in Table 5.1.

Table 5.1. Classification of externalities

Type of Externality	External Economy	External Diseconomy
Technological	No conflict Inefficiency	Conflict Inefficiency
Pecuniary	No conflict No inefficiency	No conflict No inefficiency

EXAMPLES OF EXTERNALITIES. Consider an external diseconomy in which the acting parties are farmers located in an agricultural watershed and the affected party is a hydroelectric power company located downriver from the watershed. Crop production by farmers causes excessive erosion, which generates sediment. Sediment is carried by runoff to the river, which flows into a reservoir owned by the hydroelectric power company. Some of the sediment becomes trapped in the reservoir. Trapped sediment displaces water, which reduces the water storage capacity of the reservoir and decreases electrical generating capacity. Periodically, the power company has to dredge (remove) the sediment from the reservoir to restore water holding capacity to its original level. The cost of dredging constitutes an offsite damage from soil erosion.

An external diseconomy is created because farmers ignore the offsite sediment damages to the reservoir caused by soil erosion. In terms of Figure 5.1, the private MC of production (MC) is less than the marginal social cost (MSC). If sediment deposition in the reservoir is the only offsite damage caused by crop production, then the difference between MSC and MC equals the marginal sediment damage incurred by the power company. The divergence between social and private marginal costs (MSC – MC) causes privately efficient crop production to exceed socially efficient crop production ($Q_r > Q_s$). Hence, the level of crop production is too high in the presence of an external diseconomy.

Consider an external economy between a farm operated by Mr. Jones (acting party) and a ranch operated by Mr. Davis (affected party). Mr. Jones sprays his cornfield with a herbicide to reduce yield losses from weed infestation. Mr. Davis grazes cattle on a pasture located next to the field. Herbicide use reduces the weeds not only in Mr. Jones's cornfield, but also in Mr. Davis's pasture. As long as Mr. Jones sprays his corn without regard to Mr. Davis's welfare and Mr. Davis is not indifferent to the spraying, there is an external economy. Mr. Davis benefits from improved pasture for his cattle but does not incur any additional cost.

The external economy from herbicide use is illustrated in Figure 5.4. Marginal resource cost and marginal private benefit of herbicide use for Mr. Jones are given by MRC_j and MB_j, respectively. MRC_j is the increase in herbicide application cost from using another unit of the herbicide. Because the market for herbicides is assumed to be purely competitive, MRC_j equals the market price of the herbicide (p). MB_j is the increase in corn production from using another unit of the herbicide. Due to diminishing returns, MB_j decreases as herbicide use increases. Efficient herbicide use for Mr. Jones is Q_r where $p = MB_j$. Because Mr. Jones selects the level of herbicide use based on weed control in corn production, the weed control in Mr. Davis's pasture is not likely to be as effective as in Mr. Jones's corn. For this reason, the marginal private benefit of herbicide use for Mr. Davis is below the marginal private benefit of herbicide use for Mr. Jones ($MB_d < MB_j$).

This particular external diseconomy is *undepletable*. This means that use of the herbicide by Mr. Jones does not deplete (diminish) the marginal benefit of the herbicide to Mr. Davis. In other words, both individuals benefit from use of the herbicide, although Mr. Jones receives a higher benefit than does Mr. Davis. Marginal social benefit (MSB) for an undepletable externality is determined by summing the marginal benefits to Mr. Jones and Mr. Davis. For example, when herbicide use is Q in Figure 5.4, $MSB = MB_1 + MB_2$. MB_1 is the marginal benefit to Mr. Davis and MB_2 is the marginal benefit to Mr. Jones at Q. The socially efficient use of the her-

5. Property Rights and Externalities

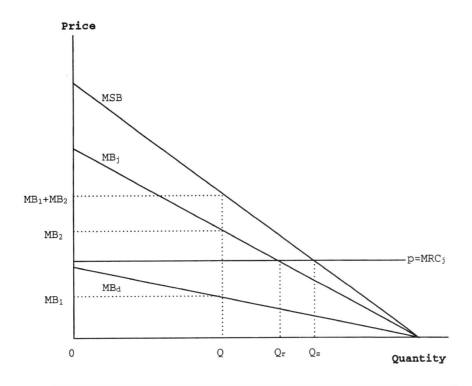

Figure 5.4. External economy from herbicide use. MB = marginal benefit; MRC = marginal resource cost; and MSB = marginal social benefit.

bicide is Q_s where p = MSB. In conclusion, when there is an external economy, the privately efficient use of a resource is less than the socially efficient use ($Q_r < Q_s$).

This example demonstrates that the inefficiency caused by an external economy has the opposite effect of the inefficiency caused by an external diseconomy. The external diseconomy results in overproduction of crops (Figure 5.1), whereas the external economy results in underutilization of the herbicide (Figure 5.4) from a social viewpoint.

Externalities can occur between households. Consider first an external diseconomy. A construction project has been proposed to widen and straighten the road between the city of Argone and the suburban community of Bellvue. Many of the residents of Bellvue use the road to commute to jobs in Argone. The project would improve road safety and reduce commuting time for Bellvue residents. Catlin is a town located between Argone and Bellvue. Residents of Catlin oppose the project because the new road would pass right by their town. This would increase road noise and traffic congestion in Catlin. The proposed road project improves the welfare of Bellvue residents but reduces the welfare of Catlin residents.

An external economy between households exists when the landscape improvements made by the Ryan family enhance the aesthetic values experienced by the next-door neighbor and/or the value of the neighbor's property.

Externalities between households and firms are quite common. Consider a rural town surrounded by large farms. To reduce pesticide costs, farmers opt for aerial application of pesticides. On windy days, some pesticide drifts over the city, increasing the exposure by and, health risk to, households. There is an external diseconomy between farms (acting party) and households (affected party). The culprit is the particular technology used to apply pesticides, not the use of pesticides. Due to the nature of the problem, a solution might be negotiated between the town's health department and farmers. One possibility is to restrict aerial application of pesticides to days when wind velocity is low.

Externalities and Property Rights

An externality results from the absence of efficient property rights. This suggests that a major way to eliminate or reduce externalities is to establish efficient property rights. If property rights can be established, then normal market transactions can be used to achieve efficient resource use. The classical treatise on this subject was developed by Ronald Coase.[2] He argued that acting and affected parties have an incentive to negotiate a reduction in external diseconomies provided:

1. The economy is decentralized, making it possible for both parties to negotiate an agreement freely and without government interference.
2. The cost of negotiating and enforcing an agreement is low.
3. Efficient property rights are established.

Consider applying these conditions to the externality caused when the rancher's cattle graze on a neighbor's property. The first condition is assumed to be satisfied. Because there are only two parties, the second condition is likely to be satisfied. What is the likelihood of the rancher and neighbor negotiating a settlement without the establishment of property rights (third condition)? This question is addressed in terms of the demand and supply curves for reducing an external diseconomy shown in Figure 5.5. D is the demand curve and S is the supply curve for externality reduction. The externality is greatest at zero externality reduction and is completely removed at R_{max}. Suppose that the neighbor is willing to bribe the rancher to keep cattle off his property. If the amount of the bribe decreases with respect to the level of externality reduction, then there are decreasing marginal benefits of externality reduction. Hence, the neighbor is willing to make a lower bribe per cow as the number of cows on his property decreases when the demand curve for externality reduction is downward sloping.

Suppose the rancher is willing to compensate the neighbor for damages from the externality and that the level of compensation per cow decreases as the number of cows on the neighbor's property increases. In other words, the rancher is willing to provide a higher compensation for the first cow than for the second cow, a higher compensation for the second cow than for the third cow, and so forth. Compensation per cow increases as the number of cows removed from the neighbor's property increases or, equivalently, as the externality is reduced. Therefore, the supply curve for externality reduction (S) is upward sloping.

5. Property Rights and Externalities

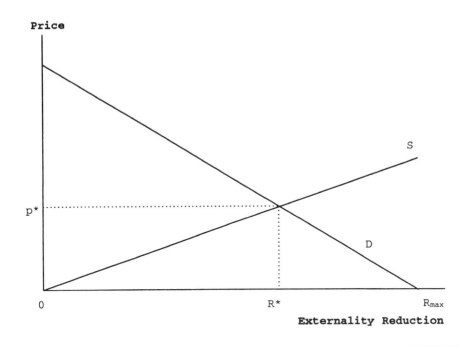

Figure 5.5. Equilibrium price and quantity for external diseconomy.
D = demand; and S = supply.

The efficient quantity and price of externality reduction occur where the demand and supply curves intersect, namely, at R* and p*, respectively. At R*, the bribe that the neighbor is willing to make equals the compensation the rancher is willing to pay. While the first and second conditions make it possible for the acting and affected parties to negotiate a settlement to the externality, negotiation is unlikely unless property rights are assigned. It will be very difficult for the rancher and neighbor to resolve the conflict over cattle grazing unless either the rancher has the right to graze cattle on the neighbor's land or the neighbor has the right to prevent cattle from grazing on his land. When all three conditions are satisfied, the two parties can negotiate an agreement. Even if property rights are established and only a few parties are involved, there needs to be agreement on the damages caused by the externality. Without such agreement, the parties are not likely to reach agreement on the bribe or compensation.

ASSIGNMENT OF PROPERTY RIGHTS. Does it make any difference how property rights are assigned to the parties? Coase showed that the efficient level of externality reduction is achieved regardless of the assignment of property rights, as long as transaction costs and income effects are zero. Transaction costs refer to all the costs of settling externality disputes, including expenses for attorney's fees and time spent negotiating a settlement. Low transaction costs mean that the cost of negotiating a settlement between the acting and affected parties is negligible. Transaction costs are likely to be low in the cattle-grazing example because there are only two parties. In the example of lake eutrophication from livestock manure, however, transaction costs are high because of the potentially large number of acting parties (polluting farms) and affected parties, and the difficulty of tracing phosphorus loading in the lake to manure application on specific farms. When the transaction costs exceed the benefits of reducing the externality, it is not efficient to reduce the externality.

Income effects occur when bribes shift the demand curve and/or compensation shifts the supply curve for externality reduction. When income effects are zero, the demand and supply curves are not affected by the assignment of property rights. For example, consider two property rules. Rule 1 legalizes open grazing (which gives the rancher the right to graze cattle on the neighbor's land), and rule 2 makes open grazing illegal. The appropriate forms of settlement are for the neighbor to bribe the rancher with rule 1 and for the rancher to compensate the neighbor with rule 2. Rule 2 reduces the income of the rancher and increases the income of the neighbor. Rule 1 does the opposite. When income effects are zero, such changes in income do not shift the demand and supply curves.

If the income effect for each party is positive and nonzero, then a lower income for the rancher reduces the amount of compensation offered for a given externality reduction and a higher income for the neighbor increases the amount of the bribe offered for a given externality reduction. Therefore, the demand curve with rule 1 is below the demand curve with rule 2 ($D_1 < D_2$) and the supply curve with rule 1 is above the supply curve with rule 2 ($S_1 > S_2$), as shown in Figure 5.6. Therefore, the equilibrium level of externality reduction is greater with rule 2 than rule 1 ($R_2 > R_1$). Equilibrium prices can be higher or lower with rule 1 than rule 2 depending on the relative magnitude of the income effects for the acting and affected parties. Positive transaction costs cause the difference between the equilibrium levels of externality reduction to be even greater than illustrated in Figure 5.6. Therefore, if transaction costs and/or income effects are not zero, which is likely to be the case, the assignment of property rights to acting and affected parties can influence the level of the external diseconomy and possibly the price of removing it. Similar statements can be made for external economies.

How difficult is it to implement property rules? It depends on the nature of the externality. In the cattle grazing example, it would be relatively easy to establish property rules because the cause and consequences of the external diseconomy are obvious: lack of fencing around the ranch or the neighbor's property allows cattle to graze on the neighbor's land. Because it would be easy to determine when a violation of the rules occurs, enforcement would be straightforward.

Property rules have their share of problems. First, they can generate perverse economic incentives. For example, allowing open grazing gives ranchers an incentive to graze their cattle on someone else's land. This behavior is rational because it reduces the amount of owned or rented land needed to graze a given number of cattle. Unfortunately, it increases the magnitude of the externality. Second, as the number of parties increases, so does the difficulty of reaching agreement. Third, when the number of parties is large, a potential *free rider problem* can arise, especially in the case of environmental externalities such as air and water pollution. A free rider problem exists when some of the parties benefiting from externality reduction do not have to bear any of the cost of achieving the reduction.

5. Property Rights and Externalities   107

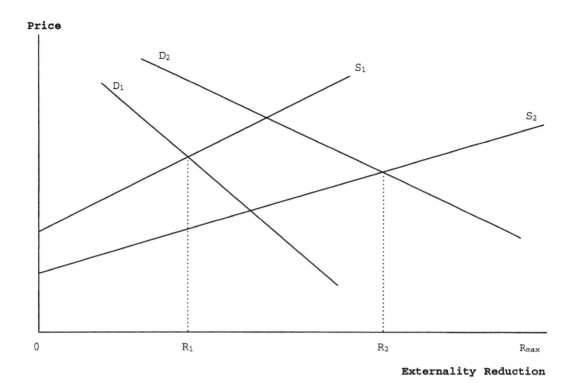

Figure 5.6. Equilibrium level of externality reduction with nonzero income effect for two grazing rights. D = demand; and S = supply.

Fourth, it is difficult to define property rules for open access resources such as air and water because many environmental interests object to giving acting parties the right to pollute or otherwise degrade the environment. Other interests object to giving affected parties the right to an unpolluted environment. Finally, society may be unwilling to establish property rules for certain resources, such as national parks, wildlife refuges and wild and scenic rivers, for fear that a market based on these rules would result in insufficient quantities and excessively high prices of the resource.

GOVERNMENT INTERVENTION. Because of the limitations of property rules, other approaches have been used to reduce external diseconomies, including liability rules, economic incentives and regulations. Some of these approaches have the same limitations as property rules. With a liability rule, the courts establish that acting parties are legally liable for the damages inflicted on affected parties. If affected parties claim damages, then the court determines the appropriate amount of compensation for the damages incurred. Economic incentives, such as taxes and subsidies, can be used to reduce externalities. Finally, regulations can be used to limit the amount of the externality to some administratively determined level. Establishment of property rules and other mechanisms for reducing externalities usually involves *government intervention.* Only the general pros and cons of this overall approach are discussed here. Chapter 9 provides a more detailed economic analysis of government intervention for reducing environmental externalities.

Proponents claim that government intervention is needed to reduce economic losses from external diseconomies. Government intervention has been the primary means of addressing externalities associated with national security, health and safety, social welfare and environmental pollution. Pigou[3] laid the foundation for such intervention when he stated: "It is the clear duty of Government, which is the trustee for unborn generations as well as for its present citizens, to watch over, and if need be, by legislative enactment, to defend the exhaustible natural resources from rash and reckless exploitation."

Mishan[4] supported government intervention for certain types of externalities: "With respect to bodies of land and water, extension of property rights may effectively internalize what would otherwise remain externalities. But the possibilities of protecting the citizen against such common environmental blights as filth, fume, stench, noise, visual distractions, etc., by a market and property rights are too remote to be taken seriously." Government intervention has been supported by the general public, as evidenced by the broad range of social, economic and environmental legislation.

Under the banners of new resource economics, new institutional economics and free market environmentalism, some economists have argued that government intervention to mediate externalities is often cost-ineffective, premature, unnecessary and misguided. Castle[5] pointed out: "Market failure in some abstract sense does not mean that a nonmarket alternative [such as government intervention] will not also fail in the same or in some other abstract sense." The new approach attempts to explain government failure in terms of the relationship between principals and agents and the influence of transaction costs on this relationship. In this approach, politicians and bureaucrats are the agents and citizens are the principals. Efforts to reach agreement (establish a social contract) are thwarted by voter ignorance, imperfect information and special interests, all of which increase the transaction costs of achieving agreement. The higher the transaction costs, the greater the likelihood of government failure. The newer approaches emphasize the formation of markets to internalize externalities by establishing effective property rights and reducing transaction costs.[6]

Summary

Centrally planned economies utilize government committees to allocate resources. Market-based economies can automatically achieve socially efficient resource use through the independent actions of consumers and producers. Markets do not function without efficient property rights. Such rights have four major attributes: exclusivity, specificity, transferability and enforceability. Exclusivity allows the resource owner to exclude others from using the resource. Specificity refers to the bundle of rights associated with a particular property. Transferability allows property rights to be conveyed to another party. Enforceability implies that property right infringements can be determined and violators can be prosecuted. Transaction costs refer to the time and money costs of establishing property rights.

Free markets will not achieve a socially efficient use of resources if a) the market is not purely competitive, b) the resource has unique physical attributes, and c) externalities exist. When monopolistic elements are present in a market, it is not purely competitive. Common property resources and public goods have unique physical attributes that do not permit these resources to be allocated by a market. Access to certain common property resources and open access resources is unrestricted due to a lack of exclusive property rights. Some common property resources, such as offshore oil and big game, have the characteristic that use by one person decreases the availability of the resource to others. The tragedy of the commons refers to the tendency for certain common property and open access resources to be overexploited. Public goods such as national parks, wildlife refuges, and wild and scenic rivers do not have exclusive property rights and, unless congestion occurs, use by one person does not diminish the benefits that the good provides to other persons.

An externality occurs when the actions of an acting party have an adverse or beneficial effect on the welfare of an affected party. When the offending action can be changed so as to increase the welfare of the affected party without decreasing the welfare of the acting party, the externality is Pareto relevant. Unlike pecuniary externalities, technological externalities influence the efficiency of resource use. Technological externalities can be either external economies, which increase the welfare, or external diseconomies, which decrease the welfare of the affected party. One way to alleviate externalities is to establish efficient property rights.

Economist Ronald Coase showed that in a decentralized economy with efficient property rights, it would be mutually advantageous for acting and affected parties to negotiate a reduction in an external diseconomy, provided transaction costs are low. He also showed that the level of externality reduction is independent of how property rules are distributed between the acting and affected parties, provided transaction costs and income effects are zero. Proponents of property rules want to make them more effective by reducing transaction costs. Critics of property rules point to their drawbacks when applied to common property resources and public goods. They support alternative approaches, such as liability rules, economic incentives and regulations, all of which entail significantly more government intervention than property rules.

Questions for Discussion

1. Conversion of eastern Europe from a communist to a capitalist government has involved implementation of market-based incentives for determining resource use. Why have the costs of converting from state-owned to privately owned property been so great?

2. Which of the four attributes of property rights appears to be most critical in terms of resolving externalities? Why?

3. Give examples of an external diseconomy and an external economy not dis-

cussed in this chapter. How might these external effects be resolved?

4. Distinguish between a common property resource, open-access resource and a public good. Illustrate with examples.

5. When is it in society's best interest to eliminate externalities?

Further Readings

Solow, Robert M. 1974. "The Economics of Resources or the Resources of Economics." *American Economic Review* 64:1–14.

Stroup, Richard L. and John A. Baden. 1983. "Property Rights: The Real Issue." *Natural Resources: Bureaucratic Myths and Environmental Management*. San Francisco, California: Pacific Institute for Public Policy Research. pp. 7–27.

Tietenberg, Tom. 1992. "Property Rights, Externalities, and Environmental Problems." Chapter 3 in *Environmental and Natural Resource Economics*, 3rd ed. New York: Harper-Collins Publishers, Inc., pp. 44–71.

Notes

1. Garrett Hardin, "Tragedy of the Commons," *Science*, December 13, 1968.

2. Ronald Coase, "The Problem of Social Cost," in *Economics of the Environment*, 2nd ed. Robert Dorfman and Nancy S. Dorfman, eds. (New York: W.W. Norton & Co., 1977), pp. 142–171.

3. A.C. Pigou, as quoted in J. W. Milliman, "Can People Be Trusted with Natural Resources?" *Land Economics* 38(1962):199–218.

4. E.J. Mishan, "A Reply to Professor Worcester," *Journal of Economic Literature* 10(1972):59–62.

5. Emery N. Castle, "The Market Mechanism, Externalities, and Land Economics," *Journal of Farm Economics* 13(1973):11–14.

6. Steven N.S. Cheung, "The Structure of a Contract and the Theory of Non-Exclusive Resource," *Journal of Law and Economics* 13(1970):49–70; Terry L. Anderson, "The New Resource Economics: Old Ideas and New Applications," *American Journal of Agricultural Economics* 64(1982):928–934; and Terry L. Anderson, "The Market Process and Environmental Amenities," in *Economics and the Environment: A Reconciliation*, Walter E. Block, ed. (Vancouver, British Columbia: The Fraser Institute), pp. 137–157.

CHAPTER 6

Natural Resource Decisions

When we try to pick anything out by itself, we find it hitched to everything else in the universe.

—JOHN MUIR, 1911

Private and public decisions regarding the use and management of natural resources and the environment are influenced by a variety of social, economic, technical and environmental factors. Resource decisions can be arrayed in a hierarchy in which the complexity of the decisions increases as more factors are considered. The physical attributes of a resource influence decisions regarding its use and management. This chapter discusses a *decision hierarchy* for natural resource management, the distinction between *exhaustible* and *renewable* natural resources, *market equilibrium,* and *static efficiency* (efficient resource use when decisions in different time periods are independent), *producer surplus, consumer surplus* and *net social benefit.*

Natural Resource Management

Resource management refers to the decisions made by resource owners, managers, interest groups and policy-makers regarding the rate, timing and method of resource depletion, conservation and management. The four major paradigms, or philosophical approaches, to resource management can be arranged in a decision hierarchy like the one illustrated in Figure 6.1.

Each paradigm corresponds to one of the four models discussed in Chapter 4: namely, circular flow, material balances, ecological economics and sustainable development. Management decisions become more diverse and complex, moving from lower to higher layers in the diagram.

The first layer consists of the simplified circular flow model of the economy. This model concentrates on decisions that govern the exchange of resources, such as land, labor and capital, between households and firms. Purchase and sale of resources by the government and other countries can be considered by adding government and import or export sectors to the model. Resource management decisions in the circular flow model are governed by market prices that are determined by demand and supply conditions. If the demand for a natural resource increases, then

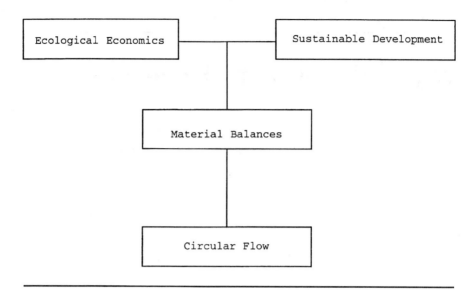

Figure 6.1. Resource decision hierarchy.

market price increases until quantity demanded equals quantity supplied, as illustrated in Figure 6.2. Market price rises from p_1 to p_2 and quantity demanded increases from Q_1 to Q_2 when demand increases from D_1 to D_2. The circular flow model represents market-based decisions quite well. Not all resource decisions are subject to market forces. Water pollution by farmers is not subject to market forces because there is no market for clean water. In some cases, markets are incomplete, as in the case of recreational hunting, which is managed primarily by the public sector. Limited recreational hunting is provided by private game preserves.

The second layer contains the material balances model, which adds three new elements to the circular flow model: consumption of environmental services, disposal of material–energy residuals and assimilative capacity of the environment. Most household, firm and government decisions related to these elements are not governed by market forces. For example, while there are markets for collection and disposal of common household and business refuse, there are no markets, or very limited markets, for disposal of residuals in air and water bodies. In the absence of a mechanism for keeping residuals below assimilative capacities, air and water pollution is likely to occur. The material balances model envisions direct public intervention to reduce environmental pollution. The third layer consists of the ecological economics model and the sustainable development model. All four models consider economic efficiency and equitable distribution of income. The material balances, ecological economics and sustainable development models address the management of environmental pollution. Achieving an optimum scale for an economy relative to the ecosystem is a unique concern of the ecological economics model. Progress in achieving an optimum scale is limited by several factors.

First, the optimum scale is subjective. Once an optimum scale has been selected, however, conventional economics can be used to determine the most efficient way of achieving it. Second, the goal of achieving an optimum scale for the economy is incompatible with economic growth, which is the most widely accepted

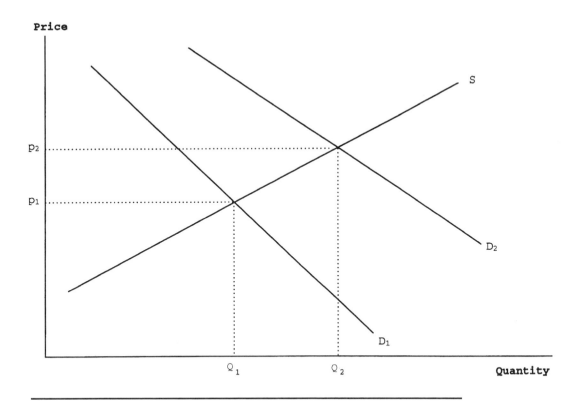

Figure 6.2. Increase in price and quantity demanded of a natural resource due to increase in demand (D). S = supply.

goal of economic development. With unlimited economic growth, the scale of the economy eventually exceeds whatever optimum scale is chosen. Third, the institutional and policy reforms proposed by ecological economists for achieving a balance between the economic subsystem and the ecosystem and an equitable distribution of income are not popular. Examples include government auctioning of depletion quotas for nonrenewable resources, limiting household size, and placing upper limits on household income.[1]

The diversity and complexity of resource management decisions, consequences of making wrong decisions and the likelihood of disagreement and conflict increase moving from the bottom to the top layers of Figure 6.1. It is no big deal when the public decides that Hollywood's latest movie release is a flop. However, it is a big deal when the hole in the atmospheric ozone layer becomes larger. In other words, the stakes generally increase from the bottom to the top of the resource decision hierarchy.

Sustainable development and ecological economics are placed in the same layer of the decision hierarchy because they have many common elements and their level of generality is comparable. Both models focus on interdependencies between the economy and the ecosystem. Sustainable development emphasizes conservation of natural resources and the environment as a means of ensuring long-term economic development. Advocates of sustainable development support the short-term goal of economic growth, especially in developing countries, and the importance of having developed countries bear the burden of reducing environmental degradation and financing environmental protection in developing countries. Since ecological economics considers the ecosystem to be the ultimate constraint on economic growth, unlimited economic growth is ruled out. Sustainable development is a philosophy about how development should proceed, whereas ecological economics is a transboundary discipline interested in characterizing the linkages between the economy and the environment.

All four paradigms in the resource decision hierarchy are useful in understanding and resolving issues related to the development and/or use of natural and environmental resources. The philosophical basis for this textbook is that a blending of the four paradigms offers a much stronger foundation for evaluating and resolving natural and environmental resource issues than any single paradigm. This eclectic approach introduces a fair bit of tension into economic-environmental analysis because of the inherent conflicts between paradigms. For example, the goal underlying the circular flow paradigm is economic growth. Resources are not limiting in this paradigm because of technological progress and substitution of manufactured capital (structures and equipment) for natural capital (natural and environmental resources). In contrast, the ecological economics paradigm is based on the goal of keeping the economic subsystem within ecological limits. Technology is not an automatic way to remove ecological limits and manufactured capital and natural capital are treated as complements rather than substitutes. Despite this tension, a blending of the four paradigms holds greater promise than any single paradigm in resolving complex natural resource and environmental issues.

How do the remaining chapters of this book relate to the four paradigms? Chapters 7 and 8 concentrate on the economic principles for determining efficient use of exhaustible and renewable resources, respectively, from the viewpoint of individual firms and households (private efficiency), and society (social efficiency). These principles are the fundamental building blocks of the circular flow and material balances models.

Chapter 9 develops economic principles for determining the efficient levels of environmental pollution and evaluates alternative policies for reducing environmental pollution. These principles and policies lie at the core of the material balances approach. Chapter 10 deals with natural resource and environmental accounting, which is an integral part of both ecological economics and sustainable development. Chapter 11 addresses benefit–cost analysis of resource investments. Benefit–cost analysis is an application of welfare economics and investment criteria, which are critical elements of all four models. Chapter 12 covers nonmarket valuation of natural and environmental resources. Nonmarket resource values are very important in determining efficient resource use, developing monetary accounts of natural resource depletion and environmental degradation, evaluating resource protection policies, and assessing the efficiency of alternative resource investments. In summary, subsequent chapters draw heavily from one or more of the four paradigms.

Types of Resources

Firm and household decisions regarding the use of natural and environmental resources are influenced by the physical and biological attributes of a resource. Trees and fish have different physical and biological attributes than do petroleum and minerals. These differences have important economic implications for the spatial and temporal use and management of natural and environmental resources. Natural and environmental resources can be classified into two broad categories: *exhaustible resources* and *renewable resources*.

EXHAUSTIBLE RESOURCES. The stock of *exhaustible resources* such as petroleum, coal and metals is fixed. Use of exhaustible resources depletes the current stock of the resource, which reduces its future availability. The greater the rate of use, the more quickly the resource is depleted. A simple model can be used to illustrate the dynamics of exhaustible resources. Let S_0 equal the initial stock of coal and U_{t-1} equal the total use of coal used through the end of period t-1. The stock of coal available at the beginning of period t is determined by the following stock equation:

$S_t = S_0 - U_{t-1}$

Because S_0 is fixed, the greater the use of coal prior to time t, the less coal is available for current and future generations. This simple stock equation overlooks differences in the availability of different qualities of coal (bituminous, subbituminous and lignite). For example, the 1990 amendments to the Clean Air Act restrict emissions of sulfur dioxide. This legislation has increased the use of western coal relative to eastern coal because the former has a lower sulfur content. As a result, mining of western coal has increased relative to mining of eastern coal and the stock of low sulfur coal is being depleted more rapidly than the stock of high sulfur coal.

An important feature of certain exhaustible resources is the uncertainty regarding their location, quantity and quality. Locations of resources such as coal and metals are generally known; however, resources such as oil and natural gas are subject to greater uncertainty because their location must be determined through exploration. If more oil is discovered, then the initial stock of oil is increased by the amount discovered. Uncertainty regarding the initial stock of oil translates into uncertainty regarding the amount of oil that will be available to future generations. Suppose initial estimate of oil resources (S_0) is 600 billion barrels and cumulative use of oil through the beginning of period t (U_{t-1}) is 300 billion barrels. If current annual use of oil is 3 billion barrels, then current reserves and the number of years of oil remaining at current use rates (*reserves-to-use ratio* or R) are:

$S_t = 600 - 300 = 300$ and $R = 300/3 = 100$.

If favorable discoveries of oil cause petroleum geologists to revise the initial estimate of oil resources upward to 1,000 billion barrels and the current use rate remains unchanged, then current reserves and the reserves-to-use ratio are:

$S_t = 1,000 - 300 = 700$ and $R = 700/3 = 233$.

In this case, a 67 percent increase in oil resources results in a 133 percent increase in the reserves-to-use ratio. If oil prices decrease in response to the higher estimate of oil resources, annual oil consumption could increase, which would have the effect of lowering the reserves-to-use ratio.

The stock of certain exhaustible resources can be extended through recycling. For example, recycling of aluminum reduces the use of bauxite, the mineral from which aluminum is manufactured. Other things equal, recycling lowers the rate of use of the resource, which decreases cumulative use (U_{t-1}) relative to what it would be without recycling. Consequently, it takes longer to exhaust the resource when recycling occurs, other things equal.

Physical exhaustion of a resource occurs when $S_t = 0$, at which point the stock of coal is depleted. *Economic exhaustion* of a resource occurs when the use of the resource falls to zero. Economic exhaustion usually occurs before physical exhaustion because extraction of a resource will be discontinued when extraction is no longer profitable. Both concepts of exhaustion are dynamic. For example, increases in the price of the resource and/or improvements in the efficiency of resource extraction improve the profitability of resource extraction. This increases resource extraction and shortens the time until the resource is physically exhausted. On the other hand, these same factors could increase resource exploration. If new discoveries are made, then the time until the resource is exhausted could be extended.

RENEWABLE RESOURCES. Soil, water, crops, fish, wildlife, forests and solar energy are renewable resources. Unlike exhaustible resources, *renewable resources* are regenerated through natural growth. The time and space requirements for regeneration vary by resource. Soil regeneration occurs at a relatively slow rate. It takes decades, and in some cases centuries, to replenish the soil lost by high rates of water and wind erosion. In arid climates, soil degradation can be irreversible. Other renewable resources, including certain plant and animal species, regenerate in a matter of hours or days. There are many interconnections among renewable resources. Crops require soil, water and sunlight (solar energy) for growth and development. Forests contain trees, plants, fish and wildlife that require soil, water and sunlight for regeneration. Deforestation (timber harvesting in excess of regeneration) not only increases the land's vulnerability to soil erosion, but decreases precipitation, which increases the risk of arid conditions.

Renewable resources typically have multiple uses. An old growth forest can be managed for commercial timber, which means that periodically timber is harvested and used to manufacture wood products. The same forest can encompass a watershed that provides drinking water for a nearby community, habitat for fish and wildlife and outdoor recreation. Because of their dependence on complex physical, biological and chemical processes, and the multiplicity of uses, renewable resources are generally more difficult to manage than exhaustible resources. Management of fish and animal populations is based on their age-sex structure, habitat and geographic distribution, all of which are influenced by economic and environmental conditions.

Consider a biological resource such as a forest. The initial stock of the forest resource is called biomass. Biomass at the beginning of period t is:

6. Natural Resource Decisions

$$S_t = S_0 - H_{t-1} + G_{t-1} - L_{t-1}$$

where S_0 is initial biomass, H_{t-1} is cumulative harvest, G_{t-1} is cumulative biomass growth, and L_{t-1} is cumulative biomass losses due to natural causes such as fire and disease. A subscript of t-1 beside a variable designates the level of the variable as of the end of period t-1. Therefore, forest biomass:

decreases when $H_{t-1} > (G_{t-1} - L_{t-1})$,

increases when $H_{t-1} < (G_{t-1} - L_{t-1})$, and

remains constant when $H_{t-1} = (G_{t-1} - L_{t-1})$,

where $G_{t-1} - L_{t-1}$ is net growth in forest biomass. A unique characteristic of a biological resource is that biomass growth depends on the current level of biomass. Chapter 8 gives a more complete explanation of the management of renewable resources.

Next, consider how the characteristics of a renewable resource, such as water, influence management. Water can be managed as a *flow resource* or a *fund resource*. It is a flow resource when the quantity available in a given period is not directly affected by human activities. A free-flowing river is a flow resource. If the river is dammed to create a water storage reservoir, then the water becomes a fund resource. Construction of the dam and reservoir allows the water to be stored for later use. The stock of water in the reservoir at the beginning of period t is:

$$S_t = F_{t-1} - W_{t-1} - L_{t-1}$$

where F_{t-1} is cumulative river flow into the reservoir, W_{t-1} is cumulative withdrawals from the reservoir and L_{t-1} is cumulative losses from the reservoir due to evaporation, seepage and other causes through the end of period t-1. Without storage capacity, withdrawals from the river cannot exceed river flows in a given period. With water storage, current withdrawals can exceed current river flows because water can be withdrawn from the reservoir. As the seven-year (1986–1992) drought in California demonstrated, development of water storage capacity can actually increase a region's vulnerability to drought when that capacity becomes the basis for rapid expansion in agricultural, urban and industrial activity. Such expansion increases current water withdrawals and increases dependence on storage capacity to maintain economic activity.

Solar energy is a unique renewable resource for three reasons. First, the accumulation and/or growth of most natural resources depends on solar energy. Formation of exhaustible resources, such as petroleum and coal, required solar energy. Second, solar energy is a flow resource that has to be used when it is available or its value in use is lost. Because the amount of solar insolation received by the earth's surface is huge compared with the amount used for photosynthesis, heating, cooling and electrical power generation, most solar energy is not utilized. Third, solar energy is free.

Static Efficiency

This section discusses economic principles for determining efficient use of natural resources with static conditions. The latter exist when resource decisions in different time periods are independent. If use resource in the current time period does not affect the amount or the price of the resource in future time periods, then resource decisions are time-independent. The normative decision criterion for determining efficient resource use in a static framework is to maximize *consumer surplus* plus *producer surplus* in each period. Consumer and producer surpluses are derived from the market demand and supply curves for the resource. The sum of consumer surplus and producer surplus is *net social benefit* (NSB). The remainder of this section discusses efficient resource use under pure competition and imperfect competition.

PURE COMPETITION. A purely competitive market contains many buyers and sellers who have perfect knowledge of technical and economic conditions. Consider how the efficient use of land is determined in a purely competitive market. The land market consists of a market demand for land as depicted in Figure 6.3. This demand is determined by the marginal productivity of land in the production of a particular commodity. Suppose the commodity is food. Marginal productivity of land in food production equals the increase in food production from using another unit of land, holding fixed all other inputs used in food production. The law of diminishing marginal productivity says that the marginal productivity of a resource decreases as more of that resource is used. For this reason, the demand curve for land is negatively sloped, which implies that the quantity demanded of land decreases as the price of land increases, and vice versa.

The market demand curve for land in Figure 6.3 shows that q_1 units of land are purchased at p_1. Total willingness to pay for land equals the area below the demand curve (D) up Q_1, which is the mp_0nQ_1 area. Total willingness to pay is the sum of the rectangular area mp_1nQ_1 and the triangular area p_1p_0n. The rectangular area equals total expenditure, and the triangular area equals *consumer surplus* for Q_1 units of land. Therefore, consumer surplus equals total willingness to pay minus total expenditure on the resource. It is a surplus value because it is the value of the resource to the buyers above and beyond what they actually spend on the resource.

The market supply curve for land is shown in Figure 6.4. It represents the marginal opportunity cost of land, or the loss in value from selling an additional unit of land. Selling land involves transferring the right to the land from the current owner to the new owner. When this occurs, the current owner relinquishes the value of the land in its current use. If the relinquished value is p_4, then the seller is not willing to sell that additional unit of land for a price less than p_4. Since the value of land in its next best use is likely to increase as less land is devoted to that use, the marginal opportunity cost of land increases as more land is sold.

Under pure competition, the relationship between the marginal opportunity cost of land and the quantity of land is the market supply curve (S) for land. The supply curve for land is upward sloping, indicating that sellers are willing to sell more land as the price of land rises. Total income from selling Q_3 units of land is the area tp_3vQ_3, and total opportunity cost is the area $tuvQ_3$ in Figure 6.4. Opportunity cost equals the income the land would have earned in its next best use. *Producer*

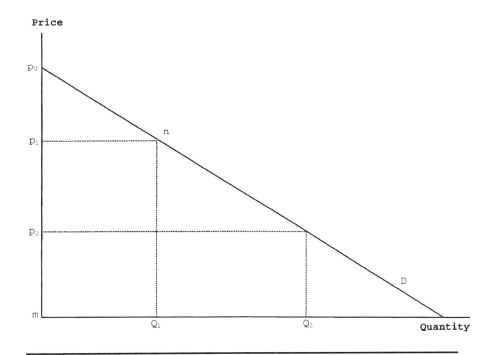

Figure 6.3. Total willingness to pay (area mp_0nQ_1) and demand curve (D) for land.

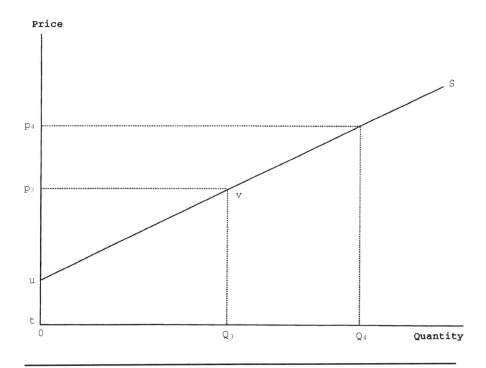

Figure 6.4. Producer surplus (up_3v) and supply curve (S) for land.

surplus equals the difference between total income and total opportunity cost, which is the area up₃v. It represents the surplus value received by the seller above and beyond the opportunity cost of the land.

The equilibrium price and quantity of land under pure competition is illustrated in Figure 6.5 for the following market demand and supply curves:

$$p_d = 50 - 0.5Q_d \quad \text{Demand}$$
$$p_s = 5 + 0.5Q_s \quad \text{Supply},$$

where p_d is the demand price, Q_d is the total quantity demanded of land in the market, p_s is the supply price and Q_s is the total quantity supplied of land in the market. The market demand curve shows the total quantity demanded at various prices, and the market supply curve gives the total quantity supplied at various prices. Market demand is determined as follows. The total quantity of land demanded in the market at a particular price (which is one point on the market demand curve) is derived by summing the quantities of land demanded by all buyers at that price. Other points on the market demand curve are derived in a similar manner. Connecting all the points gives the market demand curve. The total quantity of land supplied in the market at a particular price (which is one point on the market supply curve) is derived by summing the quantities of land supplied by all sellers at that price. Other points on the market supply curve are derived in a similar manner.

Because the market demand curve is negatively sloped and the market supply curve is positively sloped, consumer surplus increases and producer surplus decreases as price decreases. Conversely, consumer surplus decreases and producer surplus increases as price increases. For example, when price decreases from p_1 to p_2, consumer surplus increases from $p_1$tu to $p_2$tv and producer surplus decreases from $p_1$ws to $p_2$vs. Hence, there is a tradeoff between consumer surplus and producer surplus.

Land is being used efficiently when NSB (sum of producer and consumer surpluses) is maximized. Maximum NSB occurs at the market equilibrium price p_2 and equilibrium quantity Q_2; it equals the area stv. Market equilibrium price is found by equating the demand and supply prices and solving for Q:

$$50 - 0.5Q_d = 5 + 0.5Q_s \text{ and } Q_d = Q_s.$$

Equilibrium quantity is $Q_2 = 45$ and equilibrium price is $p_2 = 27.5$.

Because the area of a rectangle is the base times the height, and the area of a triangle is one half of the base times the height, in equilibrium, NSB equals consumer surplus (triangle $p_2$tv) plus producer surplus (triangle $p_2$vs). In numerical terms, consumer surplus is 418.75 {0.5[(50 − 27.5)45]} and producer surplus is 418.75 {0.5[(27.5 − 5)45)]}. In this case, consumer surplus is identical to producer surplus because the demand and supply curves have the same slope (0.5) and the supply curve has a positive intercept. NSB equals 837.5 (418.75 + 418.75).

To show that the maximum NSB occurs at Q_2, consider the NSB at a price above and below the equilibrium price. If the price of land is $p_1 = 45$, then sellers are willing to supply $Q_3 = 80$, but buyers are only willing to purchase $Q_1 = 10$. There is a surplus of 70 units of land. Ten units of land are bought and sold and NSB is 225 (stux), which is less than 837.5 (stv). If the market price is $p_3 = 10$, which is be-

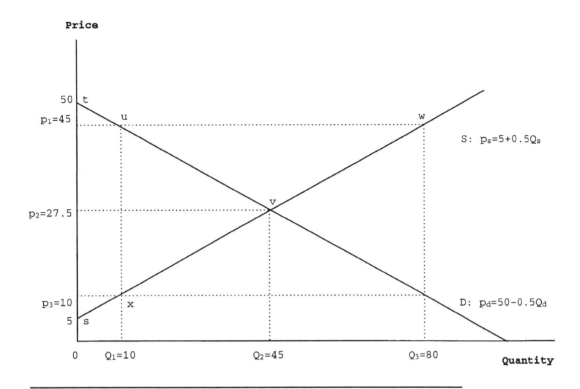

Figure 6.5. Equilibrium in land market under pure competition. D = demand; and S = supply.

low the equilibrium price, then buyers are willing to purchase $Q_3 = 80$, but sellers are only willing to supply $Q_1 = 10$. There is a shortage of 70 units of land. NSB is 225 (stux), which is less than 837.5 (stv).

It is no coincidence that NSB is maximized at the equilibrium price and quantity (p_2 and Q_2). At the equilibrium price, quantity demanded equals quantity supplied and there is no surplus or shortage of land. Because equilibrium is assured under purely competitive market conditions, maximum NSB is automatically achieved.

The use of land is said to be *Pareto efficient* when it is not possible to increase the welfare of some individuals without decreasing the welfare of at least one other individual by altering the distribution of land among users. In other words, when a Pareto-efficient use of resources is achieved, the gainers from a re-allocation are not able to compensate the losers and still be better off. In general, there can be more than one Pareto-efficient use of resources in an economy; Pareto efficiency is not unique. In the land example, the Pareto-efficient use of land changes when the distribution of land among sellers and the distribution of income among buyers is altered. Because Pareto efficiency requires purely competitive market conditions, efficient resource use is not achieved when the market has monopolistic elements and/or the demand and supply curves for resources do not include all the relevant social benefits and costs of using the resource. This is typically the case for natural and environmental resources.

IMPERFECT COMPETITION. What happens to resource use, resource price and NSB when the market is not purely competitive? Imperfect competition exists when any of the conditions for pure competition are violated. Suppose there is only one buyer of land. This single buyer is called a monopsonist. Under monopsony, the buyer has to pay successively higher prices for land to acquire additional land. The higher prices apply to all units of land purchased. Hence, the cost of purchasing an additional unit of land no longer equals the price of land but, rather, the marginal resource cost (MRC) of land. MRC is the increase in the total cost of land from buying an additional unit of land. The MRC curve, which is illustrated in Figure 6.6, has the same intercept as the supply (S) curve (50); however, the slope of the MRC curve is twice the slope of the supply curve [1 = (2)(0.05)].

The profit-maximizing quantity of land for the monopsonist is found by setting demand price equal to MRC (p_d = MRC) and solving for Q as follows:

$$50 - 0.05Q_d = 5 + 1.0Q_s \text{ and } Q_d = Q_s.$$

The solution is Q_m = 30. Compared with the purely competitive solution (Q_c), the equilibrium quantity of land with monopsony is smaller (30 < 45), the equilibrium price is larger (35 > 27.5) and NSB is lower (acd = 725 < 837.5 = abd). The loss in NSB from monopsony equals the area abc. Hence, imperfect competition is not socially efficient. Moving from pure competition to monopsony in the land market causes producer surplus to increase and consumer surplus to decrease, which makes the seller better off and buyers worse off relative to pure competition.

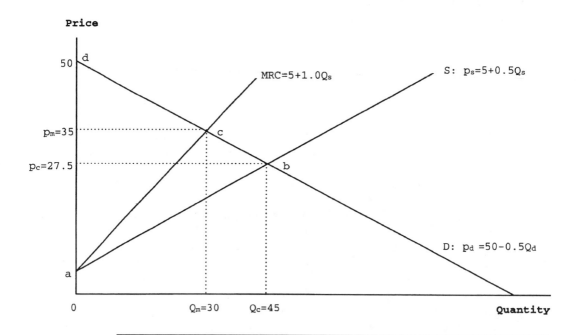

Figure 6.6. Equilibrium in land market under monopsony. D = demand; MRC = marginal resource cost; and S = supply.

Summary

Decisions regarding the management of natural and environmental resources can be approached using one of four paradigms. These paradigms consist of the circular flow model, the material balances model, the ecological economics model and the sustainable development model. The complexity and stakes of resource management decisions increase from the circular flow model to the materials balances model and from the material balances model to both the ecological economics and sustainable development models.

Natural and environmental resources are classified as exhaustible or renewable. The stock of an exhaustible resource, such as petroleum, coal and minerals, is fixed, although the time required to exhaust the stock is extended by recycling. The faster the rate of depletion of an exhaustible resource, the more rapidly the stock is exhausted. There is considerable uncertainty regarding the location, quantity and quality of certain exhaustible resources, such as crude oil and natural gas. The stock of a renewable resource, such as soil, water, crops, fish, wildlife and forests, is stabilized by ensuring that the rate of use does not exceed the rate of regeneration. Time and space requirements for regeneration vary by resource. Renewable resources are generally more difficult to understand and manage than are exhaustible resources because of their dependence on complex physical, biological and chemical processes, and their multiple and often competing uses.

Static efficiency principles are appropriate for evaluating resource management decisions when use in the current period does not affect the availability of the resource in future periods. Resource efficiency in a static framework requires maximizing NSB (producer surplus plus consumer surplus). The condition for static resource efficiency is equality between price and marginal cost under pure competition or price and marginal resource cost under monopsony.

Questions for Discussion

1. Why are economists most comfortable with the circular flow paradigm in the resource decision hierarchy depicted in Figure 6.1? What are some of the drawbacks of not advancing beyond this paradigm?

2. Conventional economics addresses the optimum scale of a firm but ignores the optimum scale of the economy. The latter is a primary concern of ecological economics. Why does this dichotomy exist?

3. What is the distinction between exhaustible resources and renewable resources? What is the implication of this difference for natural resource management?

4. How does recycling of aluminum products influence the depletion of bauxite? [Bauxite is the mineral from which aluminum is made.]

5. Suppose the demand curve for land in Figure 6.5 shifts to the right so that the new demand equation is $p_d = 60 - 0.5Q_d$. Determine the new equilibrium price and quantity of land and NSB.

Further Readings

Pearce, David, Edward Barbier and Anil Markandya. 1990. "Discounting the Future." In *Sustainable Development: Economics and Environment in the Third World*. London, England: Earthscan Publications Ltd. pp. 23–56.

Randall, Alan. 1987. "Cornucopia or Catastrophe?" Chapter 2 in *Resource Economics: An Economic Approach to Natural Resource and Environmental Policy*. New York: John Wiley & Sons, Inc. pp. 11–32.

Note

1. Herman Daly, "The Steady-State Economy: Alternative to Growthmania," *Steady-State Economics,* 2nd ed. (Cavelo, California: Island Press, 1991), pp. 180–210.

CHAPTER 7

Exhaustible Resource Use

Unless we find a way to dramatically change our civilization and our way of thinking about the relationship between humankind and the earth, our children will inherit a wasteland.

—VICE PRESIDENT AL GORE, 1992

The static condition developed in Chapter 6 for efficient resource use (price equals marginal cost or price equals marginalresource cost) must be modified to take account of market dynamics if it is to be applied to exhaustible resources. Market dynamics refers to the interactions between the extraction and stock of a resource that occur over time. *Exhaustible resources,* such as petroleum and minerals, exhibit market dynamics because the rate of extraction in the current period influences the stock of the resource available in future periods. For example, if the current oil reserve is 700 billion barrels (bbl) and 300 billion bbl are extracted in the next 10 years, then the oil reserve falls to 400 (700 − 300) billion bbl after 10 years. If higher energy prices increase energy conservation and reduce oil extraction by 100 billion bbl over the next 10 years, then the oil reserve increases from 400 billion bbl to 500 (700 − 200) billion bbl. Energy conservation increases the oil reserve by 100 billion bbl. Therefore, efficient extraction of an exhaustible resources must account for market dynamics.

This chapter examines *market dynamics* and determination of *efficient intertemporal extraction* for exhaustible resources. It also considers the effects of various factors on efficient extraction and prices of exhaustible resources.

Market Dynamics

Market dynamics for an exhaustible resource result from changes in resource prices and extraction over time due to shifts in the supply and demand for the resource. As shown in Chapter 6, market dynamics are introduced by placing a time index in the form of a subscript on stocks, extraction and prices, and other factors that shift resource demand and supply. Incorporating time indexing in the stock accounting relationship for the oil conservation example in the introduction gives:

$$S_t = S_0 - U_{t-1}$$

where S_t is oil reserves (or stock) in the beginning of period t, S_0 is initial oil reserves and U_{t-1} is total oil extraction through the end of period t−1. For the oil example, S_0 = 700 billion bbl, U_{t-1} = 200 billion bbl with conservation, and U_{t-1} = 300 billion bbl without conservation. Therefore, S_t = 500 billion bbl with conservation and S_t = 400 billion bbl without conservation.

While the stock accounting relationship indicates how oil extraction reduces oil reserves over time, it provides little insight about the efficient rates of oil extraction over time. For this and other reasons, a dynamic economic model of resource extraction is needed.

A common way to introduce dynamics into the oil market is to utilize a model that accounts for changes in demand and supply for oil over time. An example of a dynamic market model for oil is as follows:

$$p_t = g[Q_{dt}, Q_{d(t-1)}, R_t] \quad \text{Demand}$$

$$p_t = f[Q_{st}, S_t, G_t] \quad \text{Supply}$$

$$Q_{dt} = Q_{st} \quad \text{Equilibrium}$$

where:

p_t is the price of oil in period t,

Q_{dt} is the quantity demanded of oil in period t,

$Q_{d(t-1)}$ is the quantity demanded of oil in the previous period (t−1),

R_t is a demand shifter in period t which includes household preferences and income and the quantities demanded of substitutes for oil,

Q_{st} is the quantity supplied of oil in period t,

S_t is oil reserves in the beginning of period t, and

G_t is a measure of improvements in technology and other factors in period t.

The demand equation states that the price of oil in the current period depends on the quantity demanded of oil in the current and previous periods and on factors that shift the demand for oil over time, such as changes in preferences, income and the availability of oil substitutes. The supply equation states that the price of oil depends on the quantity supplied of oil in the current period, oil reserves at the beginning of the period, and factors that shift the supply curve for oil, such as changes in technology. In equilibrium, quantity demanded of oil equals quantity supplied.

The familiar two-variable demand (D) and supply (S) curves are derived from these demand and supply equations by assigning specific values to the demand and

7. Exhaustible Resource Use

supply shifters. The dynamic market model allows demand and supply functions to vary over time. The demand and supply curves for period t are illustrated in Figure 7.1. With this procedure, the intercept of the demand curve is determined by the value of the two demand shifters, $Q_{d(t-1)}$ and R_t, and the intercept of the supply curve is determined by the values of the two supply shifters, S_t and G_t. In a purely competitive industry, the equilibrium price of oil is p^*_t and equilibrium extraction of oil is Q^*_t.

Net Social Benefit

Efficient intertemporal extraction of an exhaustible resource refers to the efficient rates of resource extraction over time. These rates are determined by maximizing *net social benefit* (NSB). In the case of discrete time, NSB is:

$$NSB = PV(B_0,...,B_T) = \sum_{t=0}^{T} B_t(1+r)^{-t},$$

where $PV(B_0,...,B_T)$ denotes the present value of net benefits of resource extraction in all T + 1 periods, r is the discount rate, $(1 + r)^{-t}$ is the discount factor in period t,

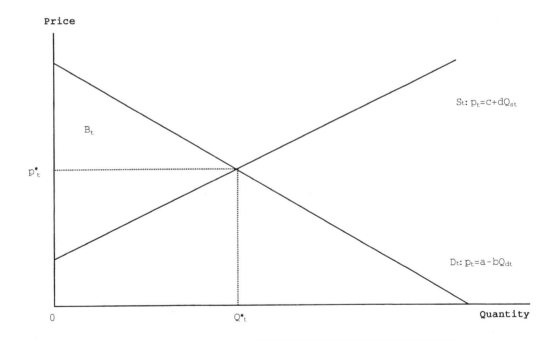

Figure 7.1. Demand (D_t) and supply (S_t) curves and market equilibrium for an exhaustible resource in period t under pure competition.

0 is the initial period and T is the last period in the planning horizon. Net benefit in period t, B_t, equals producer surplus plus consumer surplus, which is the area between the supply and demand curves for the resource in period t at the equilibrium extraction rate. $B_t(1 + r)^{-t}$ is the present value of the net benefit in period t. Hence, NSB is the present value of net benefits of extraction over the planning horizon. NSB must be maximized over all possible extraction rates for the resource in the T + 1 periods in the planning horizon.

Efficient intertemporal extraction of an exhaustible resource is evaluated for two periods in the next section and for multiple periods in the subsequent section. Both evaluations are based on two simplifying assumptions. First, the resource is used in the same period in which it is extracted. This implies that there is no storage of extracted resource from one period to the next. Second, resource markets are assumed to be purely competitive. Consequences of imperfectly competitive resource markets are discussed in the last section of this chapter. Oil is selected as the exhaustible resource.

Two-Period Dynamic Efficiency

This section determines the equilibrium conditions for efficient intertemporal extraction of oil in two periods referred to as the *current period* and the *future period*. Two demand conditions are evaluated: *constant resource demand* (Case 1) and *variable resource demand* (Case 2). Constant resource demand means demand for oil is the same in both periods. It is appropriate to consider a constant demand when oil demand is relatively stable between the current and future period. Variable resource demand means that demand for oil changes between the current and future periods. Variable resource demand is appropriate when the demand for oil is expected to change over time.

The constant and variable resource demand condition is evaluated for two supply conditions: *unrestricted supply* (Cases 1U and 2U) and *restricted supply* (Cases 1R and 2R) of oil. Oil supply is restricted (unrestricted) when the stock of oil is less than (greater than) total quantity demanded of oil in the two periods. Two cost conditions are considered: *zero marginal extraction cost* (MEC = 0) and *positive marginal extraction cost* (MEC > 0).[1] Marginal extraction cost measures the change in total cost of oil extraction when an additional unit oil is extracted. For simplicity, MEC is assumed to be constant (the same in both periods).

In summary, the economic evaluation of efficient intertemporal extraction of oil given below considers the following cases:

Case 1: Constant resource demand
    1U: Unrestricted supply
        MEC = 0 or MEC > 0
    1R: Restricted supply
        MEC = 0 or MEC > 0

Case 2: Variable resource demand
    2U: Unrestricted supply
        MEC = 0 or MEC > 0
    2R: Restricted supply
        MEC = 0 or MEC > 0

CASE 1: TWO-PERIOD EFFICIENCY WITH CONSTANT OIL DEMAND.
In this case, efficient oil extraction rates are determined for the current and future periods. The constant oil demand is as follows:

D: $p = 10 - 0.20Q_d$,

where p is the price of oil in dollars per barrel (bbl) and Q_d is the quantity demanded in bbl. The price intercept of the demand function is 10 and the slope is −0.20. Because the demand for oil is the same in both periods, there is no need to place a time index (subscript) on the price and quantity variables. This demand function is illustrated in Figure 7.2A.

Case 1U: No Supply Restriction. The constant demand function for oil with MEC = 0 is illustrated in Figure 7.2A and with MEC = $5 in Figure 7.2B.

For MEC = 0, NSB is maximized by extracting the amount of oil given by the quantity intercept of the demand function, namely, 50 bbl. Therefore, the efficient extraction rates are $Q^*_0 = Q^*_1 = 50$ bbl. Total oil extraction is 100 bbl. Equilibrium prices are $p^*_0 = p^*_1 = 0$. Extracting more than 50 bbl in each period adds nothing to NSB and, hence, is not in the best interest of society.

For the efficient extraction rates and an 8 percent discount rate, NSB is:

$$\begin{aligned} \text{NSB} &= \text{PV}(Q^*_0 = Q^*_1 = 50) = (0.50)(10)(50)(1.08^{-0}) \\ &\quad + (0.50)(10)(50)(1.08^{-1}) \\ &= \$520. \end{aligned}$$

The first term $[(0.50)(10)(50)(1.08^{-0})]$ is the present value of the entire area under the demand function in the current period, namely, B_0, or the area under the demand curve in Figure 7.1A. Since $1.08^{-0} = 1$, the discount factor is dropped in all subsequent present value terms for the current period. The second term $[(0.50)(10)(50)(1.08^{-1})]$ is the present value of the entire area under the demand function in the future period, namely, B_1. The expression $(0.50)(10)(50)$, which appears in both terms, is the formula for a triangular area.

For MEC = $5, NSB is maximized by extracting the amount of oil in each period which makes p = MEC = $5. This occurs at 25 bbl as shown in Figure 7.2B. The equilibrium extraction rates and prices of oil in each period are $Q^* = 25$ bbl and $p^* = \$5$. As long as oil demand and MEC are constant, and there is no supply restriction, the efficient extraction rates and prices are the same in both periods. Total oil extraction is less when MEC = $5 than when MEC = 0 (50 bbl vs. 100 bbl).

For the efficient extraction rates and an 8 percent discount rate, NSB is:

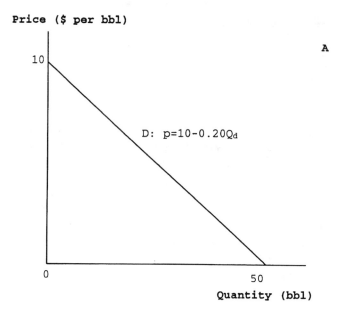
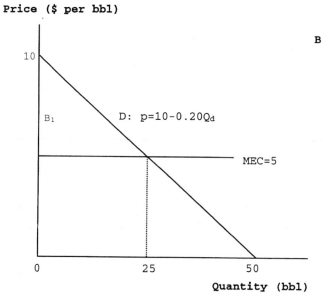

Figure 7.2. Case 1U: Efficient extraction rates for oil with constant demand (D) and unrestricted supply when marginal extraction cost (MEC) = 0 (*A*) and when MEC = 5 (*B*).

7. Exhaustible Resource Use

$$\begin{aligned}\text{NSB} &= \text{PV}(Q^*_0 = Q^*_1 = 25) = (0.50)(10 - 5)(25) \\ &\quad + (0.50)(10 - 5)(25)(1.08^{-1}) \\ &= \$120.37.\end{aligned}$$

The first term, $(0.50)(10 - 5)(25)$, is the present value of the area between the demand curve and MEC = \$5, namely, B_1. The second term, $(0.50)(10 - 5)(25)(1.08^{-1})$, is the present value of the area between the demand curve and MEC = \$5, namely, B_2.

Case 1R: Supply restriction. When there is a supply restriction, it must be taken into account when determining the efficient extraction rates. The criterion for maximizing NSB in this case is to select extraction rates that make discounted price equal to discounted MEC in both periods and that satisfy the supply restriction. Let the oil reserve be 80 bbl.

For MEC = 0, NSB is maximized by solving for the extraction rates, Q_0 and Q_1, that make discounted prices equal in the both periods and that satisfy the supply restriction. Specifically, the following equations are solved for Q_0 and Q_1:

$$p^\diamond_0 = p^\diamond_1$$

$$Q_0 + Q_1 = Q_A,$$

where $p^\diamond_t = p_t(1 + r)^{-t}$ is the discounted price of the resource and Q_A is the oil reserve.

For $r = 8$ percent, the discounted oil prices are $p^\diamond_0 = (10 - 0.20Q_0)(1.08)^{-0}$ in the current period and $p^\diamond_1 = (10 - 0.20Q_1)(1.08)^{-1} = 9.26 - 0.19Q_1$ in the future period. Substituting these values into the last two equations gives:

$$10 - 0.20Q_0 = 9.26 - 0.19Q_1$$

$$Q_0 + Q_1 = 80.$$

Solving for Q_0 and Q_1 gives the efficient extraction rates for oil, namely:

$$Q^*_0 = 40.37 \text{ bbl and } Q^*_1 = 39.63 \text{ bbl}.$$

Substituting Q^*_0 and Q^*_1 into the respective demand functions gives an equilibrium price of \$1.93 per bbl in the current period and \$2.07 per bbl in the future period. This equilibrium is illustrated in Figure 7.3.

For the efficient extraction rates and an 8 percent discount rate, NSB is:

$$\begin{aligned}\text{NSB} &= \text{PV}(Q^*_0 = 40.37, Q^*_1 = 39.63) = (0.50)(10 - 1.93)(40.37) \\ &\quad + (0.50)(10 - 2.07)(39.63)(1.08)^{-1} \\ &= \$308.39\end{aligned}$$

As expected, NSB is lower with than without a supply restriction when MEC = 0 (\$308.39 versus \$520). Net benefit in the current period is the area under the demand function up to 40.37 bbl, which is B_0 in Figure 7.3A. Net benefit in the future

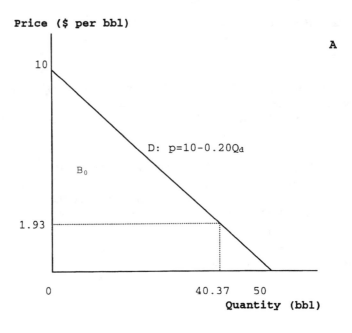

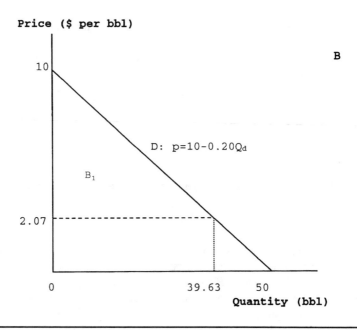

Figure 7.3. Case 1R: Efficient extraction rates for oil with constant demand (D), restricted supply (80 bbl) and marginal extraction cost (MEC) = 0 for current period (A) and future period (B).

7. Exhaustible Resource Use   133

period (B_1) is the area under the demand function for the future period up to 39.63 bbl, which is B_1 in Figure 7.3B. Any other extraction rates result in a lower NSB.

For MEC = $5, NSB is maximized at an extraction rate of 25 bbl in each period, as shown in Figure 7.2B. The supply restriction is not relevant in this case because total extraction of oil at p = MEC = $5 is less than the oil reserve (50 bbl < 80 bbl). In other words, the supply restriction is not binding. When MEC = $5, the oil reserve does not become binding until it falls below 50 bbl. For an efficient extraction rate of 25 bbl in each period, NSB is the same as it is in Case 1U with MEC = $5, namely, $120.37.

CASE 2: TWO-PERIOD EFFICIENCY WITH VARIABLE RESOURCE DEMAND. Case 2 assumes that the demand for oil increases between the current period and future period. The following two demand functions have this property:

D_0: $p_0 = 10 - 0.20Q_0$ Current demand
D_1: $p_1 = 30 - 0.50Q_1$ Future demand

These demand functions are illustrated in Figure 7.4.

Case 2U: No supply restriction. When MEC = 0, NSB is maximized by extracting 50 bbl in the current period and 60 bbl in the future period ($Q^*_0 = 50$ and $Q^*_1 = 60$) as shown in Figure 7.4. These extraction rates are the intercepts on the quantity axis of the respective demand functions. Total oil extraction is 110 bbl, and the equilibrium price (p*) is zero in both periods.

For the efficient extraction rates and an 8 percent discount rate, NSB is:

$$\text{NSB} = \text{PV}(Q^*_0 = 50, Q^*_1 = 60) = (0.50)(10)(50)$$
$$+ (0.50)(30)(60)(1.08^{-1})$$
$$= \$1,083.33.$$

B_0 is the entire area under the demand function for the current period, as shown in Figure 7.4A, and B_1 is the entire area under the demand function for the future period, as shown in Figure 7.4B. Comparing cases 1U and 2U for MEC = 0 shows that when demand increases between the current period and future period, total extraction increases from 100 bbl to 110 bbl and NSB increases from $520 to $1,083.

For MEC = $5, NSB is maximized by selecting extraction rates that make p = MEC in both periods. The condition p = MEC requires $10 - 0.20Q_0 = 5$ in the current period and $30 - 0.50Q_1 = 5$ in the future period. Solving these equations for Q_0 and Q_1 gives $Q^*_0 = 25$ bbl and $Q^*_1 = 50$ bbl as shown in Figure 7.5A and 7.5B. Total oil extraction is 75 bbl (25 + 50) and equilibrium price (p*) is $5 in both periods.

For the efficient extraction rates and an 8 percent discount rate, NSB is:

$$\text{NSB} = \text{PV}(Q^*_0 = 25, Q^*_1 = 50) = (0.50)(10 - 5)(25)$$
$$+ (0.50)(30 - 5)(50)(1.08^{-1})$$
$$= \$641.20.$$

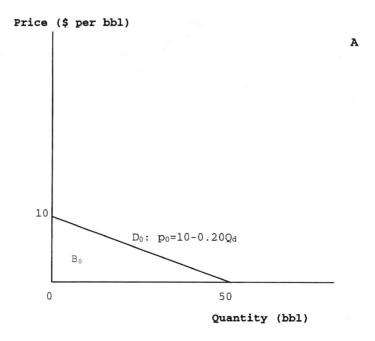

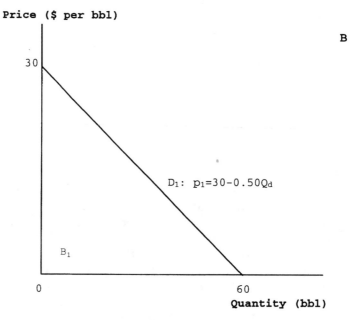

Figure 7.4. Case 2U: Efficient extraction rates for oil with increasing demand (D), unrestricted supply and marginal extraction cost (MEC) = 0 for current period (A) and future period (B).

7. Exhaustible Resource Use

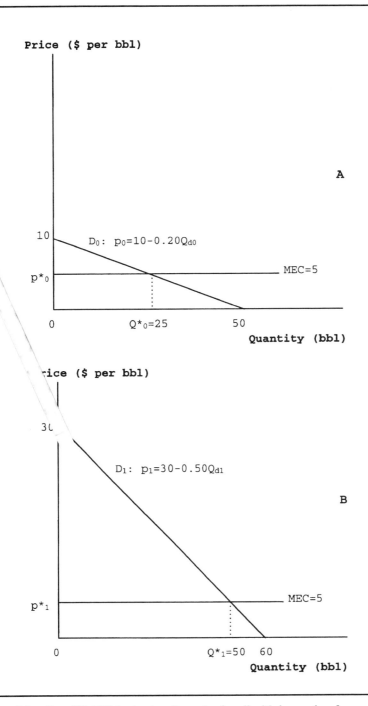

Figure 7.5. Case 2U: Efficient extraction rates for oil with increasing demand (D), unrestricted supply and marginal extraction cost (MEC) = 5 for current period (*A*) and future period (*B*).

As expected, NSB is lower when MEC = $5 than when MEC = 0 ($641.20 vs. $1,083.33).

Case 2R: Supply restriction. When total demand for oil exceeds the oil reserve, the extraction rates that maximize NSB are determined using the same procedure as in Case 1R.

For MEC = 0 and an oil reserve of 80 bbl, efficient extraction rates are determined by equating discounted prices in both periods and imposing the supply restriction. The efficient extraction rates are determined by solving the following equations for Q_0 and Q_1:

$$10 - 0.20Q_0 = 27.78 - 0.463Q_1$$

$$Q_0 + Q_1 = 80.$$

Efficient extraction rates are $Q^*_0 = 29.05$ bbl and $Q^*_1 = 50.95$ bbl. Equilibrium price is $4.19 per bbl in the current period and $4.53 per bbl in the future period. Discounted equilibrium price is $4.19 per bbl in both periods.

Another way to determine the equilibrium extraction rates and discounted price is to use a face-to-face diagram like the one shown in Figure 7.6. The diagram shows the demand function for the current period (D_0) in the conventional position. The discounted demand function for the future period (D_1) is rotated 180 degrees so that it faces D_0. Quantity extracted is measured from left to right in the current period and from right to left in the future period. The length of the quantity axis equals the supply restriction of 80 bbl. Efficient extraction rates and the discounted equilibrium price are determined by the intersection of the two demand functions.

For the efficient extraction rates and an 8 percent discount rate, NSB is:

$$\begin{aligned}NSB &= PV(Q^*_0 = 29.05, Q^*_1 = 50.95) = (0.50)(10 - 4.19)(29.05) \\ &\quad + (0.50)(30 - 4.53)(50.95) \\ &= \$733.24.\end{aligned}$$

When MEC = $5 in both periods, the efficient extraction rates are found by solving p^*_0 = MEC = $5 to obtain Q_0^* = 25 bbl and solving p^*_1 = MEC = $5 to obtain Q_1^* = 50 bbl, as shown in Figure 7.7. Total oil extraction is 75 bbl, which is less than 80 bbl. Hence, an oil reserve of 80 bbl is not a binding supply restriction when MEC = $5. For the demand functions used in case 2R and MEC = $5, the oil reserve becomes binding when it is less than 75 bbl.

When the oil reserve is a binding constraint, the efficient oil extraction rates have to be determined by equating p − MEC in both periods and imposing the restriction $Q_0 + Q_1 = Q$. Consider the efficient extraction rates for an oil reserve of 60 bbl and MEC = $5. This oil reserve is binding because it is less than 75 bbl. Extracting more than 10 bbl in the current period causes the amount of oil available in the future period to fall below 50 bbl, which decreases the future period's net benefit. The present value loss in the future period's net benefit from extracting an additional barrel of oil in the current period is called *marginal user cost* (MUC).

How is MUC calculated? Let oil extraction in the current period be increased from 10 bbl to 11 bbl. Since the oil reserve is 60 bbl, a 1 bbl increase in current ex-

7. Exhaustible Resource Use

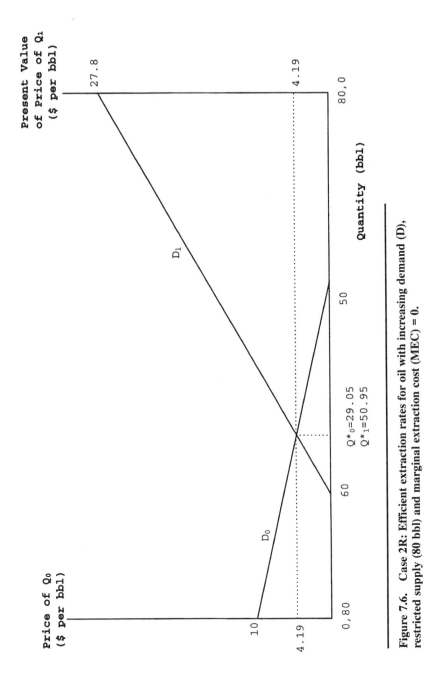

Figure 7.6. Case 2R: Efficient extraction rates for oil with increasing demand (D), restricted supply (80 bbl) and marginal extraction cost (MEC) = 0.

traction reduces future extraction from 50 to 49 bbl. The MUC of increasing oil extraction from 10 bbl to 11 bbl in the current period is the present value of triangle abc in Figure 7.7B, which is $0.231 [(0.50)(50 − 49)($5.50 − $5.00)(1.08^{-1})]$. This value is the slope of the MUC curve in Figure 7.7A. The efficient extraction rate in the current period (Q^*_0) is determined by solving p_0 = MEC + MUC for the equilibrium quantity to obtain $Q^*_0 = 14.53$ bbl. The efficient extraction rate in the future period is $Q^*_1 = 60$ bbl − 14.53 bbl = 45.47 bbl.

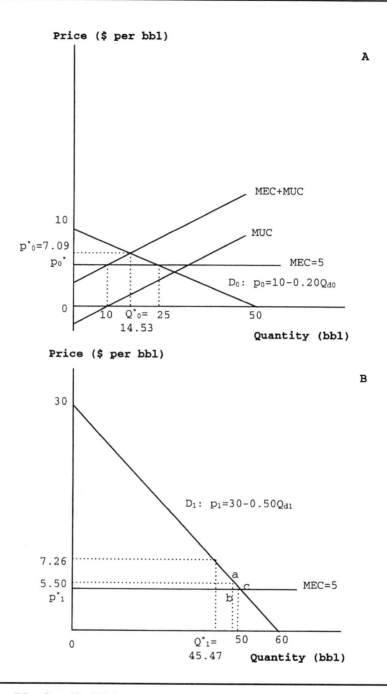

Figure 7.7. Case 2R: Efficient extraction rates for oil with increasing demand (D), restricted supply (60 bbl) and marginal extraction cost (MEC) = 5 for current period (A) and future period (B).

Efficient extraction rates can be determined directly by solving the following equations for Q_0 and Q_1:

$$p^0_0 - MEC^0_0 = p^0_1 - MEC^0_1$$

$$Q_0 + Q_1 = Q_A.$$

Substituting MEC = $5 and Q_A = 60 bbl into these equations gives:

$$10 - 0.20Q_0 - 5 = 27.78 - 0.463Q_1 - 4.63$$

$$Q_0 + Q_1 = 60.$$

Solving the two equations for Q_0 and Q_1 gives $Q^*_0 = 14.53$ bbl and $Q^*_1 = 45.47$ bbl.

Substituting Q^*_0 into the demand function for the current period ($p_0 = 10 - 0.20Q_0$) and Q^*_1 into the demand function for the future period ($p_1 = 30 - 0.50Q_1$) gives the following equilibrium prices:

$$p^*_0 = \$7.09 \text{ and } p^*_1 = \$7.26.$$

For the efficient extraction rates and an 8 percent discount rate, NSB is:

$$\begin{aligned}NSB &= PV(Q^*_0 = 14.53, Q^*_1 = 45.47) = [(7.09 - 5) + 0.5(10 - 7.09)]14.47 \\ &\quad + (1.08)^{-1}[(7.26 - 5) + 0.5(30 - 7.26)]45.47 \\ &= \$625.14.\end{aligned}$$

Multiple-Period Efficiency

The two-period analysis used above is limiting because it allows the stock of the resource to exceed total quantity demanded of the resource. When this occurs, some of the resource is not extracted, which reduces its value to society. This limitation is easily overcome in a multiple-period analysis by requiring total extraction of the resource to equal the stock of the resource. Given that efficient resource extraction can be determined for an infinitely long planning horizon ($T = \infty$), this requirement does not imply that the resource is unavailable to future generations.

EQUILIBRIUM CONDITIONS. Efficient extraction of an exhaustible resource over multiple periods requires that two equilibrium conditions be satisfied. The *first equilibrium condition* is:

$$p_0 - MEC_0 = p^0_t - MEC^0_t \text{ for } t = 1,...,T,$$

where p is the market price of the resource and MEC is marginal extraction cost. This condition states that the difference between discounted price and discounted MEC must be equal in all periods. This condition applies even when demand and/or marginal extraction cost vary over time. For Case 2R with MEC > 0:

$p_0 - MEC = \$7.09 - \$5 = \$2.09$ and

$p^0_1 - MEC^0_1 = (\$7.26 - \$5)/1.08 = \$6.72 - \$4.63 = \$2.09$.

Because p − MEC = MUC in pure competition, the first equilibrium condition for efficient intertemporal extraction of an exhaustible resource can be written as:

$MUC_0 = MUC_1 = ... = MUC_T$.

In equilibrium, MUC must be the same in all periods. As shown in the two-period analysis, MUC is positive when the total demand for an exhaustible resource exceeds the available supply. If $MUC_t > MUC_{t'}$ ($t \neq t'$), then NSB is increased by raising extraction in period t and lowering extraction in period t′ until MUC is equal in all periods.

When the first equilibrium condition is satisfied, MUC increases at a compound rate of r over time:

$MUC_t = MUC_0(1 + r)^t$ for $t = 1,...,T$,

where r is the discount rate.

The *second equilibrium condition* requires that total extraction of the resource equals total resource stock. To satisfy this condition, resource price must increase at such a rate that quantity demanded of the resource becomes zero in the same period as the stock of the resource is exhausted. The second equilibrium condition ensures there is neither a surplus nor a shortage of the resource at the end of the planning horizon (T).

TIME PATH OF PRICES AND EXTRACTION. The two equilibrium conditions have implications for the time path of resource prices and extraction. In the special case where resource demand is constant and MEC = 0 over time, the path of resource prices is described by *Hotelling's condition*:

$p_t = p_0(1 + r)^t$ for $t = 1,...,T$.

This condition requires:

$(p_{t+1} - p_t)/p_t = r$,

which implies that resource price increases smoothly and exponentially over time by the rate of discount. When prices increase in this manner, resource extraction decreases smoothly and exponentially over time. The time paths for resource price and extraction rates implied by Hotelling's condition are illustrated in Figure 7.8.

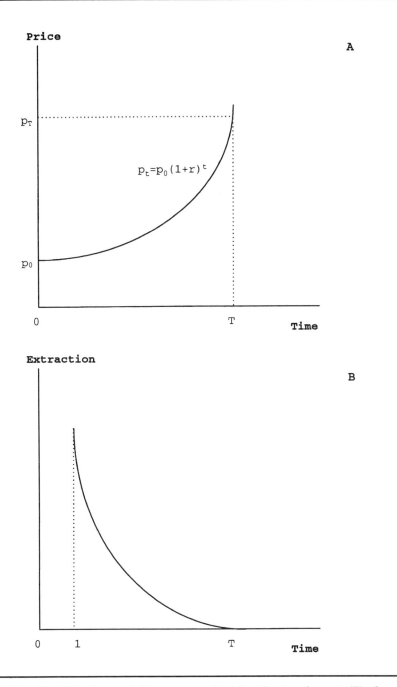

Figure 7.8. Hotelling time path for resource price (A) and extraction rate (B) when resource demand is constant and marginal extraction cost (MEC) = 0 in all periods.

When MEC = 0, equilibrium requires p = MUC, which requires prices to increase in accordance with increases in MUC. When MEC is greater than zero and constant over time, the time path of prices is displaced upward by MEC. In period T, the extraction rate is zero and resource price reaches its highest level, namely, p_T.

Changes in resource demand, resource supply and the discount rate influence the equilibrium time path of resource extraction and prices. Demand can increase due to increases in population and income and can decrease due to lower prices for substitutes. MEC can increase over time when the resource stock is not homogenous (stock effects) and per unit taxes on resource extraction (severance taxes) are increased. MEC can decrease over time due to technological improvements that increase the efficiency of resource exploration and extraction and lower severance taxes. Other things equal, a higher discount rate accelerates resource extraction and a lower discount rate decelerates resource extraction. The direction of change in near-term resource prices and extraction for different demand and supply conditions is summarized in Table 7.1.

Consider the implications of an increasing demand for oil demand with constant MEC. Increases in the demand for oil imply that oil prices increase over time. This causes the present value loss in returns from current oil extraction to increase, which means MUC is higher in later periods than in earlier periods. Equalization of MUC across periods, which is required by the first equilibrium condition, necessitates shifting oil extraction from the present to the future.

How do oil companies react to an increase in resource demand? If oil companies do not adjust intertemporal resource extraction rates, then oil shortages would occur in the future and overall profit would be lower. By shifting extraction from the present to the future, companies can eliminate shortages and increase profit. Oil companies would continue to shift extraction from the present to the future until profit margins on oil extraction (p − MEC) are equal in all periods.

As shown in Table 7.1, certain combinations of changes in resource demand and supply result in a particular change (up or down) in resource price and extraction and other combinations result in indeterminate changes in resource price and extraction. Because an exhaustible resource is finite, higher extraction rates shorten the period required to exhaust the stock of the resource. Improvements in technology can lengthen this period.

Table 7.1. Directional movement in near-term prices and extraction of an exhaustible resource for various resource demands and MEC

Future Change		Near-Term Change	
In Demand	In MEC	In Price	In Extraction
Constant	Constant	Smooth increase	Smooth decrease
Increase	Constant	Increase	Decrease
Increase	Decrease	Indeterminate	Decrease
Increase	Increase	Indeterminate	Indeterminate
Decrease	Constant	Decrease	Increase
Decrease	Decrease	Decrease	Indeterminate
Decrease	Increase	Decrease	Increase
Constant	Decrease	Increase	Decrease
Constant	Increase	Decrease	Increase

MEC = marginal extraction cost.

Underlying Factors

This section describes how six factors influence the efficient rate of resource extraction and resource prices: a) technological progress, b) imperfect competition, c) external costs, d) discount rate, (e) recycling and (f) exploration and development. Conclusions are based on the assumption that only one factor at a time is being varied.

TECHNOLOGICAL PROGRESS. Technological progress can influence resource extraction and prices two ways. First, it can increase the efficiency of resource extraction. Second, technological progress can lead to development of substitutes for the resource. Suppose a developing technology has the potential to reduce future MEC relative to current MEC for oil. This causes future MUCs to exceed current MUCs, which violates the first equilibrium condition. To restore equilibrium, oil extraction must be shifted from the present to the future. How do oil companies react to the new technology? Part of the decrease in future extraction costs will be passed on to consumers in the form of lower prices. If demand for oil is elastic, then lower future oil prices would increase future oil consumption and revenues. The expected increase in future revenues would prompt oil companies to shift oil extraction from the present to the future.

Suppose technological progress lowers the cost of photovoltaic cells (electricity from solar energy) and makes electricity production from photovoltaic cells competitive with electricity production from oil-powered sources. This change in the relative cost of electricity might cause some households to substitute photovoltaic cells for oil furnaces, which would reduce the demand for and price of oil. If the decline in oil prices were appreciable, then oil extraction would likely decline.

A major concern with technological progress is the uncertainty regarding its timing and impact. Because technological progress is usually the result of engineering and management innovations, the rate of development of new extraction technologies and substitutes is affected by nonmarket conditions. The history of many industrialized countries indicates that technological progress is accelerated when there is a greater need for the technology. Development of synthetic rubber during World War I was the direct result of channeling natural rubber into the production of tires for airplanes and vehicles used in the war effort. As pointed out in Chapter 1, technological progress often increases the use of natural resources and degrades the environment. Development of inorganic fertilizer decreased the amount of land needed for crop production (fertilizer was substituted for land) but increased the use of natural gas (which is used to produce inorganic fertilizer); hence, extraction of natural gas increased.

IMPERFECT COMPETITION. Most markets for exhaustible resources are not purely competitive. Except for publicly sanctioned monopolies such as those established for the local and regional distribution of oil, gas and water, a monopoly is illegal in the United States. Nevertheless, exploration, development and marketing of many exhaustible resources, especially energy and minerals, are done by a few, large companies. Hence, most natural resource markets have elements of imperfect competition.

How does imperfect competition influence efficient intertemporal extraction of an exhaustible resource? This question is answered by comparing efficient extraction of an exhaustible resource under pure competition and imperfect competition. Let the oil reserve be 60 bbl and assume that the markets for all other inputs used in extraction of oil are purely competitive. As shown in the previous section, efficient extraction in purely competitive markets requires $p^0_t - MEC^0_t$ to be the same in all periods (ignoring environmental costs) and exhaustion of the resource. Efficient extraction with pure competition is 14.53 bbl and equilibrium price is $7.09 per bbl in the current period, and 45.47 bbl and $7.26 per bbl in the future period, as shown in Figure 7.9. This is the same equilibrium illustrated in Figure 7.7. Hence, with pure competition, an oil reserve of 60 bbl constitutes a supply restriction.

Under imperfect competition, profits are maximized by selecting resource extraction rates such that MR − (MEC + MUC) is equal in all periods on a discounted basis, and total extraction equals the stock of the resource. The abbreviation MR stands for marginal revenue, which is the change in total revenue from a one-unit change in resource extraction. When the market demand for a resource is negatively sloped, marginal revenue equals one half of the price. Hence, the slope of the MR curve is one half of the slope of the demand curve.

For the demand and marginal extraction costs illustrated in Figure 7.9, efficient resource extraction and price in an imperfectly competitive market is $Q^*_{0m} = 12.5$ bbl and $p^*_{0m} = \$7.50$ per bbl in the current period and $Q^*_{1m} = 25$ bbl and $p^*_{1m} = \$17.50$ per bbl in the future period. Efficient extraction is slightly lower (12.5 bbl < 14.53 bbl) and equilibrium price is slightly higher ($7.50 > $7.09) in the current period with imperfect competition than with pure competition. In the future period, efficient extraction is substantially lower (25 bbl < 45.53 bbl) and price is much higher ($17.50 > $7.26) with imperfect competition than with pure competition. Because total oil extraction is less than the oil reserve, 22.5 bbl of oil (60 − 137.5 bbl) are unused at the end of the future period.

The time paths for resource price when MEC is constant over time and the same discount rate is used for imperfect competition and pure competition are illustrated in Figure 7.10. When the price of oil exceeds the price of an oil substitute (p_S), consumers of oil switch to the less expensive substitute, and the quantity demanded of oil becomes zero. Therefore, the upper limit on oil price is p_S. Because the initial price of oil is higher with imperfect competition than with pure competition ($p_{0m} > p_{0c}$), more time is required to exhaust the resource under imperfect competition than under pure competition ($T_m > T_c$). In other words, extraction of the resource is spread out over a longer period with imperfect competition. When the discount rate is positive, shifting oil extraction from the present to the future reduces NSB. Therefore, NSB is lower with imperfect competition than with pure competition. Different results can occur when MEC and the discount rate are not the same for pure competition and imperfect competition.

EXTERNAL COSTS. The first equilibrium condition for efficient resource extraction requires that marginal user cost be the same in all periods. Extraction of many exhaustible resources generates an external diseconomy, which imposes a cost on affected parties. Surface mining of coal severely disturbs the landscape. Mining of precious metals, like gold and silver, result in slag piles. Rain leaches

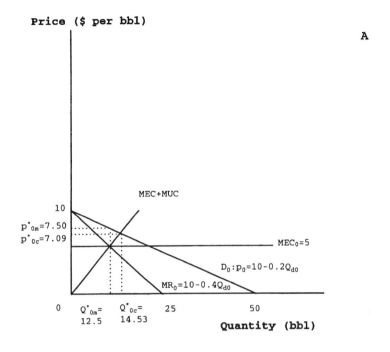

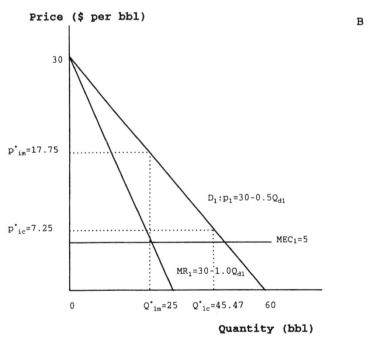

Figure 7.9. Efficient extraction rates and prices for oil under pure competition (c) and imperfect competition (m) with a supply restriction of 60 bbl in current period (A) and future period (B). MEC = marginal extraction cost; MUC = marginal user cost; and MR = marginal revenue.

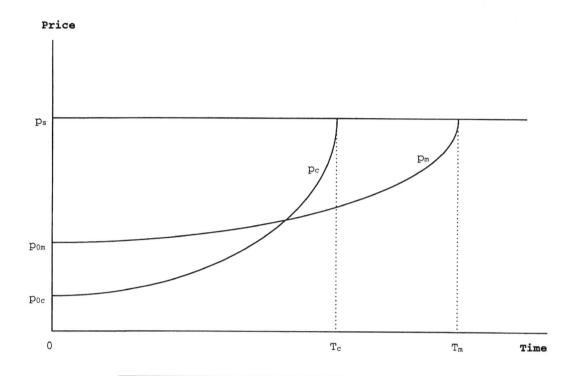

Figure 7.10. Time path of resource price and time to exhaustion under pure competition (p_c, T_c) and imperfect competition (p_m, T_m) when p_s is price of a substitute resource.

heavy metals from slag piles. The metals are transported to streams and lakes in runoff, causing water pollution. Burning fossil fuels contributes to global warming and acid precipitation. Environmental pollution can diminish human health and reduce the capacity of the environment to assimilate wastes and perform vital ecological services. Losses in environmental quality from the extraction and use of exhaustible resources are an external cost. External costs should be taken into account when selecting the socially efficient extraction rates for exhaustible resources.

Suppose the marginal environmental cost of oil extraction in period t (MNC_t) is positive when oil extraction exceeds a threshold extraction rate of Q_h, as indicated in Figure 7.11. When the extraction rate is below Q_h, the environment assimilates all the disturbances and wastes associated with resource extraction. Therefore, $MNC_t = 0$ for $Q < Q_h$. When the rate of extraction exceeds assimilative capacity ($Q > Q_h$), MNC_t increases with the rate of extraction. Threshold extraction rates and the rate at which MNC_t increases with extraction are likely to vary over resources and time. For example, coal generates about 75 percent more carbon dioxide per 1,000 BTUs of heat energy than does natural gas.[2] Therefore, burning coal produces a greater risk of global warming than does burning natural gas. Underground mining of coal usually causes less environmental degradation than does surface mining of coal. Temporal differences in environmental degradation can also be significant. Once degradation of an ecological system exceeds a certain level, the loss in biodiversity increases rapidly.

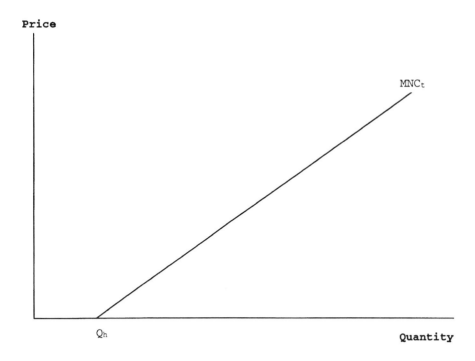

Figure 7.11. Marginal environmental cost (MNC) curve for an exhaustible resource in period t.

The efficient extraction rates for an exhaustible resource when marginal environmental cost are considered are illustrated in Figure 7.12. When MNC is ignored, the efficient extraction rate is Q_0 where p = MEC + MUC. When MNC is nonzero, the efficient extraction rate is Q'_0 where p'_0 = MEC + MUC + MNC. Because $Q'_0 < Q_0$ and $p'_0 > p_0$, the efficient extraction rate is lower, and the efficient price is higher, with than without environmental cost. In general, the condition for efficient intertemporal extraction of an exhaustible resource is that the present value of p − MEC − MUC − MNC be the same in all periods and that total resource extraction equals the stock of the resource.

DISCOUNT RATE. How do changes in the discount rate influence the efficient intertemporal extraction of an exhaustible resource? A lower discount rate causes oil prices to rise more slowly, according to Hotelling's condition, $p_t = p_0(1 + r)^t$. For a fixed demand, slower growth in oil prices means faster growth in quantity demanded of oil. As a result, the oil reserve is exhausted before the quantity demanded of oil reaches zero. There is excess demand for oil and the second equilibrium condition is violated.

To re-establish equilibrium, the initial price of oil (p_0) must be increased until once again the oil reserve equals total extraction. When p_0 is raised, oil prices are higher in earlier periods than later periods, which allows oil companies to increase

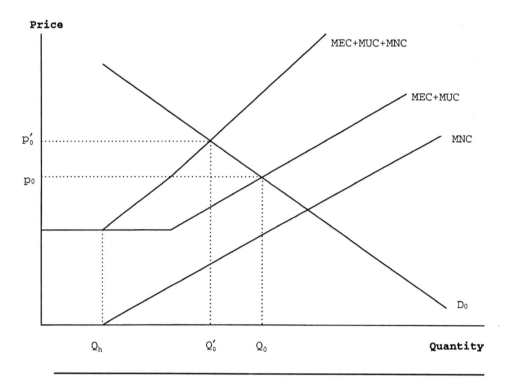

Figure 7.12. Comparison of equilibrium extraction rate (Q) and price (p) in current period considering (Q'_0) and ignoring (Q_0) environmental costs. MEC = marginal extraction cost; MUC = marginal user cost; MNC = marginal environmental cost; and D = demand.

profit by shifting extraction from later to earlier periods. A higher initial price, combined with a lower discount rate, increases the time required to exhaust the resource. A higher discount rate has the opposite effect. The initial price must be reduced so that the resource can be exhausted in a shorter period of time. This effect can be illustrated in Figure 7.10 by letting p_c be the price path before and p_m the price path after the discount rate increases. The higher discount rate increases the time to exhaustion from T_c to T_m.

How do oil companies respond to a lower discount rate? A lower discount rate increases the present value of future income from selling oil. Therefore, oil revenues can be increased by shifting extraction from the present to the future, provided marginal extraction costs do not rise over time. In contrast, a higher discount rate reduces the present value of future income from selling oil, which causes oil companies to shift extraction from the future to the present. Consequently, a change in the discount rate gives oil companies a profit incentive to make the same adjustment in extraction rates as required by the equilibrium conditions.

What are the intergenerational implications of changing the discount rate? When the amount of resource extracted is fixed, a lower discount rate shifts extraction from the present to the future, which increases the amount of the resource available for future generations. Conversely, a higher discount rate shifts extraction from the future to the present, which depletes the resource more rapidly and reduces the amount of the resource available for future generations. This intergenerational effect of the discount rate on resource extraction has sparked interest in other criteria for allocating exhaustible resources over time. Following the line of thinking proposed by Howarth and Norgaard,[3] intertemporal extraction can be determined by assigning future generations a property right to a portion of the current resource base. In essence, this reserves a portion of the current stock of the resource for future generations.

RECYCLING. Recycling converts residuals from production and consumption into useful products. Many materials are recycled, including newspaper, glass, metal, cardboard, plastic, motor oil and by-products of agricultural and industrial production. Products can be made from primary (virgin) material and/or secondary (recycled) material. Consider aluminum products, which include containers for a variety of consumer products, equipment, structures and motor vehicles parts. One of the major aluminum products is beverage containers. Aluminum beverage containers can be made from primary and/or secondary aluminum. Primary aluminum is manufactured from bauxite ore, which is mined. Secondary aluminum comes from recycled aluminum containers. Most consumers cannot tell the difference between aluminum products made from primary or secondary aluminum. This is not true of all products. Recycled paper can often be distinguished from virgin paper.

There are three types of recycling that apply to both exhaustible and renewable resources. *Closed-loop recycling* involves converting a residual into a new consumer product that has form and properties similar to the residual. Using recycled aluminum in aluminum containers or using old paper fiber in new paper products is each an example of closed-loop recycling. *Open-loop recycling* utilizes residuals in a product that is different in form and/or properties from the residual. Using the residuals from timber harvesting (bark and branches) to make a mulch for ornamental beds or utilizing sawdust from a lumber mill for animal bedding are examples of open-loop recycling. *Energy recycling* recovers energy from residuals. Using methane gas recovered from animal manure to generate electricity or burning solid waste to produce heat are examples of energy recycling. Because recycling is not 100 percent efficient, it generates its own wastes.

Recycling has benefits and costs that depend on the type of recycling and other factors. Consider the benefits and costs of closed-loop recycling of aluminum. There are several benefits. First, it decreases the demand for bauxite, which reduces the extraction of bauxite and extends the life of the bauxite resource. Second, environmental degradation from bauxite mining declines. Third, recycling aluminum decreases the energy required to mine, transport and process bauxite. Making aluminum products from recycled aluminum requires only 6 percent as much energy as producing these products from bauxite.[4]

Fourth, aluminum recycling reduces the amount of household and industrial trash, which increases the longevity of existing landfills and decreases the demand for new landfills. The longevity of a landfill is the time required for the landfill to reach its capacity. Increasing the longevity of landfills is a significant benefit, especially in major cities where landfills have reached or are near capacity and suitable space for new landfills is very limited. Landfills also involve an opportunity cost equal to what the land would earn in its next best use. Fifth, aluminum recycling reduces litter, which has aesthetic benefits.

Recycling of aluminum entails several costs. First, the household or business must separate aluminum residuals from other types of residuals. Second, the aluminum has to be transported to a recycling facility. Recycling costs are influenced by the technologies for gathering and processing residuals; the geographic concentration of recyclable materials; and government policies, such as tariffs, discriminatory freight rates, tax credits, severance taxes and depletion allowances.

Benefits and costs of recycling determine the extent to which households and businesses recycle their residuals. Consider a household that is deciding whether or not to recycle a residual and, if so, how much of the residual to recycle. The decision depends on the marginal private benefit (MPB) and marginal cost (MC) of recycling, as illustrated in Figure 7.13. MPB reflects the private benefit to the household of recycling. If the household is refunded a deposit when the recyclable material is returned, then MPB includes the refund. The figure also shows the marginal social benefit (MSB) of recycling. MPB is generally less than MSB because the household does not consider the benefits to society of recycling, such as extending the life of the primary resource and the longevity of landfills, decreasing waste loads and aesthetic benefits of reduced litter. MC include the marginal cost to the household of separating residuals (aluminum, paper, glass, etc.) and transporting them to a recycling center. The transport cost is negligible in cities that have curbside recycling.

When MC is greater than MPB, as shown in Figure 7.13A, the privately efficient amount of recycling is zero ($Q_r = 0$). The socially efficient amount of recycling is positive ($Q_s > 0$). When MC and MPB intersect, as depicted in Figure 7.13B, the privately efficient amount of recycling is positive ($Q_r > 0$) but less than the socially efficient amount ($Q_r < Q'_s$).

Next, consider a firm's decision whether or not to use secondary (recycled) materials in its production activity, and if so, how much secondary material to use. When primary and secondary materials are perfect substitutes, the isoquant is a straight line with a 45-degree angle, as illustrated in Figure 7.14A. In this case, the decision about how much primary and/or secondary material to use depends on the relative prices of the materials. If the price of the primary material exceeds the price of the secondary material, then the isocost line has a slope greater than 45 degrees and production cost is minimized by using only the secondary material (point a). Conversely, if the price of the primary material is less than the price of the secondary material, then the isocost line has a slope less than 45 degrees and production cost is minimized by using only the primary material (point b). In the event the primary and secondary materials have the same price, the isocost line is congruent with the isoquant and production cost is the same regardless of which combination of primary and secondary materials is used.

When the primary and secondary materials are less than perfect substitutes, production cost is minimized by using a combination of the two materials, as illustrated in Figure 7.14B. The higher the price of the primary (secondary) material relative to the price of the secondary (primary) material, the greater the use of the secondary (primary) material.

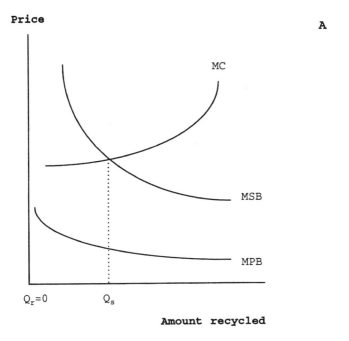

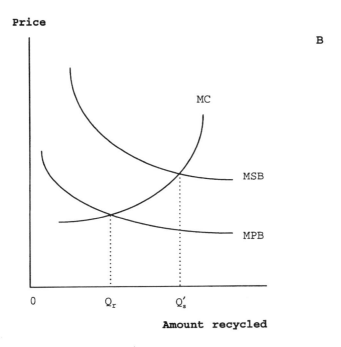

Figure 7.13. Privately (Q_r) and socially (Q_s) efficient amount of recycling showing when it is inefficient (A) and efficient (B) for a household to recycle. MC = marginal cost; MPB = marginal private benefit; and MSB = marginal social benefit.

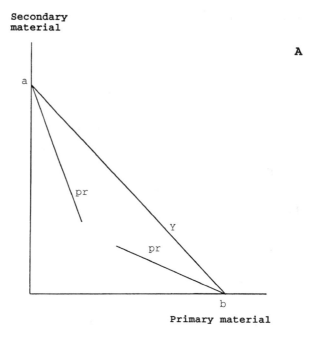

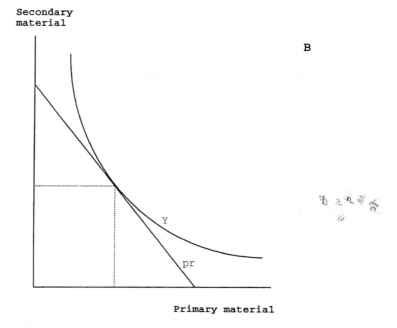

Figure 7.14. Efficient use of primary and secondary materials when they are perfect substitutes (*A*) and less than perfect substitutes (*B*). Y = isoquant; and pr = ratio of price of primary material to price of secondary material.

7. Exhaustible Resource Use 153

EXPLORATION AND DEVELOPMENT. While the stock of an exhaustible resource is fixed, the size and location of specific resource deposits are uncertain. The degree of uncertainty is higher for certain exhaustible resources (petroleum) than others (coal). *Exploration* is the process of determining the location and size of resource deposits. If the search results in a discovery, then exploration continues until the extent and size of the deposit are confirmed. If a deposit is economical to produce, then development follows. *Development* involves the activities leading up to extraction of the resource. If resource deposits are not found in favorable areas, then estimates of the resource stock are usually revised downward. Significant discoveries can lead to upward revision in the estimated stock of the resource, which exerts downward pressure on resource prices. Downward (upward) revision in the resource stock decreases (increases) the time to exhaustion, other things equal.

Summary

Exhaustible resources have the fundamental property that current extraction decreases the stock of the resource, which reduces the amount of the resource available in future periods. This property of exhaustible resources makes resource extraction and prices dynamic or time-dependent. Market dynamics can be introduced by placing a time index on stocks, extraction, prices and demand/supply shifters such as population, income, availability of substitutes and technology. Efficient intertemporal extraction of an exhaustible resource is determined by maximizing net social benefit, which equals the present value of the sum of producer surplus and consumer surplus over all periods.

In general, efficient intertemporal extraction of an exhaustible resource requires the present value of resource price minus marginal extraction cost minus marginal user cost minus marginal environmental cost to be the same in all periods, and all of the resource to be extracted. The time path of resource extraction and prices is influenced by extraction costs, resource demand and the discount rate. In the special case where resource demand is constant and MEC is zero in all periods, resource price increases over time by the discount rate.

Efficient extraction rates and prices for an exhaustible resource are influenced by several factors, including technological progress, imperfect competition, external cost, discount rate, recycling and exploration and development. Technological progress can dampen and even reduce resource prices. Relative to pure competition, imperfect competition typically extends the period needed to exhaust the resource, which reduces net social benefit from the resource. Consideration of external environmental costs of resource extraction reduces the socially efficient extraction rates and increases resource price. Higher (lower) discount rates increase (decrease) the rate of extraction in the near term and shorten (lengthen) the period over which the resource is exhausted. Recycling of residuals generated by consumption or production of products made using exhaustible resources has several benefits and costs, including conserving the resource, decreased energy use, extending the longevity of landfills, reduced littering and others. Resource exploration is undertaken to find additional resource deposits and to reduce uncertainty regarding the location and size of resource deposits.

Questions for Discussion

1. Modify the stock accounting relationship for an exhaustible resource ($S_t = S_0 - U_{t-1}$) to take account of recycling.

2. It is often argued that the discount rate for a private company is higher than the discount rate for society. What is the basis for this argument? How does the discrepancy between private and social discount rates influence the efficient intertemporal extraction of an exhaustible resource?

3. Show that Hotelling's condition ($p_t = p_0(1 + r)^t$ for $t = 1,...,T$) is consistent with the equilibrium condition for efficient intertemporal resource extraction ($p_t = MEC_t + MUC_t$ for $t = 1,...,T$) when resource demand is constant and $MEC = 0$.

4. Suppose the demand equations for oil are $p_0 = 10 - 0.20Q_0$ in the current period and $p_1 = 30 - 0.50Q_1$ in the future period. Marginal extraction cost is \$1 per bbl in the current period and \$2 per bbl in the future period. Determine the efficient prices and extraction in both periods and net social benefit when the oil reserve is 80 bbl and the discount rate is 10 percent.

5. A backstop technology is a technology that becomes economical to use when the price of an exhaustible resource exceeds a certain level. For example, electricity generated from photovoltaic cells is a backstop technology for electricity generated from fossil fuels. How is the efficient intertemporal extraction of an exhaustible resource likely to be influenced by backstop technology? Does it make any difference whether the backstop technology for an exhaustible resource involves extraction of another exhaustible resource or exploitation of a renewable resource?

6. The rate of recycling in St. Lucia is low. The St. Lucia City Council is reviewing a policy recommendation from the Resource Recovery Commission to initiate curbside recycling. The cost of providing this service would be paid by the city. Use Figure 7.13 to illustrate the likely effects of this recommendation on household recycling in St. Lucia. Do the effects of this policy depend on who pays the cost of providing curbside recycling?

Further Readings

Common, Michael. 1988. "Natural Resource Exploitation." Chapter 7 in *Environmental and Resource Economics: An Introduction*. New York: Longman Inc., pp. 198–268.

Dasgupta, P. S. and G. M. Heal. 1979. *Economic Theory and Exhaustible Resources*. Cambridge: Cambridge University Press.

Hotelling, H. 1931. "The Economics of Exhaustible Resources." *Journal of Political Economy* 39:137–175.

Notes

1. While MEC = 0 is highly unlikely, it is evaluated for two reasons. First, it provides a stepping stone to the more complex case of MEC > 0. Second, MEC = 0 is one of the assumptions underlying Hotelling's condition, which explains the behavior of prices and extraction over time.

2. Christopher Flavin, "The Bridge to Clean Energy," *World Watch* (Washington, D.C.: Worldwatch Institute, July/August 1992), pp. 10–18.

3. Richard B Howarth and Richard B. Norgaard, "Intergenerational Resource Rights, Efficiency, and Social Optimality," *Land Economics* 66(1990):1–11.

4. John E. Young, "Aluminum's Real Tab," *World Watch* (Washington, D.C.: Worldwatch Institute, March/April 1992), pp. 26–33.

CHAPTER 8

Renewable Resource Management

Destruction of tropical forests drives the extinction of countless unique plants and animals. It also adds to Earth's greenhouse effect, spurs erosion of valuable topsoil, and exacerbates deadly flooding.

—RICHARD MONASTERSKY, 1990

Renewable resources have a capacity to renew or regenerate themselves. There are two types of renewable resources: *conditionally renewable* or *nondegradable*. *Conditionally renewable resources,* such as soil, water, fish, wildlife and forests, have a regenerative capacity, which is influenced by a variety of natural processes and human activity. Soil regeneration is influenced by chemical, geological, hydrological and biological processes. Water regeneration is controlled by the hydrologic cycle, which is affected by solar energy, climate and topography. Regeneration of fish, wildlife and forests is influenced primarily by biochemical and hydrologic processes. The rates at which conditionally renewable resources are harvested affects their capacity to regenerate particularly when harvest rates exceed rates of regeneration.

A conditionally renewable resource is degraded when use rates exceed regenerative capacity. Excessive degradation can lead to total exhaustion or extinction of the resource. Huge irrigation diversions from the Aral Sea in Soviet Central Asia have caused what was once the world's fourth-largest inland sea to be dramatically reduced in size and decimated a highly productive fishery.[1] Use rates for certain exhaustible resources can influence the regenerative capacity of conditionally renewable resources. Extensive use of coal to generate electricity in the Midwestern United States produces large amounts of sulfur dioxide, which, in combination with rain water, produces acid precipitation. Deposition of acid precipitation in forests and lakes in the northeastern United States and eastern Canada has decreased the productivity of these ecosystems. Sustainable use of conditionally renewable resources requires that use rates be kept below regenerative rates. When this occurs, the flow of human and environmental services provided by these resources can be maintained indefinitely, which benefits current and future generations.

Nondegradable renewable resources, such as energy from the sun, wind and tides, have a regenerative capacity that is not influenced by human activity. Only a small fraction of the energy contained in nondegradable renewable resources is directly utilized in economic activities. For example, only a small amount of the solar energy that reaches the earth's surface is used by plants in photosynthesis. While human activities do not influence the regenerative capacity of nondegradable resources, they influence other environmental conditions. For example, atmospheric testing of nuclear devices introduces highly toxic substances into the atmosphere and the burning of fossil fuels contributes to global warming.

This chapter focuses on the management and use of conditionally renewable resources in general and biological resources in particular. The bioeconomic implications of three management regimes are evaluated: *private property, common property with limited access* and *common property with unlimited access*. Private and socially efficient harvest rates and optimal resource stocks are evaluated under static and dynamic conditions. Harvest rates that result in *maximum sustainable yield, species extinction* and *maximum net social benefit* are compared.

Simplifying Assumptions

The economic analysis of conditionally renewable resources presented here is based on a combination of the following assumptions:

Assumption 1. The resource is private property, which implies limited access. There is either a single owner or manager or multiple owners with a single manager. The resource manager's objective is to maximize profit from use of the resource. An example of this assumption is a privately owned ranch. Only the owners of the ranch and other individuals designated by the owners have a legal right to use the property.

Assumption 2. The resource is common property with limited access. The resource manager's objective is to maximize total profit from use of the resource. An example of this assumption is when a tribe rations the use of tribal grazing land among its members.

Assumption 3. The resource is common property with unlimited access. Each resource user attempts to maximize his or her own profit from use of the resource. An example of this assumption is when fishers have unlimited access to an ocean fishery.

Assumption 4. Biological growth processes, demand for the resource and cost of harvest are known. This assumption implies an absence of biological and economic uncertainty.

Assumption 5. The resource is managed for a single use and/or for a single species. Single-use management implies that a forest is managed for its commercial timber or recreational value, or that an ocean is exploited for its commercial fishery. Single species management ignores interdependencies among species in an ecosystem such as the potential negative effects on small mammals of managing a forest ecosystem for grizzly bear habitat. Managing for multiple species or uses is more complex than managing for a single use or species.

8. Renewable Resource Management

Assumption 6. The resource is managed for multiple uses and/or multiple species. By law, many publicly owned resources in the United States such as the national forest and lands managed by the Bureau of Land Management must be managed for multiple uses. Multiple-use management entails both competitive and complementary uses of the resource. Outdoor recreation and commercial timber production are likely to be competitive uses of a forest when timber harvesting rates are very high and/or there is clearcutting. Enhancing fish and wildlife habitat and improving fishing and hunting conditions are likely to be complementary uses of the forest.

Assumption 7. Markets for natural resources and inputs used in conjunction with natural resources are purely competitive. Pure competition requires complete and accurate knowledge of markets and technology, large numbers of buyers and sellers, and homogeneous and divisible inputs and outputs.

Natural Growth

The basic accounting relationship for a biological resource is:

$$S_t = S_0 + R_t - H_t - L_t.$$

This equation states that *resource stock* or biomass (S_t) equals initial biomass (S_0), plus cumulative biomass growth (R_t), minus cumulative harvest of the resource (H_t), minus cumulative biomass losses (L_t), all measured at the end of the current period (t). For example, in the case of an ocean fishery, S_t is the stock of fish (biomass) at the end of the current period, S_0 is the initial fish biomass, R_t is cumulative growth in fish biomass through the end of period t, H_t is cumulative harvest of fish through the end of period t, and L_t is cumulative fish losses through the end of period t. Losses are due to natural causes such as disease, tsunami and changes in ocean temperatures (El Nino), floods and earthquakes.

While this simple accounting relationship gives a good historical accounting of the resource stock, it has several deficiencies in terms of explaining changes in the stock and determining economically efficient harvest rates for the resource. First, it simplifies the dynamic nature of renewable resources by ignoring important interactions between harvest and growth over time. For example, when the rate of harvest exceeds the threshold rate of growth needed to preserve the species, the species can become *extinct* at which point the resource stock (S_t) is zero.

Second, it does not consider how changes in technology and resource prices influence harvest. For example, improvements in technology that increase harvest rate per unit of effort have important implications for the profitability of resource extraction, resource price and resource conditions. A more comprehensive model of renewable resource management is needed that takes into account the interaction between growth and harvest over time.

Natural growth is the growth in excess of biomass losses from natural causes. Natural growth in Beluga whales equals additions to the population from births minus losses due to mortality. It describes the change in biomass in the absence of commercial harvesting and can be discrete or continuous.

Discrete growth occurs in discrete time periods (week, month or year) over a finite or infinite number of periods as indicated by the following equation:

$$G_t = S_t - S_{t-1} \qquad (t = 0,...,T).$$

This equation states that growth in period t (G_t) equals the difference between biomass at the end of the current period (S_t) and biomass at the end of previous period (S_{t-1}). Discrete time is indicated by placing a t subscript on a variable. T is the number of periods in the planning horizon and is referred to as the length of the planning horizon.

Continuous growth occurs continuously over a finite or infinite number of time periods:

$$G(t) = dS(t)/dt \qquad (t = 0,...,T).$$

This equation says that growth at time t equals the instantaneous change in S(t), or the change in S(t) for a small change in t, which is written as dS(t)/dt. Continuous time is denoted by placing a parenthesized t on a variable. When biomass growth depends on the level of biomass, as in the last two equations, growth is said to be *density dependent*. To indicate that growth is dependent on biomass, growth functions are sometimes written as $G(S_t)$ for discrete growth or G[S(t)] for continuous growth.

Natural growth rates for biological resources differ over space and time due to natural variation in nutrients, precipitation, temperature, topography, wind and other biogeophysical conditions. While no single mathematical equation best describes natural growth in biomass, the *logistic curve* has considerable appeal. The logistic growth rate in continuous time is:

$$G[S(t)] = iS(t)[1 - S(t)/S_{max}],$$

where i is the intrinsic growth rate and S_{max} is the maximum stock. A logistic growth curve is depicted in Figure 8.1. Growth rate, G(S), increases at a decreasing rate as the stock increases from S_{min} to S_{msy}. S_{min} is the minimum or threshold biomass. When $S_{min} = 0$, as in Figure 8.1, there is *exhaustion of the resource* or *species extinction*. S_{msy} is the stock that achieves *maximum sustainable yield* (q_3), which is the maximum growth that can be sustained by the resource. As the stock increases from S_{msy} to S_{max}, the growth rate declines and eventually reaches zero at S_{max}. The magnitude of S_{max} is determined by the environmental carrying capacity.

Purely compensatory logistic growth requires $S_{min} = 0$ and that $G(S_t)$ or G[S(t)] be strictly concave (∩) as illustrated in Figure 8.1. For any initial biomass, if S(t) approaches S_{max} as t approaches infinity, then growth in biomass over time can be described by a logistic curve as illustrated in Figure 8.2.

The general *harvest function* for a renewable resource without the time designation is:

$$q = H(S, I)$$

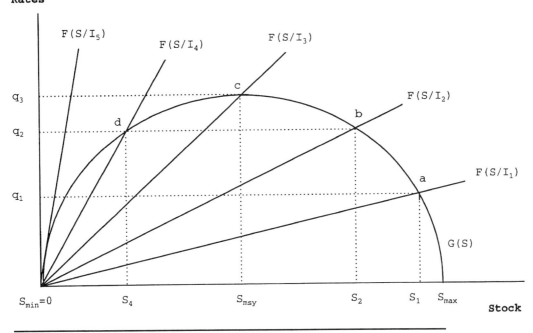

Figure 8.1. Purely compensatory logistic growth in stock of renewable resource, G(S), conditional harvest functions, $F(S/I_j)$ and efficient harvest rates (q).

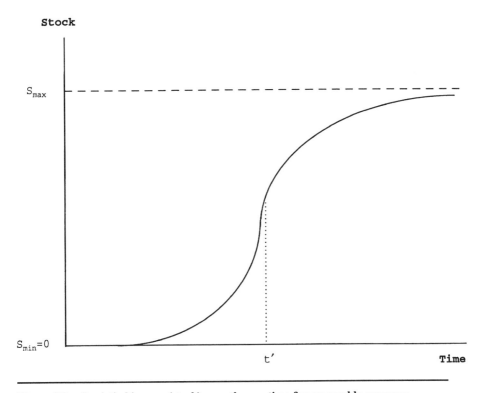

Figure 8.2. Logistic biomass (stock) growth over time for renewable resource.

where q is the rate of harvest, S is the stock and I stands for the inputs used in harvesting the resource. For the moment, I includes only variable inputs (fixed inputs ignored). The harvest function indicates that the harvest rate depends on both input use (I) and stock level (S). In general, q is an increasing function of I and S.

The *conditional harvest function* is:

$$q = F(S / I = I_j).$$

This function expresses the relationship between harvest rate (q) and the stock (S) when I is fixed at I_j. When input use is fixed, harvest rate increases as the stock increases, and vice versa. In other words, q is an increasing function of S for fixed I. Conditional harvest functions for input uses I_1, I_2, I_3, I_4 and I_5 are shown in Figure 8.1. Each conditional harvest function is a ray or straight line through the origin. This implies that harvest is zero when stock is zero and harvest per unit of stock is constant when input use is constant. Harvest increases as input use increases from I_1 to I_5. Therefore, the conditional harvest function rotates counterclockwise around the origin as I increases.

Biologically efficient harvest rates occur at q_1, q_2 and q_3 where the conditional harvest function intersects the growth curve. Why are these rates biologically efficient? q_1 is the most biologically efficient harvest rate for I_1 because increasing input use ($I > I_1$) causes the harvest rate to exceed the natural growth, which decreases the stock. Conversely, at a lower input use ($I < I_1$), natural growth exceeds the harvest rate causing the stock to increase. Biologically efficient harvest rates increase with higher input use as long as $I < I_3$, namely, $q_3 > q_2 > q_1$.

What happens when input use exceeds I_3? Consider I_4, which corresponds to the conditional harvest function labeled $F(S/I_4)$. Input use is so high relative to natural growth at I_4 that the stock falls below the level that achieves maximum sustainable yield, $S_4 < S_{msy}$, and harvest is less than at I_3 ($q_2 < q_3$).

Plotting the biologically efficient harvest rates in Figure 8.1, namely, q_1, q_2 and q_3, with the corresponding input use gives the yield function depicted in Figure 8.3. The *yield function* relates biologically efficient harvest rates to the corresponding input use. Input use greater than I_3 is biologically inefficient because harvest rates decrease for $I > I_3$. For example, I_4 and I_2 give the same harvest rate, namely, q_2, but I_4 is substantially greater than I_2.

Static Efficiency

PRIVATE PROPERTY. This section examines the management of privately owned renewable resources (assumption 1) when there is no uncertainty (assumption 4), a single use for the resource or a single species being managed (assumption 5) and pure competition (assumption 7). In addition, decisions are assumed to be made in a static setting, which makes single-period analysis appropriate. When a resource is privately owned, only the resource owner and those having permission from the owner are allowed to use the resource. Others are excluded from using the resource, which makes access limited.

8. Renewable Resource Management 163

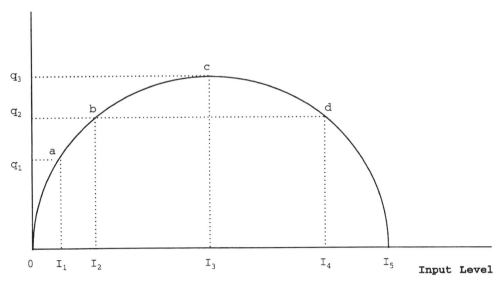

Figure 8.3. Yield function for renewable resource with purely compensatory growth.

Total profit or *economic rent* for a privately owned resource is the difference between total revenue (TR) and total cost (TC):

$\pi(q) = TR(q) - TC(q)$.

Maximizing $\pi(q)$ with respect to q gives the profit maximizing rate of harvest. Total profit can also be defined in terms of input use as follows:

$\pi(I) = TR(I) - TC(I)$.

Maximizing $\pi(q)$ with respect to I gives the profit maximizing input use. Both approaches yield the same efficient harvest rates and input use. In some applications it is easier to work with $\pi(q)$ than $\pi(I)$, and vice versa. In the following explanation, the efficient harvest rate is determined by maximizing $\pi(q)$. In the second example of profit maximization given in the next section, efficient input use is determined by maximizing $\pi(I)$.

When the resource market is purely competitive, the price of the resource (p) does not depend on the harvest rate (q). Therefore, TR = pq, which indicates that total revenue is a linear function of q as shown in Figure 8.4A. The total cost curve (TC) is derived from the yield function given in Figure 8.3. For simplicity, all inputs are assumed to be variable, which represents economic conditions in the long run. Multiplying the amount of input from the yield curve by its price gives total cost. When input markets are purely competitive, input prices do not depend on the quantity of inputs purchased. If input use I_2 requires i_1 units of labor, i_2 units of fuel and i_3 units of equipment and the market prices of these inputs are c_1, c_2 and c_3, respectively, then total cost for I_2 is:

$$TC(I_2) = TC_2 = c_1 i_1 + c_2 i_2 + c_3 i_3.$$

Other values of TC are derived by repeating this calculation for other input levels.

Plotting total cost for different input uses with the corresponding harvest rates gives the total cost curve in Figure 8.4A. Harvest rate q_3 and total cost TC_3 correspond to maximum sustainable yield. Only the segment of the total cost curve between 0 and f contains economically efficient harvest rates. Harvest rates on the backward bending segment of the total cost curve (f to h) are economically inefficient because these harvest rates can be achieved at lower total cost by decreasing input use. For example, I_4 is greater than I_2, yet both levels of input use provide the same harvest rate (q_2). Therefore, TC is greater with I_4 than with I_2 ($TC_4 > TC_2$). Total profit is maximized at a harvest rate of q_2 where the difference between TR and TC, namely, $TR_2 - TC_2$, is a maximum. Harvest rates above or below q_2 are less profitable. Because $q_2 < q_3$, the profit maximizing harvest rate is less than maximum sustainable yield.

Figure 8.4B depicts the profit maximizing harvest rate in terms of marginal revenue (MR) and marginal harvest cost (MHC). MR equals the change in total revenue with respect to harvest. Because the resource is sold in a purely competitive market, MR equals resource price, which makes the MR curve horizontal. MHC is the change in TC with respect to a change in the harvest rate, which equals the slope of the TC curve. Because the slope of the TC curve increases as harvest increases, MHC increases at an increasing rate as harvest rises. For a harvest rate greater than q_2, MHC > p, which implies cost increases more than revenue. Conversely, for a harvest rate less than q_2, p > MHC, which implies revenue increases more than cost. Profit is a maximum at q_2 where p = MHC. As resource price increases and/or harvest cost decreases, the harvest rate that maximizes profit approaches maximum sustainable yield ($q_2 \rightarrow q_3$).

In summary, when a privately owned, conditionally renewable resource is managed for maximum profit under static conditions:

1. The profit maximizing harvest rate is generally less than maximum sustainable yield ($q_2 < q_3$).
2. The profit maximizing harvest rate approaches maximum sustainable yield ($q_2 \rightarrow q_3$) as resource price increases and/or marginal harvest cost decreases.
3. In the unlikely event that marginal harvest cost is zero, the profit maximizing harvest rate equals maximum sustainable yield ($q_2 = q_3$).
4. The resource is not likely to become extinct because the input use that maximizes profit is below the input use that leads to extinction ($I_2 < I_5$).

EXAMPLES. Two examples are used to illustrate how the profit maximizing harvest rate for a privately owned renewable resource is determined. The first example does not specify the mathematical form of the yield function, whereas the second example does.

8. Renewable Resource Management

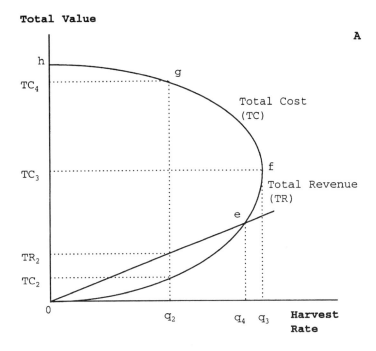

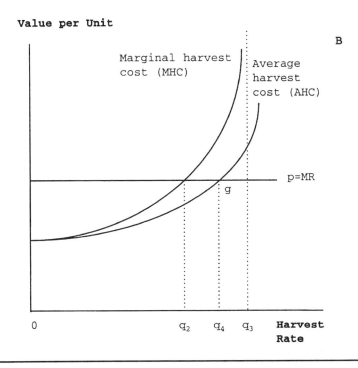

Figure 8.4. Profit maximizing harvest rate based on total (*A*) and marginal (*B*) values.

The first example is based on input use, harvest rates, total cost, total revenue and profit given in Table 8.1. The table includes only the efficient portion of the yield function. Input price is $0.50 and resource price is $3. Input use is in column 1 and harvest rate is in column 2. Total cost equals input use (column 1) times the price of the input. Total revenue is harvest rate (column 2) times the price of the resource. Profit is total revenue (column 4) minus total cost (column 3). Profit reaches a maximum of $2 at a harvest rate of 2 and an input use of 8. Maximum sustainable yield is greater than the profit maximizing harvest rate (5 > 2). In terms of Figure 8.4A: $q_2 = 2$ and $I_2 = 8$; $q_3 = 5$ and $I_3 = 35$; and $q_4 = 4$ and $I_4 = 24$. Note that profit is negative (there is a monetary loss) at maximum sustainable yield.

In the second example, growth in biomass is logistic and the *harvest function* is $q = \alpha SI$ where α is a constant. For this harvest function, harvest per unit of input approaches zero as the stock approaches zero ($q/I \to 0$ as $S \to 0$). The resulting *yield function* is:

$$q = \alpha NI(1 - \alpha I/\beta),$$

where N is environmental carrying capacity and β is the intrinsic rate of growth in the resource. For this yield function, harvest is a maximum when input use is[2]:

$$I = \beta/2\alpha.$$

Harvest rates (q) corresponding to different input use (I) for $\alpha = 0.002$, $N = 150$ and $\beta = 0.40$ are given in Table 8.2. q increases as I increases from 0 to 100 and decreases when I exceeds 100. Maximum sustainable yield occurs at $I = 100$.

As stated earlier, the profit function in terms of I is:

$$\pi(I) = TR(I) - TC(I).$$

Substituting pq(I) for TR(I) and cI for TC(I), where p is the price of the resource and c is unit input cost, gives the following profit function:

$$\pi(I) = pq(I) - CI.$$

Substituting the yield function for q(I) in the last expression gives:

$$\pi(I) = p\alpha NI(1 - \alpha I/\beta) - CI.$$

Substituting the numerical values for α, N and β used in Table 8.2 into the last ex-

Table 8.1. Example of harvest rates for different input levels and corresponding total cost, total revenue and profit

Input Level	Harvest Rate	Total Cost ($)	Total Revenue ($)	Profit ($)
3	1	1.50	3.00	1.50
8	2	4.00	6.00	2.00
15	3	7.50	9.00	1.50
24	4	12.00	12.00	0
35	5	17.50	15.00	−2.50

Profit is a maximum at $I = 8$.

8. Renewable Resource Management

Table 8.2. Harvest rates and input levels for $q = \alpha N\, I(1 - \alpha I/\beta)^a$

Input Level	Harvest Rate
20	5.4
40	9.6
60	12.6
80	14.4
100	15.0
120	14.4
140	12.6
160	9.6

[a] q is the rate of harvest, N is environmental carrying capacity, I is input, α is a constant, and β is the intrinsic rate of growth in the resource. $\alpha = 0.002$, $N = 150$, and $\beta = 0.40$.

pression, setting the derivative of the resulting expression with respect to I equal to zero and solving for I gives the following profit maximizing input use:

$$I^* = \frac{0.3p - c}{0.003p}.$$

For $p = \$20$ and $c = \$4$, $I^* = 33.33$. Substituting I^* into the yield function gives a profit maximizing harvest rate of $q^* = 8.33$. Substituting q^* and I^* into the harvest function and solving for S gives an optimal stock of $S^* = 124.96$. Finally, substituting p, c, q^* and I^* into the profit function gives a maximum profit of $33.28. Note that the profit maximizing harvest rate is less than maximum sustainable yield (8.33 < 15). For $p = \$25$ (higher price) and $c = \$3$ (lower cost of input), the profit maximizing harvest rate increases from 8.3 to 12.6. Therefore, as p increases and/or c decreases, q^* approaches q_{max}.

COMMON PROPERTY. Many renewable resources such as national forests, big game herds and ocean fisheries are common property resources. A *common property resource* is a resource owned by a large group of individuals whose access to the resource can be limited or unlimited. National forests in the United States are the common property of the general public and are managed for multiple uses by the United States Forest Service. Access to national forests is essentially unlimited although use of certain portions of the forest, such as campgrounds, usually require payment of a fee.

Access to common property resources can be limited by a variety of institutional arrangements. With pastoral grazing, such as in the developing countries of Africa, grazing land is the common property of a tribe. The tribe limits access to commonly owned grazing land by specifying the period of time each member is allowed to graze cattle. Grazing on tribal land is not permitted by nonmembers of the tribe. Grazing on publicly owned land in the western United States is managed by the Bureau of Land Management and the Forest Service. Both agencies issue grazing permits to private ranchers. A permit gives the rancher an exclusive right to graze a certain number of animals in a particular area. In return for grazing rights, ranchers pay an annual grazing fee to the agency, which is based on the fee per animal times the number of animals grazed in the area.

This section examines the use of renewable common property resources with limited access (assumption 2) or unlimited access (assumption 3). There is assumed to be no uncertainty (assumption 4), single-use or species management (assumption 5) and pure competition (assumption 7).

Limited Access. Suppose access to and harvesting of a renewable common property resource is controlled by a single resource manager. If the resource manager's objective is to maximize profit, then the manager selects the same input use and harvest rate as a private owner of the resource, namely, I_2 and q_2, respectively. There is typically no difference between the profit maximizing input use and harvest rate for a privately owned renewable resource and a commonly owned renewable resource with limited access when both resources are managed for maximum profit.

Unlimited Access. When access to a renewable common property resource is unlimited, use of the resource is quite different even when the objective of each resource user is to maximize profit. Total revenue exceeds total cost (there is excess profit) at the harvest rate that maximizes profit when access is limited (q_2 in Figure 8.4). When there is unlimited access, excess profit causes resource users to increase input use and harvest rates in order to capture some of the excess profit. In addition, excess profit is likely to attract new resource users. Input use and harvest increase until excess profit is zero at q_4. In terms of Table 8.1, input use increases from 8 with limited access to 24 with unlimited access. At a harvest rate of 24, profit is zero and there is no incentive for firms to increase harvest.

A harvest rate in excess of q_4 generates economic losses that stimulates resource users to reduce input use and harvest rates. In particular, for $q > q_4$, TC > TR or AHC > p, which results in economic losses. In an effort to reduce losses, resource users decrease harvest rates until q_4 is reached. For the revenue and cost relationships depicted in Figure 8.4, the profit maximizing harvest rate is greater with unlimited access than with limited access ($q_4 > q_2$) and both rates are less than maximum sustainable yield ($q_2 < q_4 < q_3$).

The profit maximizing harvest rate can also be determined using marginal and average cost curves as shown in Figure 8.4B. For a privately owned resource, profit is a maximum at the harvest rate that equates price to marginal harvest cost (p = MHC), namely, q_2. At q_2, p > AHC, which implies excess profit per unit of harvest; the same result obtained with the total revenue and total cost curves. When access to the common property resource is unlimited, excess profit causes harvest to increase until price equals average harvest cost (p = AHC at point g), which occurs at q_4 where profit per unit of harvest is zero. Point e in Figure 8.4A, at which there is zero economic profit, corresponds to point g in Figure 8.4B.

In general, there is greater likelihood of resource overexploitation and extinction for a renewable common property resource when access is unlimited than when it is limited. This occurs for several reasons. First, increases in resource price generally lead to higher input use when access to the resource is unlimited. Suppose resource price increases such that total revenue equals total cost at point h on the backward bending portion of the total cost curve in Figure 8.5. Excess profit is zero at h. Because point h corresponds to point d in Figures 8.1 and 8.3, the input use at point h is I_4 and the stock is S_4. Because $S_4 < S_{msy}$, the resource is being overexploited at

8. Renewable Resource Management

an input use of I_4. In general, when there is unlimited access, increases in resource price bring about increases in input use that cause the optimal stock of the resource to approach S_{min}. Therefore, the likelihood of overexploitation and extinction is much greater with unlimited access than with limited access, other things equal.

A second source of overexploitation is optimism regarding the productivity of the resource. Suppose resource users in the industry overestimate the yield for the resource. In this case, the estimated yield function lies above the true yield function as shown in Figure 8.6A. Based on the estimated yield function, resource users believe I_6 results in maximum sustainable yield; however, the true maximum sustainable yield occurs at I_7, which is substantially less than I_6. Because the true yield function is below the estimated yield function, the true total cost curve (TC′) is to the left of the estimated total cost curve (TC) as shown in Figure 8.6B.

When access to the resource is unlimited, resource use increases until profit is zero, which occurs when total revenue equals estimated total cost. This occurs at I_5 where TR = TC(I_5). At I_5, resource users expect a yield of q_5 but only achieve a yield of q_6. Because I_5 is greater than the input use that achieves true maximum sustainable yield (I_7), the stock falls below the level that achieves true maximum sustainable yield and the resource is overexploited. The greater the overestimation of yield, the greater the overexploitation and the likelihood of extinction. High resource prices in combination with optimism regarding resource productivity result in an

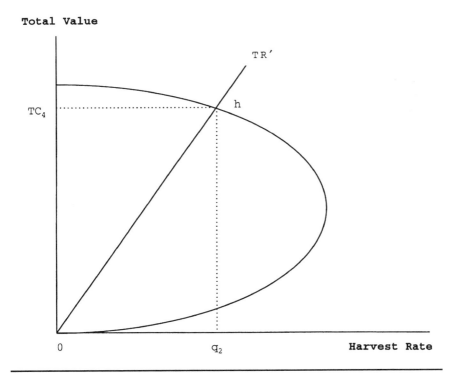

Figure 8.5. Overexploitation of common property renewable resource when access is unlimited and resource price is high. TC = total cost; TR = total revenue.

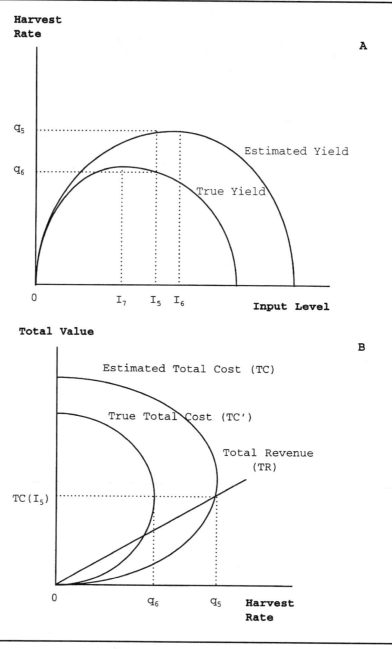

Figure 8.6. Effects of biological uncertainty on input levels (*A*) and profit maximizing harvest rates (*B*).

even greater likelihood of overexploitation and extinction than either factor taken alone.

For unlimited access to a common property resource:

1. Harvest rate is higher and the stock is lower than with limited access.
2. Harvest rate is likely to exceed maximum sustainable yield.
3. Resource stock can drop to levels that eventually lead to extinction.

MULTIPLE OBJECTIVE MANAGEMENT. Multiple objective management is desirable when the resource has more than one use and/or harvest rates for one species influence the productivity of other species. Conflicts in multispecies management is exemplified by the tuna-dolphin controversy that occurred between the United States and Mexico. In early 1991, the United States placed an embargo on tuna caught using harvesting methods (purse-seine nets) that cause incidental death of dolphins. The embargo was motivated by amendments to the United States' Marine Mammals Protection Act that limited dolphin killed during tuna harvesting. Mexico, which relies heavily on the objectionable harvesting method, charged that the Act violated the General Agreement on Tariffs and Trade's (GATT's) free trade rules because it constituted a barrier to trade.

A panel of the GATT ruled that the tuna sanctions imposed by the Act did in fact violate free trade rules. This ruling brought into question not only the validity of the Act but the whole issue of inconsistencies between domestic environmental policies and international trade agreements.[3] Mexico decided not to use the ruling to oppose the United States ban because it did not want the tuna–dolphin controversy to jeopardize the North American Free Trade Agreement among the United States, Canada and Mexico.

Resource conflicts from multiple-species management occur in the protection of endangered species. In a landmark case based on the Endangered Specie Act, the United States Forest Service reduced harvest rates in old growth forests in the northwestern United States to protect the habitat of the northern spotted owl. These forests are the only habitat for this species of owl.

Effects of multiple use management on resource use depend on whether the resource uses are competitive or complementary and how alternative management strategies influence use. Consider multiple use management in which nonconsumptive use of a renewable resource decreases as consumptive use increases (competition in use). For example, the roads needed for commercial logging operations are a major source of water pollution in streams and lakes. Water pollution impairs the ecological services provided by the forest. Suppose the loss in ecological services per unit increase in harvest rate, called marginal ecological loss (MEL), increases exponentially when the harvest rate exceeds some threshold rate of q_0 as shown in Figure 8.7.

Because most ecological services provided by a forest are unpriced and do not influence the value of commercial timber, they are usually ignored by timber companies in determining harvest rates. The privately efficient harvest rate is q_2 where p = MHC (p is the price of timber) and the socially efficient harvest rate is q_1 where p = MHC + MEL. Because $q_2 > q_1$, the privately efficient harvest rate exceeds the socially efficient harvest rate. Therefore, harvest rates are higher and the stock is smaller when marginal ecological losses from timber harvesting are ignored.

Not all uses of renewable resources are competing. For example, the clearcutting of forest patches allows sunlight to reach the forest floor, which improves feeding habitat for deer and other grazing animals. In this case, there is a complementary relationship between the two uses of the forest up to some rate of clearcutting. What are the implications of complementary uses? Suppose harvesting of timber generates positive ecological benefits for wildlife that graze but that the marginal ecological benefit (MEB) decreases as harvesting increases and eventually becomes zero at q_4 as shown in Figure 8.8. When the resource is managed to maximize profit, the efficient rate of harvest is q_2 where p = MHC.

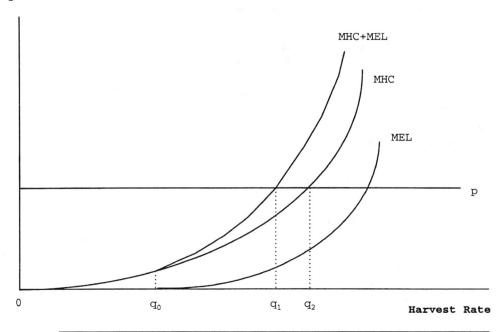

Figure 8.7. Efficient timber harvest rates in a forest with (q_1) and without (q_2) competition between timber production and ecological services. MEL = marginal ecological loss; MHC = marginal harvest cost; and p = resource price.

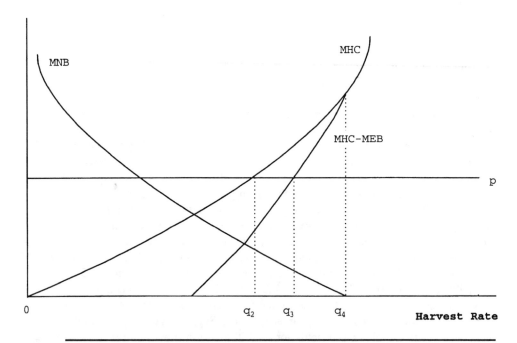

Figure 8.8. Efficient harvest rates in a forest with (q_3) and without (q_2) complementarity between timber production and ecological services. MEB = marginal ecological benefit; MHC = marginal harvest cost; MNB = marginal environmental benefit; and p = resource price.

When the resource is managed to maximize net social benefit, the efficient rate of harvest is q_3 where p = MHC − MEB. Because $q_2 < q_3$, the privately efficient harvest rate is less than the socially efficient harvest rate when resource uses are complementary. The relationship between private and socially efficient harvest rates with complementary uses is just the opposite of what it is for competitive uses (compare Figures 8.7 and 8.8).

Static efficiency conditions for a conditionally renewable resource have limited value in evaluating renewable resource issues because they ignore the time-dependent interactions between physical factors (natural growth and stock) and economic factors (prices and costs). The next section examines efficient use of a conditionally renewable resource using a dynamic framework that explicitly recognizes these interactions.

Dynamic Efficiency

A renewable resource generates direct benefits to those who produce or use it and indirect benefits to society. Direct benefits include: income from production of the resource, such as revenue from the sale of timber and fish, employment and tax revenues. Indirect benefits may or may not involve use of the resource (see Chapter 12). An indirect benefit that does not involve use of the resource is preservation value. Renewable resources have preservation value when society wants to preserve such resources for the benefit of future generations. When certain species of whales became threatened by excessive harvesting, the International Whaling Commission announced voluntary harvest quotas. These restrictions were supported by most countries that harvested whales and by consumer and special interest groups concerned about preserving whales. The Endangered Species Act and Marine Mammals Protection Act in the United States address the preservation of plant and animal species threatened or endangered by human activities (preserving biodiversity). More recently, the objective of resource management has shifted from preserving species to preserving ecosystems. Generally speaking, preservation of an ecosystem dominated by renewable resources, such as a forest or coral reef, is higher (lower) when the stock of the resource is higher (lower).

In *continuous time,* social benefit derived from a renewable resource can be represented by the following function:

$B(t) = B[q(t), S(t)]$.

This function states that social benefit, $B(t)$, depends on the rate of harvest, $q(t)$, and the stock of the resource, $S(t)$. How do $q(t)$ and $S(t)$ influence $B(t)$? $B(t)$ is positively related to $q(t)$ and $S(t)$. Increasing (decreasing) the harvest rate of commercial timber in a rain forest increases (decreases) the commercial value of the forest, but decreases (increases) the stock of trees and related biota. A lower (higher) forest biomass decreases (increases) preservation value (biodiversity), which lowers (raises) social benefit. Therefore, increased (decreased) timber harvesting in the rain forest increases (decreases) direct benefits but decreases (increases) indirect benefits. In other words, enhancing direct and indirect benefits of a rain forest are competing objectives. Such competition causes conflict between commercial timber and environmental interests.

The social benefit function contains a *threshold effect* when decreases in the stock of the resource do not affect social benefit until the stock drops below a threshold level. When there is a threshold effect, the benefit function is:

$$B(t) = B[q(t)] \text{ if } S(t) \geq S' \text{ or } B[q(t), S(t)] \text{ if } S(t) < S'.$$

The first function states that social benefit is not influenced by the resource stock as long as the stock exceeds a threshold level of S'. The second function states that social benefit is affected when the stock falls below the threshold level. Applying the benefit function to the spotted owl example indicates that the preservation value of the forest as habitat for the spotted owl is not adversely affected as long as forest biomass is above the threshold level. However, if forest biomass falls below the threshold level, then additional harvesting degrades owl habitat, which endangers the owl. For stock levels below the threshold level, preservation value of the forest decreases as forest biomass decreases, which reduces social benefit. This implies that decreases in $S(t)$ decrease social benefit when $S(t) < S'$.

In *discrete time,* the benefit function is expressed by replacing the parenthesized t in the continuous version by a subscript t, namely:

$$B_t = B(q_t) \text{ if } S_t \geq S' \text{ or } B(q_t, S_t) \text{ if } S_t < S'.$$

The efficient intertemporal use of a renewable resource is determined by selecting harvest rates and stock levels that maximize net social benefit (NSB), which equals:

$$\int_0^T B(t)e^{-rt}dt \text{ in continuous time, and}$$

$$\sum_{t=0}^T B_t(1 + r)^{-t} \text{ in discrete time,}$$

where $B(t)$ or B_t is the undiscounted social benefit in period t, r is the discount rate and $(1 + r)^{-t}$ or e^{-rt} is the discount factor and e is the base for the natural logarithm. Benefit in each period equals consumer surplus plus producer surplus. Both of the above expressions indicate that NSB equals the present value of benefits over the entire planning horizon.

The remainder of this section discusses a simplified dynamic model of efficient intertemporal use for a conditionally renewable resource. The model invokes assumptions 3 (no uncertainty), 4 (single use or species management) and 6 (pure competition).

OBJECTIVES AND CONSTRAINTS. Let the demand function for a conditionally renewable resource be:

$$p(t) = D[q(t)],$$

where $dp(t)/dq(t) < 0$. This condition states that resource price in period t, $p(t)$, is inversely related to quantity demanded of the resource in period t, q(t), when demand shifters are held constant. Hence, the demand curve is negatively sloped as illustrated in Figure 8.9A.

Total economic value from using a given amount of the resource, say $q'(t)$, is the shaded area under the demand curve between 0 and $q'(t)$ in Figure 8.9A. Mathematically, this area is:

$$V[q'(t)] = \int_0^{q'(t)} D(m)dm,$$

where D(m) is the demand function. This area is also known as total willingness to pay.

Let the supply function for the resource be:

$$p(t) = S[q(t)],$$

where $dp(t)/dq(t) > 0$. This condition states that resource price in period t, p(t), is directly related to quantity supplied of the resource in period t, q(t), when supply shifters are held constant. Hence, the supply curve is positively sloped as illustrated in Figure 8.9B.

Total cost of producing $q'(t)$, designated $C[q'(t)]$, is the area under the supply curve between 0 and $q'(t)$, which is the shaded area in Figure 8.9B. Mathematically, total cost of $q'(t)$ is:

$$C[q'(t)] = \int_0^{q'(t)} S(m)dm,$$

where S(m) is the supply function.

The benefit of $q'(t)$ is:

$$B[q'(t)] = V[q'(t)] - C[q'(t)].$$

$B[q'(t)]$ equals consumer surplus plus producer surplus at $q'(t)$. It is the shaded area in Figure 8.10.

Efficient intertemporal harvest rates for a conditionally renewable resource maximize NSB subject to two conditions. The *first condition* requires that changes in the stock over time equal growth minus harvest rate and the *second condition* requires that the stock of the resource in period 0 (initial stock) is known. Other conditions can be imposed. For example, sustainable use of a renewable resource might be achieved by requiring the stock of the resource to be above some minimum level, specifically $S(t) > S_{min}$.

NECESSARY CONDITIONS. Certain necessary conditions must be satisfied to achieve efficient intertemporal harvest rates and stocks for a renewable resource. A necessary condition does not guarantee the desired result; however, the result cannot be achieved without satisfying the necessary condition. With rare exceptions, to

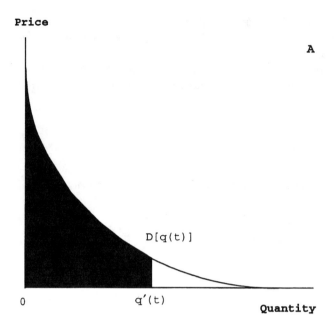

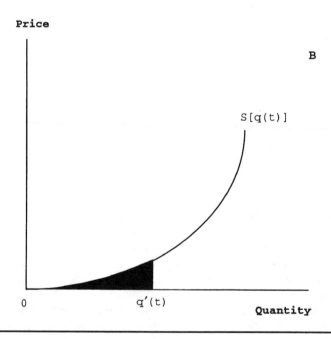

Figure 8.9. (A) Demand curve. *Shaded area* indicates total economic value. (B) Supply curve. *Shaded area* indicates total cost of production.

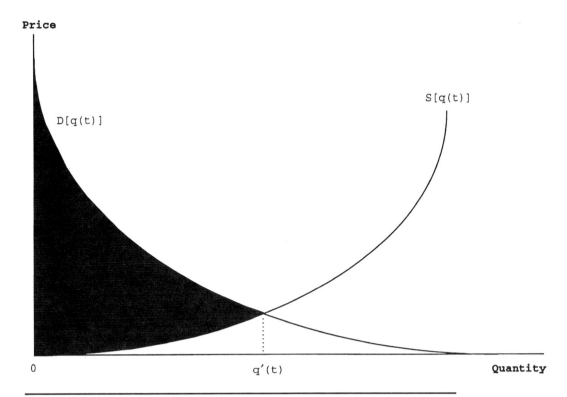

Figure 8.10. Net social benefit of q'(t) or consumer surplus plus producer surplus (*shaded area*). D = demand; S = supply.

enter college you must have a high school, or equivalent, diploma. Having a high school diploma is a necessary condition for entering college; however, entering college requires more than a high school diploma. The person must apply and be accepted for admission, and acquire the necessary financial resources to pay tuition, fees and living expenses. This section derives and discusses the necessary conditions for efficient intertemporal harvest rates for a conditionally renewable resource.

To simplify the derivation and interpretation of the necessary conditions, the demand curve for the resource is assumed to be perfectly elastic, input markets are assumed to be purely competitive and all costs are assumed to be variable. A perfectly elastic demand curve implies that quantity demanded is not a function of price, which means the demand curve is horizontal. Purely competitive input markets imply that the supply curves for all inputs used in conjunction with the resource are perfectly elastic. In the case of timber production, this assumption implies that the demand curve for timber and the supply curves for labor and manufactured capital are perfectly elastic. The assumption that all costs are variable implies a long-run situation. Under these assumptions, total economic value and total cost are, respectively:

$$V[q(t)] = p(t)q(t)$$

$$C[q(t)] = c[q(t)]q(t),$$

where p(t) is resource price and c[q(t)] is cost per unit of harvest. Social benefit in period t is:

$$B(t) = p(t)q(t) - c[q(t)]q(t).$$

Factoring out q(t) gives:

$$B(t) = \{p(t) - c[q(t)]\}q(t) = n(t)q(t),$$

where n(t) is net return per unit of resource harvested.

The first necessary condition for efficient intertemporal harvest is:

$$q(t) = 0 \text{ if } n(t) < \mu(t) \text{ or } q_{max} \text{ if } n(t) > \mu(t),$$

where q_{max} is the maximum harvest rate and $\mu(t)$ is the shadow price of the resource in period t. The shadow price of the resource is the increase in social benefit from allowing the stock of the resource to increase by one additional unit. The latter could be accomplished by reducing the harvest rate by one unit. Therefore, the first necessary condition requires the harvest rate to equal zero when net return per unit of harvest is less than the shadow price or to equal the maximum harvest rate when net return per unit of harvest exceeds the shadow price. For example, if n(t) is $4 and $\mu(t)$ is $5, then the harvest rate should equal zero because not harvesting the resource (allowing the stock to increase) increases social benefit. Conversely, if n(t) is $5 and $\mu(t)$ is $4, then as much of the resource should be harvested as possible because it increases social benefit. The remaining necessary conditions (not presented here) place limits on changes in the shadow price and stock.

The optimal stock depends on whether or not the cost of harvesting the resource is dependent (stock effect) or independent (no stock effect) of the stock. If there is a stock effect, then the natural rate of return on the resource is less than the discount rate. When there is no stock effect, the natural rate of return on the resource equals the discount rate. As a result, the optimal stock is greater with than without a stock effect. The stock that achieves maximum sustainable yield, S_{msy}, is the efficient stock if and only if there is no stock effect and the discount rate is zero.

When the necessary conditions for efficient intertemporal use of a renewable resource are satisfied, NSB is a maximum, the harvest rate equals the natural growth rate, and the stock achieves an optimal steady-state level of S_{opt}. Except in the unlikely event that the initial stock equals the optimal steady-state stock ($S_0 = S_{opt}$), the harvest rate needs to be adjusted until the optimal steady-state stock is achieved.

Under certain conditions, the steady-state stock can be achieved by adjusting the harvest rate according to the following rule:

$$q(t) = 0 \text{ if } S(t) < S_{opt}, \text{ or } G[S(t)] \text{ if } S(t) = S_{opt}, \text{ or } q_{max} \text{ if } S(t) > S_{opt}.$$

This condition states that the harvest rate in period t should equal: a) zero when the stock falls below the optimum; b) the growth rate, G[S(t)], when the stock equals

8. Renewable Resource Management

the optimum; and c) the maximum harvest rate when the stock exceeds the optimum. In special circumstances, there is a unique solution to this dynamic optimization problem.

While resource users do not like a policy of zero harvest, such a policy has been implemented in cases where the stock of the resource is very low. For example, fishing for ocean salmon in the northwestern United States was banned in the mid-1990s to allow stock of these fish to recover. The season for Atlantic cod on Georges Bank off the coast of the northeastern United States has been temporarily closed to allow stocks of cod to recover. In addition, fisheries management agencies routinely control harvest by imposing quotas and/or limiting the length of the harvest season.

When steady-state conditions are achieved, the harvest rate equals the growth rate in all periods and the change in harvest equals the change in growth over time. The equilibrium stock condition in steady state is:

$$\frac{dNSB/dS}{r} = p - c(q),$$

where $dNSB/dS$ is the change in NSB with respect to a change in the stock, r is the discount rate, p is the resource price and $c(q)$ is the unit cost of harvest. When there is a stock effect, $c(q)$ depends on the level of the stock.

For an infinite planning horizon ($T = \infty$), the left-hand side of this equation equals the loss in NSB from reducing the stock by one unit or equivalently by increasing current harvest by one unit. This loss equals marginal user cost (MUC) of current harvest. Hence, in steady-state:

$$p - c(q) = MUC.$$

This condition states that current net return per unit of resource harvested equals marginal user cost of harvesting, which is analogous to the condition required for efficient intertemporal extraction of an exhaustible resource. The equilibrium steady-state condition for a renewable resource stock is illustrated in Figure 8.11. Net price, $p - c(q)$, is an increasing function of S because unit harvest cost generally decreases as the stock increases. MUC is a decreasing function of S because MUC decreases as the stock increases.

The effects of changes in resource price, unit harvest cost and the discount rate on the optimal stock are as follows:

1. Increases (decreases) in resource price (p) holding unit harvest cost and MUC constant cause S_{opt} to decrease (increase). When resource price increases from p_1 to p_2, net return increases from $p_1 - c(q)$ to $p_2 - c(q)$, which causes S_{opt} to decrease from $S^{(1)}_{opt}$ to $S^{(2)}_{opt}$, and vice versa, as shown in Figure 8.12A.

2. Decreases (increases) in unit harvest cost (c) holding resource price and MUC constant cause S_{opt} to decrease (increase). A decrease in unit harvest cost from $c_2(q)$ to $c_1(q)$ increases net return from $p - c_2(q)$ to $p - c_1(q)$, which causes the optimal stock to decrease from $S^{(2)}_{opt}$ to $S^{(1)}_{opt}$, and vice versa, as shown in Figure 8.12B.

3. Increases (decreases) in the discount rate r holding resource price and unit cost constant cause S_{opt} to decrease (increase). When the discount rate increases from r_1 to r_2, MUC decreases from $MUC(r_1)$ to $MUC(r_2)$, which causes S_{opt} to decrease from $S^{(1)}_{opt}$ to $S^{(2)}_{opt}$, and vice versa, as shown in Figure 8.12C.

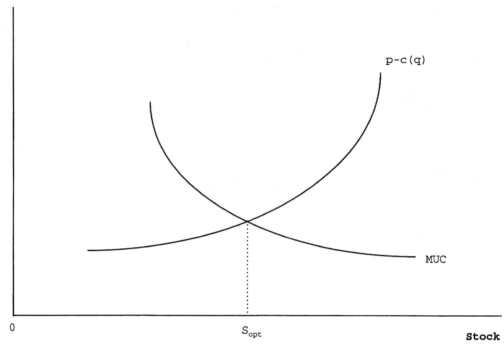

Figure 8.11. Equilibrium steady-state stock (S_{opt}) for a renewable resource. MUC = marginal user cost; p = resource price; and c = unit cost of harvest.

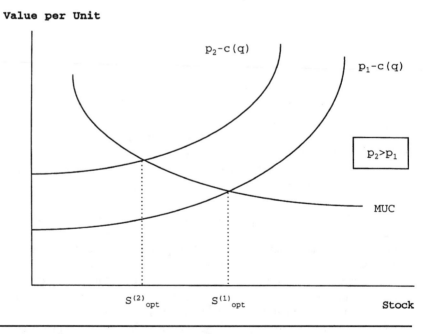

Figure 8.12. Effects of changes in resource price (*A*), unit cost (*B*) and discount rate (*C*) on optimal steady-state stock. MUC = marginal user cost; p = resource price, c = unit harvest cost; and r = discount rate.

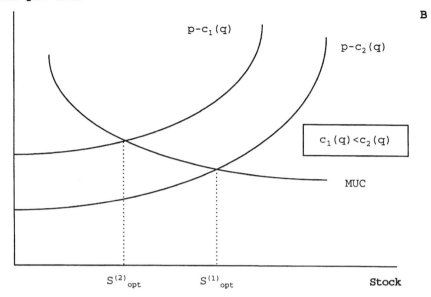

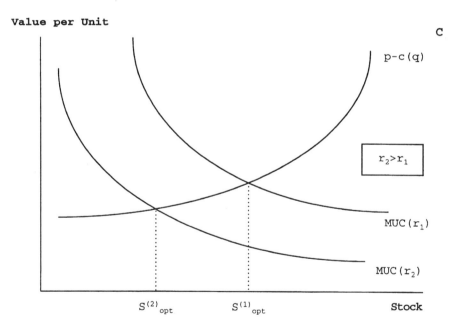

Figure 8.12. *continued.*

Renewable Resource Policies

Efficient intertemporal use of renewable resources is based on the economic principle of maximizing net social benefit. Socially efficient resource use is typically not achieved for a variety of reasons. In the case of privately owned renewable resources, socially efficient harvest rates are not achieved because of external diseconomies. For example, hydroelectric power development in the northwestern United States was done with little regard for the negative impacts on spawning and migration of anadromous fish. As a result, populations of certain species of anadromous fish have become extinct and others have been seriously depleted. As discussed earlier, overexploitation of a renewable resource can also occur when access to the resource is unlimited. Major declines in ocean fisheries and the near extinction of certain species of whales are prime examples of this phenomenon. Overexploitation occurs because reducing harvest to its socially efficient rate is not profitable. This causes the stock to fall to levels that can lead to extinction of the resource.

External diseconomies can occur for common property resources even when access is limited. An example of this is the ecological damage caused by timber harvesting in national forests. Even though private access to timber harvesting in national forests is controlled by the United States Forest Service, ecological damage still occurs. If this damage becomes serious, then harvest rates in national forests are too high from a social viewpoint (see Figure 8.7).

Potential social losses from overexploitation of renewable resources have been or can be reduced by various public policies including: a tax on harvest; access fees; and limitations on access to and harvest of the resource and subsidies. These policies are discussed below.

TAX ON HARVEST. A tax on harvest attempts to bring the privately efficient rate of harvest in line with the socially efficient rate of harvest. Consider an optimal tax to reduce overexploitation resulting from unlimited access to a renewable common property resource. Suppose each firm harvesting the resource has the same cost function:

$TC[q(t)] = c[q(t)]q(t),$

where $c[q(t)]$ is the unit cost of harvest and $q(t)$ is the harvest rate. Assuming perfectly elastic resource demand and purely competitive conditions in input markets, social benefit in period t is:

$B[q(t)] = \{p(t) - c[q(t)]\}q(t),$

where $p(t)$ is the price of the resource in period t.

Socially efficient rates of harvest are determined as follows. Let $q_j(t)$ be the harvest rate for resource user j (j = 1,...,J, where J is the number of users) in period t (t = 0,...,T, where T is the number of periods over which the resource is harvested). Socially efficient harvest rates are determined by solving the following optimization problem for $q_j(t)$:

8. Renewable Resource Management

Maximize $\int_0^T \sum_{j=1}^{J} B_j[q_j(t)]e^{-rt}dt$

with respect to $q_j(t)$ subject to $dS(t)/dt = G[S(t)] - \sum_{j=1}^{J} q(t)$. $B_j[q_j(t)] = \{p(t) - c[q_j(t)]\}q_j(t)$ is the social benefit generated by user j, dS/dt is the change in resource stock over time, $G[S(t)]$ is natural growth of the resource in period t and $\Sigma q_j(t)$ is total harvest of the resource in period t. The first equation requires choosing the harvest rates for each user in all periods that maximize the user's contribution to NSB. The second equation requires that changes over time in the stock equal the difference between natural growth and harvest.

The necessary condition for a solution to this problem is:

$p(t) = MHC(t) + \mu(t)$ for all t,

where $MHC(t)$ is the marginal harvest cost and $\mu(t)$ is the shadow price of the resource in period t. The latter equals the increase in net return from increasing the stock by one unit in period t. The necessary condition for socially efficient harvest rates is $p(t) = MHC(t) + \mu(t)$ in all periods, which occurs at $q_s(t)$ as shown in Figure 8.13. When access to the resource is unlimited, resource users select harvest rates that drive the price of the resource down to average harvest cost, namely: $p(t) = AHC[q(t)]$, where $AHC[q(t)]$ is the average cost of harvesting $q(t)$. Hence, the privately efficient harvest rate with unlimited access is $q_u(t)$ in Figure 8.13. Because $MHC(t) + \mu(t) > AHC(t)$, the socially efficient harvest rate is less than the privately efficient harvest rate with unlimited access, $q_s(t) < q_u(t)$.

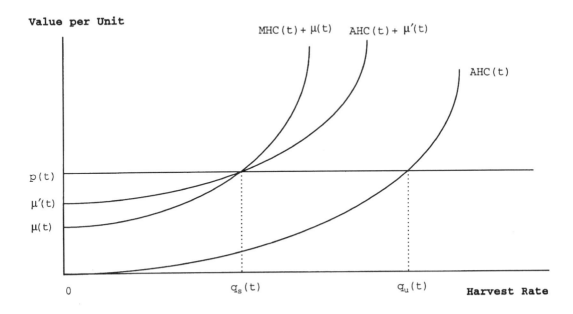

Figure 8.13. Comparison of privately efficient harvest rate with unlimited access (q_u) and socially efficient harvest rate (q_s) for a renewable resource. AHC = average harvest cost; MHC = marginal harvest cost; $\mu(t)$ = shadow price of resource; and p = resource price.

Imposing a Pigouvian tax on harvest equivalent to $\mu'(t)$ shifts the AHC(t) upward until it intersects MHC(t) + $\mu(t)$ at $q_s(t)$. With a tax of $\mu'(t)$, the privately efficient harvest rate with unlimited access equals the socially efficient harvest rate. Because $\mu'(t) > \mu(t)$, the tax exceeds the shadow price.

Determination of the Pigouvian tax is difficult because it requires information on the marginal harvest cost for all users in all periods. Unless, a regulatory agency has the authority to require resource users to reveal such information, the Pigouvian tax cannot be determined ahead of time. In the absence of this cost information, the regulatory agency could adjust the tax until the desired harvest rates are achieved. For example, an initial tax can be selected. If the resulting harvest rates still cause overexploitation of the resource, then the tax is increased until the desired rate of harvest is achieved. If the resulting harvest rates are below the desired harvest rates, then the tax is reduced.

ACCESS FEES, EFFORT RESTRICTIONS AND OTHER APPROACHES.

Another public policy for achieving a more socially efficient use of renewable resources is to require each resource user to pay an *access fee*. For example, an access fee for an ocean fishery requires every fishing vessel to pay a set fee for each day of fishing. An access fee represents a variable cost because it depends on the level of fishing effort. In contrast, a license fee is a fixed cost per season, the payment of which gives a resource user a legal right to harvest the resource during that season. Daily access fees decrease the privately efficient rate of harvest by reducing the number of days harvest effort occurs. Access fees can be adjusted upward or downward to reduce or increase total harvest effort.

Because an access fee is based on the level of harvest effort and a tax is based on actual harvest, it is somewhat more difficult to achieve the socially efficient rate of harvest with an access fee than with a tax. Revenues generated by access fees can be used to offset the administrative cost of enforcing and collecting the fee. As with any fee, the incentive to cheat (not pay the fee) increases as the fee increases. Cheating is greater for more widely dispersed resources such as fish in an ocean or rhinoceros in a large game reserve.

A common way to limit overharvesting of a renewable resource is to control when and where access to the resource is permitted, the type of resource that can be harvested and/or the maximum daily harvest. In the case of fish and game, access is limited by restricting the length of the harvest season, sex, age, size and number of fish or game that can be harvested. In the case of timber harvesting on public land, access is limited by controlling the timing and volume of timber sales to private companies. While controlling access does not generate revenue, it reduces overexploitation of the resource. To be effective, management policies need to be revised periodically to reflect changes in harvesting technology.

A novel way to control access to a renewable resource is to pay resource users for not harvesting the resource. This is a subsidy. In 1993, the Secretary of the Interior for the United States announced a plan to restore seriously depleted stock of salmon in the Atlantic Ocean off the northeastern coast of the United States. Because the depletion was caused by overfishing, the plan paid $800,000 to fishermen to stop fishing for Atlantic salmon for a period of two years. A more extreme example is the total ban on salmon harvesting in the Pacific Ocean off the northwestern United States, which was imposed in the mid-1990s.

A combination of resource policies typically affords greater flexibility in achieving the desired stock and harvest rates than a single policy. For example, in the United States, rights to hunt and bag certain big game species (big horn sheep and mountain goat) are typically rationed by a lottery that limits the number of hunting permits issued for these species. Each season, hunters randomly drawn from an eligible list of hunters are issued permits that allow them to hunt. If an animal is not bagged, then the hunter may be allowed to apply for another special permit, perhaps after a delayed period of time. In some states, bagging an animal means you can never again apply for a special permit. Additional flexibility is achieved by allowing special permit holders to sell their permits to hunters whose names are not selected during the lottery. This makes the hunting permits tradable.

Summary

Renewable resources are either conditionally renewable or nondegradable. Growth in conditionally renewable resources, such as fish, wildlife and forests, allows these resources to be managed on a sustainable basis for an indefinite period of time. The stock of a conditionally renewable resource is influenced by harvesting and natural growth (regeneration). Management objectives for and property rights to conditionally renewable resources influence the rate of harvest and the likelihood of overexploitation. Availability of nondegradable renewable resources, such as the sun, wind and tides, is independent of human activities.

Overexploitation of conditionally renewable resources occurs when harvest rates exceed natural growth rates causing the stock to decline. Failure to control overexploitation of conditionally renewable resources increases the likelihood of species extinction and loss of biodiversity. Conditionally renewable resources can be managed under different property regimes: profit maximization by a single owner or manager of private property, profit maximization by multiple users of a common property resource who limit access to the resource, and profit maximization by multiple users of a common property resource who have unlimited access to the resource. Renewable resources can be managed for a single objective or species or for multiple objectives or species.

Determining the efficient intertemporal harvest of a renewable resource requires knowledge of the growth function, harvest function, yield function, resource prices and harvest costs. The growth function explains the relationship between changes in the stock of the resource in the absence of harvesting. The harvest function describes how the harvest rate changes with respect to the stock and input use. The yield function relates efficient harvest rates to input use. Harvesting a renewable resource involves expenditures on inputs such as labor, fuel and equipment. If per unit cost of harvesting depends on the stock, then there is a stock effect.

Privately efficient harvest rates for conditionally renewable resources under static conditions (no time-related interactions between biological and economic factors) require that resource price equals either marginal harvest cost with limited access or average harvest cost with unlimited access. Compared to limited access, the harvest rate with unlimited access is generally greater and more likely to exceed maximum sustainable yield. In addition, the stock is lower with unlimited access that increases the risk of species extinction and loss of biodiversity. The socially efficient rate of harvest for a conditionally renewable resource for which economic and ecological uses are competing occurs where resource price equals marginal harvest cost plus marginal ecological loss. When economic and ecological uses are complementary, the socially efficient harvest rate occurs where resource price equals marginal harvest cost minus marginal ecological benefit.

Evaluating efficient intertemporal resource harvest in a static framework ignores the time-dependent interactions between biological and economic factors. Determining harvest rates and stocks in a dynamic framework entails maximizing net social benefit over some planning horizon subject to conditions regarding the initial stock and changes in the stock. While the mathematical derivation of efficient harvest rates is tedious, the resulting necessary conditions for efficient intertemporal use have a relatively straightforward bioeconomic interpretation. The necessary condition for efficient intertemporal harvest is that the harvest rate equals zero (the maximum harvest rate) when net return per unit of harvest is less (greater) than the value of allowing the resource stock to increase by one unit. When the stock achieves a steady state (harvest equals growth), the loss in net social benefit from reducing the stock by one unit, which equals marginal user cost, must equal the current net return on the resource. Under steady-state conditions, increases in resource price, decreases in cost of harvest and increases in the discount rate cause efficient harvest rates to increase and the stock to decrease. Opposite changes in these variables cause efficient harvest rates to decrease and the stock to rise.

External diseconomies in the use of renewable resources cause the privately efficient harvest rate to exceed the socially efficient harvest rate that reduces net social benefit from use of the resource. A variety of public policies can be used to reduce external diseconomies. These include a tax on harvest; access fees; restrictions on when, where and how the resource is accessed and harvested; and subsidies for not harvesting the resource. A combination of policies is generally superior to any single policy because it affords greater flexibility in achieving resource management objectives and generates the revenue needed to fund resource management activities.

Questions for Discussion

1. Explain how the degree of access to a renewable common property resource influences the rate of harvest and net social benefit compared to a situation where the resource is privately owned.

2. What are the likely economic consequences of eroding soil at a rate that exceeds the rate of regeneration? What public policies might be used to alleviate these economic consequences?

3. Compare the efficient harvest rates for a renewable resource under static conditions when the demand function for the resource is downward sloping and the resource is managed to maximize net social benefit.

4. The yield curve for a nonrenewable resource is $q = \alpha NI(1 - \alpha I/\beta)$, where $\alpha = 0.002$, $N = 150$ and $\beta = 0.40$. Determine efficient input use, efficient harvest rate and optimal stock when profit is maximized, $p = \$25$ and $c = \$4$.

Further Readings

Clark, C. W. 1976. *Mathematical Bioeconomics: The Optimal Management of Renewable Resources.* New York: Wiley-Interscience.

Conrad, Jon M., and Colin W. Clark. 1987. Renewable Resources. Chapter 2 in *Natural Resource Economics: Notes and Problems.* Cambridge: England: Cambridge University Press, pp. 62–116.

Howe, Charles W. 1979. The Management of Fisheries: A Case of Renewable but Destructible Common Property Resources. Chapter 13 in *Natural Resource Economics.* New York: John Wiley & Sons, pp. 256–275.

Pearce, David W., and R. Kerry Turner. 1990. Renewable Resources (Chap. 16). *Economics of Natural Resources and the Environment.* Baltimore, Maryland: The Johns Hopkins University Press, pp. 242–261.

Notes

1. Al Gore. 1993. *Earth in the Balance: Ecology and the Human Spirit* (New York: Penguin Books USA Inc.), pp. 19–20.

2. This expression for optimal input use is obtained by setting the derivative of q with respect to I equal to zero and solving for I.

3. Hilary F. French, "The Tuna Test: GATT and the Environment," *World Watch* (Washington, D.C.: Worldwatch Institute, March/April 1992), p. 9.

CHAPTER 9

Economics of Environmental Pollution

I finally closed my mouth and began looking at the planet in more detail. Most people don't get to see how widespread some of the environmental destruction is. From up there, you look around and see that it's a worldwide rampage.

—MARIO RUNCO, JR., U.S. ASTRONAUT, 1991

The material balances approach discussed in Chapter 4 indicates that the environment provides three services to the economy: resource extraction, assimilation of residuals, and amenities. Extractive services refer to the exhaustible and renewable resources that are used in the production of commodities. Residual assimilation services refer to the assimilation of production and consumption residuals by various environmental receptors (air, land and water). Amenity services include the natural beauty, solitude and recreation provided by the environment. Chapters 7 and 8 focused on efficient use of exhaustible and renewable resources in economic production, which is equivalent to efficient use of the extractive services provided by the environment. Chapter 12 deals with the valuation of amenity services provided by the environment that are not priced in regular markets. This chapter addresses efficient use of the residual assimilation services provided by the environment, which constitutes the environmental sector of the material balances model.

Two major issues arise regarding the efficient use of the environment's capacity to assimilate residuals. First, the residual assimilation services provided by the environment tend to be overexploited because they are unpriced. When the environment's capacity to assimilate residuals is exceeded, pollution occurs. Second, pollution reduces the environment's capacity to provide exhaustible resources, such as coal, timber and precious metals; amenity services, such as outdoor recreation; and ecological services, such as air and water purification. Pollution damages are an external diseconomy that can have a negative effect on both humans and ecological systems. A major goal of environmental economics is to determine efficient levels of environmental pollution or pollution abatement and the effectiveness of alternative public policies in controlling pollution.

This chapter expands the environmental segment of the material balances model to account more fully for the relationships among *residual loads, residual emissions, pollution* and *pollution damages*. The expanded model is combined with static and dynamic efficiency criteria to derive the private and socially *efficient lev-*

els of pollution and *pollution abatement*. In addition, efficient allocation of pollution reduction to pollution generators and the *role of government* in reducing pollution-related externalities are addressed. Finally, specific public policies for reducing point and/or nonpoint sources of pollution are evaluated: namely, *emission charges, input taxes, input restrictions, emission standards, mandatory production methods, cost sharing, tradable emission permits* and *environmental liability.*

Residual Emissions, Pollution and Pollution Damages

Emissions of residuals to the environment come from multiple sources. Pollution and pollution damages resulting from residual emissions have important spatial and temporal aspects. Consider global warming, which results from the accumulation of carbon dioxide, methane and other greenhouse gases in the upper atmosphere (multiple sources). Carbon dioxide, which is the largest source of greenhouse gases, is produced when fossil fuels are burned. Because the bulk of past carbon dioxide emissions were generated by developed countries (see Figure 1.1), most of the global warming that has already taken place is attributed to developed countries. The greatest growth in future carbon dioxide emissions is expected to occur in developing countries, and advanced technology for reducing carbon dioxide emissions resides in developed countries (spatial and temporal aspects).

Besides country-to-country differences, global warming has different impacts within a country. For example, global warming is expected to lengthen the growing season in northern latitudes and increase drought conditions in southern latitudes in the United States. Changes in growing conditions are likely to cause major shifts in the location of crop production with some areas experiencing increases and others decreases in production of specific crops. Warmer temperatures are also expected to increase the melting of major ice formations causing rises in sea level and flooding in coastal areas. A group of New Zealand scientists found that the ice in Antarctica melted completely three million years ago when average global temperature increased by just a few degrees above the current average.

If emissions of carbon dioxide and other greenhouse gases continue their current trend, global temperatures could increase by 3 to 8 degrees Farenheit (1.7 to 4.4 degrees Celcius) over the next 100 years.[1] If the increased temperature melts the Antarctic ice, the world's oceans could rise by as much as 215 feet (65.6 meters).[2] Changes in temperature and rainfall patterns caused by global warming is expected to alter ecological conditions and the distribution of plant and animal species. Certain species and ecosystems will be adversely affected and others will be enhanced by global warming.

Changes in temperature and rainfall from global warming are likely to reduce the quality and productivity of certain land- and water-based resources and reduce the extraction and amenity services provided by the environment. Other areas might experience improvements in resource productivity and environmental services. International agreements to curb global warming are difficult to negotiate because of intercountry differences in emissions of greenhouse gases, the high degree of uncertainty regarding the extent and timing of benefits from reducing global warming, and the huge investments needed to reduce greenhouse gas emissions.

Unlike global warming, localized pollution has a much smaller sphere of influence. There are two types of localized pollution: *point source* and *nonpoint source*. Point source pollution is pollution that can be traced to a single, well-identified source, such as a paper mill that disposes of its effluent in a pipeline that empties into a river. A pesticide holding tank that is rinsed near a wellhead is a point source of pollution for groundwater. Nonpoint source pollution is pollution from diffuse sources such as agricultural fields in a watershed and streets and parking lots in urban areas. The U.S. Environmental Protection Agency's (EPA) National Water Quality Inventory report for 1994 indicates that nonpoint source pollution is the major cause of water pollution in all types of water bodies. Of the data submitted to the U.S. EPA covering assessment of 17 percent of the nation's rivers, 42 percent of lakes and 78 percent of estuaries, agriculture is the primary source of water pollution in rivers and lakes.[3]

It is difficult to trace nonpoint source pollution to specific sources. Runoff from cropland treated with herbicides can pollute receiving water bodies such as streams and lakes. The pollution is difficult to trace to an individual field or farm without taking extensive measurements or using a biophysical simulation model.

Environmental economics does not provide a comprehensive framework for dealing with environmental pollution and damages. For the most part, the economic theory of pollution control deals primarily with damages from a single point source of pollution. In addition, most of the legislation aimed at reducing environmental pollution addresses individual pollutants. Even within the same category, such as water pollution, separate legislation has been passed to reduce point sources of water pollution versus nonpoint sources of water pollution. There is growing awareness of the need to address the multiple sources and impacts of pollution. For example, Carol M. Browner, Administrator of the U.S. EPA, stated that: "We need to approach water quality not pollutant by pollutant, not county by county, but by looking at entire watersheds, the entire natural system, as single units."[4] This approach, referred to as total watershed management or ecosystem management, has been endorsed by several resource agencies in the United States including the Natural Resources Conservation Service, the Forest Service and the Fish and Wildlife Service.

EXPANDED MATERIAL BALANCES MODEL. The environmental component of the material balances model presented in Chapter 4 (Figure 4.2) is expanded in Figure 9.1. The upper left corner of the figure shows the *residual loads* generated by economic production and consumption. Residuals are either recycled (A) or emitted to the environment (B). Recycling of residuals has a double benefit. It decreases the demand for natural resources (D) and reduces residual emissions to the environment (B); however, recycling of residuals generates its own residuals, which are emitted to the environment (C). Overall, recycling results in a net loss in residual emissions. Because aluminum recycling decreases the demand for bauxite, the mineral used to produce aluminum, it extends the life of bauxite resources and reduces the amount of aluminum emitted to the environment. Recycling in the United States has dramatically reduced primary consumption of iron and aluminum.

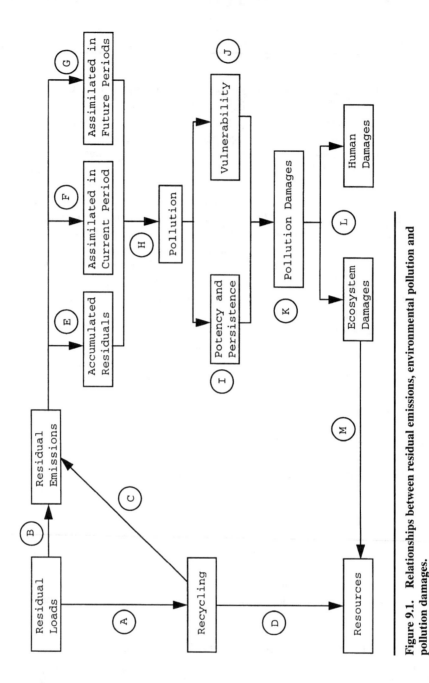

Figure 9.1. Relationships between residual emissions, environmental pollution and pollution damages.

Several things can happen to residuals. When residuals are not assimilated or degraded by the environment, they become *accumulated residuals* (E). Unrecycled glass and metal containers; toxic residuals such as dioxin, which is a degradation product of creosote; PCBs, which leak from discarded electrical capacitors; plutonium residuals from nuclear power plants and facilities that manufacture nuclear weapons; and heavy metals leached from the tailings of abandoned mines accumulate in the environment. Accumulated residuals can be harmful to humans and ecosystems.

Residuals that are not accumulated in the environment are either *assimilated in the current period* (F) or *assimilated in future periods* (G). Carbon dioxide and sulfur dioxide emissions from electrical power facilities that burn fossil fuels and that portion of applied nitrogen fertilizer not used by plants are examples of residuals that are assimilated by the environment in the current period or temporarily stored in the environment for assimilation in future periods. Human settlement and resource development patterns, such as urban and commercial development in floodplains, loss of prime farmland to urban development, conversion of prairie, wetlands and forests to crop and livestock production, overgrazing of rangeland and others, decrease the capacity of the environment to assimilate residuals and to temporarily store residuals.

In residual assimilation, biological, chemical and hydrological processes driven by carbon, nitrogen, water and other natural cycles transform residuals into derivative products that can be harmful to humans and other living organisms. A portion of the nitrogen fertilizer applied to agricultural crops is used by the crops for growth. That portion not used by crops is assimilated by or stored in the environment. Some of the nitrogen fertilizer is converted to nitrate–nitrogen, which is highly soluble. Water having nitrate–nitrogen concentrations that exceed 10 parts per million is considered potentially harmful to humans, especially infants and pregnant women.

Pollution (H) results from accumulated residuals (E) and/or residual emissions not assimilated until future periods (G). *Pollution damages* to humans and ecosystems (K) depend on the *potency* and *persistence* of the pollutant (I) and the *vulnerability* of human and biological systems to that pollutant (J). Potency refers to the toxicity of the pollutant and persistence refers to the time it takes for one half of the original mass of a pollutant to degrade (half-life). Plutonium is a very hazardous substance because it is extremely harmful to living organisms, even in minute doses (high toxicity), and takes thousands of years to degrade (high persistence). Other highly persistent substances, such as DDT, are especially detrimental to wildlife. Pollutants with high potency and persistence typically cause the greatest damage. Vulnerability refers to the degree to which humans and ecosystems are at risk from pollution. For example, there is a health risk to humans who drink water with concentrations of nitrate–nitrogen in excess of 10 ppm. There is, however, no health risk when the water is used for purposes other than human consumption such as hydroelectric power generation and recreation.

Humans and ecosystems are adversely affected by pollution damages (L). Ozone depletion occurs when total emissions of chlorofluorocarbons (CFCs) exceed the capacity of the upper atmosphere to assimilate them. Ozone depletion reduces the stratosphere's capacity to screen out ultraviolet radiation from the sun. Higher levels of ultraviolet radiation at the earth's surface increase the risk of skin cancer. Air pollution, water pollution and hazardous residuals can impair human health and, in extreme cases, cause death. Major sources of urban air pollution include particulates (smoke and soot), sulfur dioxide, nitrogen oxides, ozone, carbon monoxide and lead.[5] Sulfur dioxide and particulate matter impair respiratory functions and increase the risk of lung disease. Nitrate contamination of water in excess of the drinking water standard can result in methemoglobinemia (blue-baby syndrome). When pollution impairs human health, worker productivity declines and cost of production rises. Pollution can also damage humans by reducing the amenity services provided by the environment. Pollution of lakes, reservoirs and coastal areas can diminish the quantity and quality of recreational activities such as swimming, boating, fishing and wildlife viewing.

Pollution can damage ecosystems in several ways. Acid rain, which is caused by sulfur dioxide emissions from coal-fired power plants, has damaged aquatic and forest ecosystems in the northwestern United States and eastern Canada. In several parts of the world, extensive irrigation and groundwater mining in coastal areas lead to excessive salt buildup in topsoil (salinization). Excessive salinization decreases soil productivity and, in extreme cases, leads to abandonment of agricultural land. The World Bank has estimated that approximately 5 million acres (2 million hectares) of irrigated land are lost each year due to irrigation-induced salinization of soils.[6] Ecosystem damages can reduce the productivity and hence the quantity and quality of natural resources (M). Timber operations, such as road building and clearcutting in mountainous terrain, often cause sediment pollution in nearby streams, which reduces biological productivity and diversity. Sediment pollution can also occur when farming operations and urban and commercial developments do not incorporate soil conservation practices.

Efficient Reduction in Environmental Pollution

Pollution of environmental resources by human activities is an external diseconomy. The efficient level of an external diseconomy is determined using externality theory, which is discussed in Chapter 5. This theory indicates that the privately efficient rate of production exceeds the socially efficient rate of production when there is an external diseconomy (see Figure 5.1). Because it is usually not feasible to establish complete property rights for environmental resources, especially common property resources such as air and water, and the transaction costs of negotiating a settlement between acting and affected parties is high, public policy is often needed to reduce the loss in social welfare from pollution externalities.

If pollution damages in the current period depend only on residual emissions in the current period, then the efficient level of pollution externality can be determined using static efficiency criteria. If, however, current emissions influence pollution damages in current and future periods, then the efficient level of pollution externality should be based on dynamic efficiency criteria. A similar argument is made in Chapter 7 to justify the use of dynamic efficiency criteria to determine the efficient intertemporal depletion of an exhaustible resource. The next section covers static principles and the following section covers dynamic principles for controlling pollution.

Numerous pollution control policies have been employed including restrictions on the use of polluting inputs, maximum emission levels or pollutant concentrations, taxing pollution, subsidies for production methods that reduce pollution, best available control technologies, tradable emission permits and environmental liability. This chapter evaluates specific policies for controlling point and nonpoint sources of pollution. Implications of pure and imperfect competition for pollution control are also considered.

STATIC EFFICIENCY WITH PURE COMPETITION. Static efficiency criteria are used to determine the efficient level of pollution under pure competition when residual emissions in the current period are capable of causing pollution damages in the current period. Several sources of pollution satisfy this requirement. For example, particulate emissions from a coal-fired power generates soot, which falls on the surrounding communities. Households and businesses in that community are likely to incur pollution damages in the form of higher cleaning bills that occur shortly after the particulates are emitted. Reducing particulate emissions from the plant will result in an immediate reduction in pollution damages.

Production Restrictions. The efficient level of pollution can be achieved by restricting the production of commodities that cause pollution when there is a fixed, proportional relationship between residual emissions and production rates. This implies that emissions cannot be reduced by utilizing other inputs. While this assumption of the production-restricted model is quite restrictive, the model is helpful in demonstrating the basic relationships among consumption, production, emissions and pollution. The model is as follows:

Utility function: $U = U(Q, L)$

Production function: $Q = F(X)$

Emission function: $R = R(Q)$

Pollution function: $L = L(R)$

The utility function states that a household's satisfaction or utility (U) depends on consumption (Q) and pollution (L). Utility is assumed to be an increasing function of Q and a decreasing function of L. In other words, U increases when Q increases or L decreases, or conversely, U decreases when Q decreases or L increases. The production function implies that each firm's production rate (Q) is an increasing function of input use (X). Firms are allowed to dispose of their residuals to the environment freely, which implies that the residual assimilation capacity of the environment is unpriced. The emission function states that emission of residuals to the environment (R) is an increasing function of production (Q). Because the production-restricted model does not allow substitution between inputs and emission, X does not appear in the emission function. The pollution function indicates that pollution (L) is an increasing function of emission (R). Taken collectively, these functions imply that a) firms are the acting parties (polluters) and households are the affected parties (pollution victims), b) there is only one pollutant, and c) pollution impairs humans and/or ecosystems.

The external cost of pollution depends on the relationship between residual emissions (R) and production (Q) as shown in Figure 9.2A and between pollution (L) and residual emissions as depicted in Figure 9.2B. Let residual emissions be related to production in the following manner (see Figure 9.2A):

$R = aQ$ $(a > 0)$.

This relationship states that residual emissions (R) increase in fixed proportion to the production rate (Q), which implies that firms are unable to reduce emissions by using different inputs and/or technologies.

The relationship between pollution and residual emissions is expected to have a threshold effect. This means that for $R \leq R'$, residuals are assimilated by the environment and pollution is zero. For $R > R'$, however, pollution is expected to increase at an increasing rate as R increases. Therefore:

$$L = 0 \text{ for } R \leq R' \text{ and } L = bR^\lambda \text{ for } R > R' \text{ } (b > 0, \lambda > 1).$$

This relationship is consistent with a decrease in the environment's capacity to assimilate additional residuals as residual emissions increase. The more rapidly assimilative capacity decreases with increased emissions, the higher the value of λ. The b term captures local environmental conditions that influence the extent to which residual emission causes pollution. For example, pesticide residues from surface application of pesticides to cropland is more likely to result in pesticide pollution of surface water when application is followed by a major rainstorm than when rain follows application by several weeks.

Combining the residual emission-production relationship (Figure 9.2A) and the pollution-residual emission relationship (Figure 9.2B) indicates that Q' is the production rate above which pollution occurs. That is, for $Q > Q'$, $R > R'$ and $L > 0$.

The relationship between total pollution damage (LD) and emission of residuals is shown in Figure 9.3A, and that between pollution damage and production in Figure 9.3B. LD equals damage per unit of pollution (d) times the level of pollution (L), namely:

$$LD = dL = dbR^\lambda \text{ for } R > R'.$$

If per unit damage (d) does not depend on the level of pollution, then LD is an exponential function of R for $R > R'$ as shown in Figure 9.3a. When d increases with pollution, LD increases more rapidly as R rises for $R > R'$. The pollution damage function is more complex when damage is caused by more than one pollutant and/or lifestyle factors influence vulnerability to pollution. For example, exposure to two pollutants at levels below the maximum contaminant levels could result in human damages because of interactions between the two pollutants. In this case, d is a function of emission levels for both pollutants. Lifestyle choices like smoking are likely to increase the risk of getting emphysema from air pollution. In this case, the damage function is steeper for a person who smokes.

Quite often, pollution damage can be reduced by averting behavior. A potential decline in recreational swimming in a lake due to pollution might be partially averted by building a community swimming pool. Similarly, damage from pollution of well water might be reduced by substituting bottled water or water obtained from a rural water district. In some cases, averting behavior is less costly than reducing pollution. For example, using bottled water may be less costly than attempting to reduce the source of pollution particularly when the pollution is from nonpoint sources. While bottled water reduces the human health risk of drinking polluted well water, it does nothing to reduce the ecological damages caused by water pollution. Therefore, averting behavior does not necessarily eliminate pollution damage.

9. Economics of Environmental Pollution 197

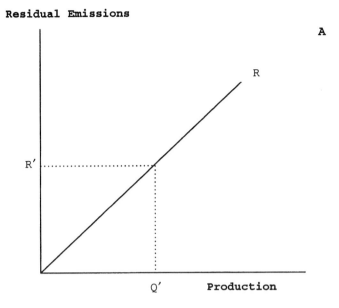

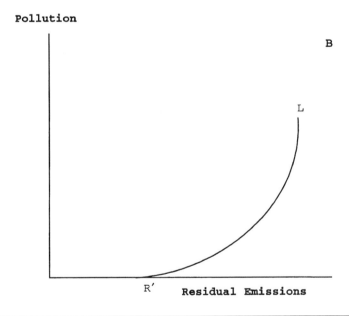

Figure 9.2. Relationships between residual emissions (R) and production (Q) and between pollution (L) and residual emissions. (A) Residual emissions as a function of production. (B) Pollution as a function of residual emissions.

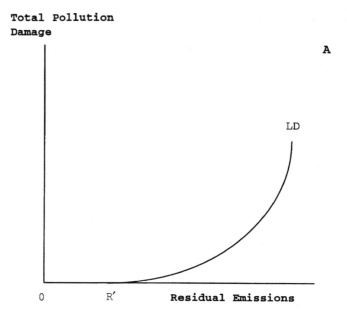

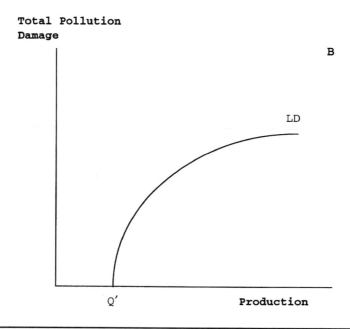

Figure 9.3. Relationships between total pollution damage (LD) and residual emissions (R) and between total pollution damage and production (Q). (*A*) Total pollution damage as a function of residual emissions with constant per unit pollution damage. (*B*) Total pollution damage as a function of production.

9. Economics of Environmental Pollution

The relationship between LD and Q, as depicted in Figure 9.3B, is determined as follows. For $Q \leq Q'$, $R \leq R'$ and $L = 0$, which implies $LD = dL = 0$. For $Q > Q'$, $R > R'$, which implies $LD = dbR^\lambda$. For $Q > Q'$, $R = aQ$, which implies $LD = dba^\lambda Q^\lambda$. Summarizing:

$LD = 0$ for $Q \leq Q'$ and $LD = dba^\lambda Q^\lambda$ for $Q > Q'$.

Marginal pollution damages (MLD) equal the derivative of LD with respect to Q, namely:

$MLD = 0$ for $Q \leq Q'$ or $KQ^{\lambda-1}$ for $Q > Q'$,

where $K = \lambda dba^\lambda$. MLD is illustrated in Figure 9.4A for different values of λ.

The privately efficient production rate occurs where output price equals marginal production cost (p = MPC), which occurs at Q_r in Figure 9.4B. When price exceeds marginal production cost (p > MPC), profit is increased by raising production. Conversely, when price is less than marginal production cost (p < MPC), profit is increased by reducing production. Therefore, in a purely competitive industry, profit is maximized at the output where price equals private marginal production cost (p = MPC) which occurs at Q_r in Figure 9.4B.

The socially efficient production rate is determined by equating p to MLD + MPC, which occurs at Q_s in Figure 9.4B. Increasing production beyond Q_s causes total pollution damages to increase more than net private benefit, which decreases net social benefit. Conversely, decreasing production below Q_s causes net private benefit to decrease more than pollution damages, which reduces net social benefit. At Q_s, net social benefit is area 1 and pollution damage is area 2.

An example of the privately and socially efficient rates of production is given in Table 9.1. The figures in the table are based on the following MLD, MPC and demand (D) functions, respectively:

$MLD = 2.5Q^{1.5}$ for $Q \geq 3$

$MPC = 0.25Q^2$

D: $p = 44 - 1.9Q$.

Table 9.1. Example of privately and socially efficient production rates

Q	MLD	MPC	MLD+MPC	p
3	12.99	2.25	15.24	38.3
4	20.00	4.00	24.00	36.4
5	27.95	6.25	34.20	34.5
6	36.74	9.00	45.74	32.6
7	46.30	12.25	58.55	30.7
8	56.57	16.00	72.57	28.8
9	67.50	20.25	87.75	26.9
10	79.06	25.00	104.06	25.0

Q = production rate, MLD = marginal pollution damage; MPC = marginal production cost; and p = price. Q = 5 is the socially efficient production rate and Q = 10 is the privately efficient production rate.

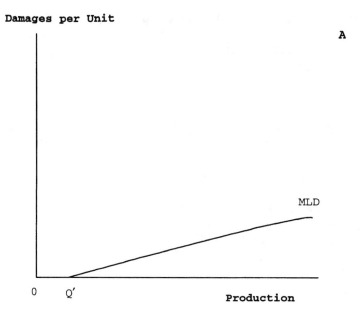

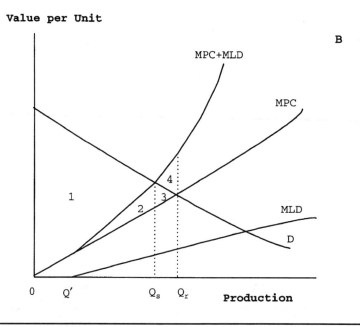

Figure 9.4. Marginal pollution damage (*A*) and privately (Q_r) and socially (Q_s) efficient production (*B*) under pure competition. D = demand; MLD = marginal pollution damage; and MPC = marginal production cost.

The privately efficient production rate is $Q_r = 10$ where p = MPC (25 = 25) and the socially efficient production rate is at $Q_s = 5$ where p = MPC + MLD (34.5 = 34.2). Hence, the privately efficient production rate is greater than the socially efficient production rate (9 > 6).

As shown earlier, total net private benefit to the industry is maximized at Q_r where p = MPC. At Q_r, net private benefit equals areas 1 + 2 + 3 and pollution damage equal areas 2 + 3 in Figure 9.4B. Because pollution is an external diseconomy, there is no incentive for firms to reduce production below Q_r, or equivalently, to reduce the externality. Net social benefit at Q_r is area 1 minus area 4 (1 + 2 + 3 − 2 − 3 − 4). Of the total externality occurring at Q_r (2 + 3 + 4), removing the externality given by 3 + 4 is Pareto relevant. Removal of this amount of externality increases net social benefit from area 1 minus area 4 to area 1. By reducing production from Q_r to Q_s, the Pareto-relevant externality (3 + 4) is completely removed. The externality remaining at Q_s (2) is not Pareto relevant because its removal decreases net private benefit more than it decreases pollution damage.

For the demand, MPC and MLD relationships depicted in Figure 9.4B, the socially efficient level of pollution damage is greater than zero. This can be seen as follows. Because $Q_s > Q'$, there is pollution damage at the socially efficient rate of production. If, however, p = MPC at a production rate less than or equal to Q', as illustrated in Figure 9.5, then the profit-maximizing production rate is Q_r and there is no pollution. Zero pollution is likely to be socially efficient when a) product prices are low and/or marginal production cost is high and/or b) MLD is zero at a relatively high rate of production.

In summary:

1. The privately efficient production rate exceeds the socially efficient production rate ($Q_r > Q_s$).
2. The socially efficient production rate approaches the privately efficient production rate as the threshold production rate increases, which causes pollution damage (Q') to decrease.
3. Zero pollution is generally not socially efficient.

Emission Restrictions. Controlling emissions by restricting production is valid only when emission is a fixed proportion of production. In the emission-restricted model, firms can reduce emissions by employing more inputs or changing technologies. For example, sulfur dioxide emissions from power plants can be reduced by substituting low sulfur coal for high sulfur coal or natural gas for coal. Alternatively, a cleaner technology could be used, such as fluidized-bed combustion, which removes the sulfur from coal as it is burned. The firm's production function in the emission-restricted model is:

$$Q = G(X, R),$$

where X represents input and R represents emission of residuals. The production function in the production-restricted model includes only X, whereas the production function in the emission-restricted model includes both X and R. This allows a given production rate, say Q_0, to be achieved with various combinations of X and R as shown in Figure 9.6. When X and R are substitutes, the current production rate

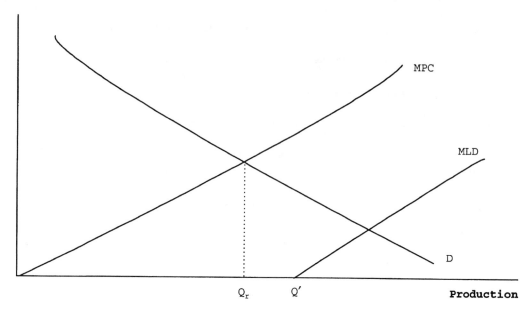

Figure 9.5. Efficient pollution equals zero when $Q_r \leq Q'$. D = demand; MLD = marginal pollution damage; and MPC = marginal production cost.

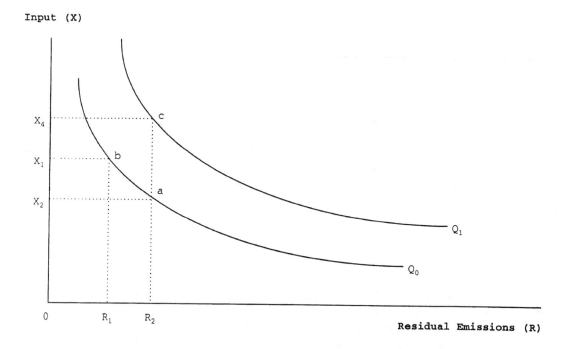

Figure 9.6. Combinations of input (X) and residual emissions (R) that give equal production (Q).

can be maintained while reducing emissions by employing more X. Moving from a to b reduces emissions from R_2 to R_1 but requires increasing input levels from X_2 to X_1. For example, electrical power production can be maintained, and particulate emissions reduced, by installing electrostatic precipitators that remove particulates from smoke before it leaves the smokestack.

Alternatively, the production rate can be increased without increasing emissions. Increasing the production from Q_0 to Q_1 while maintaining emission at R_2 can be achieved by increasing the input level from X_2 to X_4. Installation of precipitators entails an investment in manufactured capital and labor that increases the cost of production. Because R is a by-product of production, it might seem inconsistent with the material balances model to allow substitution between X and R. It is not. R depends on Q, but Q is also a function of R.

If pollution reduces not only the general welfare of consumers but also the firm's production, then L enters the production function:

$$Q = H(X, R, L).$$

Q is a decreasing function of L, other things constant. Suppose wastewater emissions from a shrimp pond result in pollution (L) of the water supply for the pond (Q). This is not uncommon in countries like Thailand. Then, L enters the production function for shrimp. Reducing water pollution is beneficial to the firm because it improves shrimp health and hence the productivity of the pond.

Just as producers can undertake investments to reduce residual emissions and pollution, households can make defensive expenditures to reduce the adverse effects of pollution. For example, households might relocate to a less polluted area to reduce water pollution–related illnesses caused by wastewater emissions from the shrimp pond. Alternatively, the household can install a water filtration system to purify the water. Defensive behavior by households is incorporated in the utility function as follows:

$$U = V[Q, F(L, Y)].$$

In this modified utility function, L is replaced by a subfunction F(L, Y), which represents the household's exposure to pollution. Exposure depends on the level of pollution (L) and the level of defensive expenditures (Y). Because F is a decreasing function of L, the household can reduce pollution damages by undertaking defensive expenditures.

Determination of the efficient level of pollution abatement for a single polluting firm in a purely competitive market is illustrated in Figure 9.7. It is assumed that additional units of pollution abatement can only be achieved by employing increasingly expensive abatement technologies. Therefore, the marginal cost of pollution abatement (MCA) is an exponentially increasing function of pollution abatement as shown in Figure 9.7. When the firm receives no benefit from pollution abatement (L does not enter the production function), the marginal private benefit of pollution abatement (MPBA) is zero and the privately efficient level of pollution abatement (A_r) is zero as shown in Figure 9.7A.

When pollution abatement enhances the firm's own production (L enters the production function), MPBA is nonzero as shown in Figure 9.7B. For example,

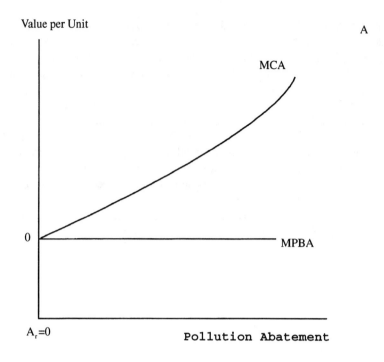

A

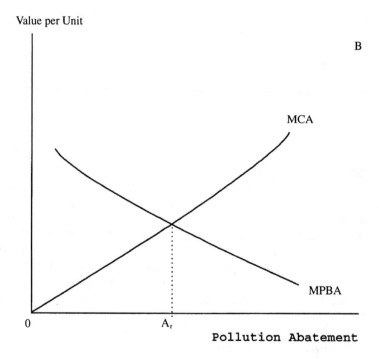

B

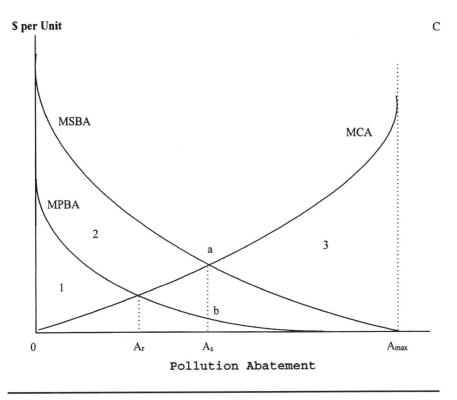

Figure 9.7. Efficient pollution abatement for a single polluting source under pure competition with zero marginal private benefit of abatement (MPBA) (*A*); nonzero, decreasing MPBA (*B*); and nonzero, decreasing marginal social benefit of abatement (MSBA) and a Pigouvian tax of ab (*C*). MCA = marginal cost of abatement.

since wastewater emissions from the shrimp pond pollute the water supply for the pond, reducing water pollution improves shrimp production. Reducing pollution of the pond increases income from shrimp production, but the increments in income become successively smaller as pollution abatement increases. In this case, MPBA is an exponentially decreasing function of pollution abatement as illustrated in Figure 9.7B. The privately efficient level of pollution abatement is A_r where MPBA = MCA.

If the wastewater from shrimp ponds degrades the quality of water used by nearby households, then water pollution abatement results in an external community benefit. Let the marginal community benefit of pollution abatement (MCBA) be an exponentially decreasing function of abatement. This means that equal increments in pollution abatement result in smaller and smaller increments in community benefits. Marginal social benefit of pollution abatement (MSBA = MPBA + MCBA) as shown in Figure 9.7C. Because MPBA and MCBA are exponentially decreasing functions of abatement, so is MSBA. Equating MCA to MSBA gives the socially efficient level of pollution abatement of A_s.

Because $A_r < A_s$, there is a Pareto-relevant externality at A_r. In particular, the loss in net social benefit from reducing pollution to A_r instead of A_s is area 2 in Figure 9.7C. Area 2 is the size of the Pareto-relevant externality. Reducing pollution beyond A_s is Pareto-inferior because the cost of abatement exceeds the social benefit. For example, reducing pollution to A_{max} (zero pollution) generates a net social benefit equal to areas 1 + 2 minus area 3, which is less than areas 1 + 2.

An example of the privately and socially efficient levels of pollution abatement is given in Table 9.2. The figures in this table are based on the following MCA, MPBA and MSBA functions:

$$MCA = 5A^{1.5}$$

$$MPBA = 2200e^{-A}$$

$$MSBA = 8200e^{-A}.$$

In this example, the privately efficient level of pollution abatement occurs at $A_r = 4$ where MCA = MPBA ($40 \cong 40.29$) and the socially efficient level of pollution abatement occurs at $A_s = 5$ where MCA = MSBA ($55.90 \cong 55.25$).

The production-restricted and emission-restricted models both indicate that a Pareto-relevant externality is present. The production-restricted model (Figure 9.4B) shows that the privately efficient production rate exceeds the socially efficient production rate ($Q_r > Q_s$) and the externality generated by Q_r is excessive from a social viewpoint (2 + 3 + 4 > 2). The emission-restricted model (Figure 9.7C) shows that the privately efficient level is less than the socially efficient level of pollution abatement ($A_r < A_s$), or equivalently that the privately efficient rate exceeds the socially efficient rate of production.

STATIC EFFICIENCY WITH IMPERFECT COMPETITION. The production- and emission-restricted models have been used to derive the efficient rates of production and pollution abatement for a purely competitive industry. How are efficient rates of production and pollution affected by imperfect competition? The answer is examined by comparing production and pollution rates for imperfectly and purely competitive industries. For simplicity, it is assumed that the demand curve, marginal production cost (MPC), marginal pollution damages (MLD) and marginal

Table 9.2. Example of privately and socially efficient levels of pollution abatement

A	MCA	MPBA	MSBA
1	5.00	809.33	3,016.61
2	14.14	297.74	1,109.75
3	25.98	109.53	408.25
4	40.00	40.29	150.19
5	55.90	14.82	55.25
6	73.48	5.45	20.33
7	92.60	2.01	7.48
8	113.14	0.74	2.75

A is the level of pollution abatement; MCA = marginal cost of abatement; MPBA = marginal private benefit of abatement; MSBA = marginal social benefit of abatement. A = 4 is the privately efficient level of pollution abatement and A = 5 is the socially efficient level of pollution abatement.

cost of pollution abatement (MCA) are the same under pure and imperfect competition. Consider using the production-restricted model to compare the production rates under pure and imperfect competition. Under pure competition, profit is maximized at the production rate where p = MPC. p is determined from the demand function (D). Under imperfect competition, profit is maximized at the production rate that makes MR = MPC.

Because MR < p with imperfect competition, the efficient rate of production is lower with imperfect competition than with pure competition ($Q_m < Q_c$) as shown in Figure 9.8A. Because the production-restricted model implies that pollution is a continuous function of production above some threshold rate of production, the privately efficient level of pollution is also less under imperfect competition. The magnitude of the difference in production and pollution between pure and imperfect competition depends on the relative slopes of the underlying marginal revenue and marginal production cost curves. Likewise, the socially efficient level of pollution, which equates marginal social benefit of abatement to marginal pollution damage, is less with imperfect competition than with pure competition.

Next, consider the emission-restricted model. Let MPB be the marginal private benefit to the industry for a one unit change in pollution abatement. It can be shown that:

$$MPBA_m > MPBA_c.$$

The last relationship is depicted in Figure 9.8B. Efficient pollution abatement is less with pure competition than with imperfect competition. Hence, pollution is less under imperfect competition than under pure competition. In summary, both the privately efficient rate of production and level of pollution are greater under pure competition than under imperfect competition. While the lower pollution level achieved by imperfect competition is preferable from a social viewpoint, the lower rate of production is not. It is coincidental that the privately efficient levels of production and pollution abatement are closer to the corresponding socially efficient levels with imperfect competition than with pure competition.

DYNAMIC EFFICIENCY. When residual emission in the current period influences both current and future pollution damages, the efficient level of pollution abatement should be determined in a dynamic framework. This framework considers the intertemporal relationship between emissions and pollution damages. For example, pollution of groundwater in some agricultural areas has been traced to excessive application of animal manure and/or storage of animal manure in barnyard areas that occurred several years earlier. Nutrients in animal manure are stored in the soil and released over time, which can cause nutrient contamination of groundwater.

Dynamic efficiency analysis is considerably more complex than static analysis of pollution externalities. A simple dynamic model is used here, which assumes that a) some of the residuals generated by production are not assimilated by the environment, b) pollution and pollution damages are proportional to emissions of residuals, and c) there is only one pollutant. The model is in discrete time. Following the same logic as used in the static efficiency analysis of pollution externalities, pollu-

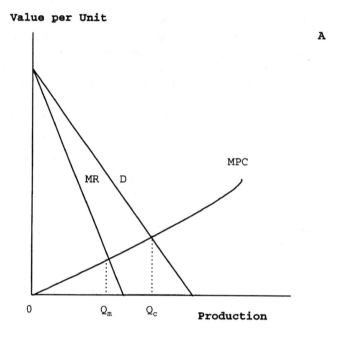

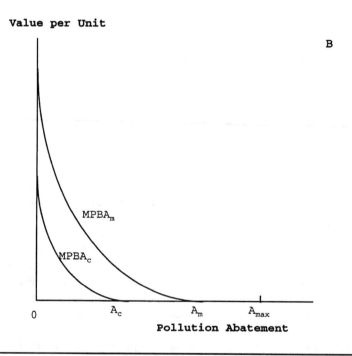

Figure 9.8. Comparison of efficient production under pure (Q_c) and imperfect (Q_m) competition for production-restricted (A) and emission-restricted (B) model. Efficient abatement is A_c for pure competition and A_m for imperfect competition, as shown in (B). D = demand; MPBA = marginal private benefit of abatement; MPC = marginal production costs; and MR = marginal revenue.

tion associated with current production (L_t) is assumed to be a power function of current production (Q_t):

$$L_t = a_t Q_t^\lambda \quad (t = 1,\ldots,T,\ a_t > 0 \text{ and } \lambda > 1).$$

L_t can be decreased by investing in pollution control technologies that reduce a_t.

Accumulated pollution at time t (AL_t) is the sum of pollution flows through period t:

$$AL_t = \sum_{\tau=1}^{t} L_\tau.$$

Therefore:

$$AL_t = \sum_{\tau=1}^{t} a_\tau Q_\tau^\lambda.$$

Pollution damage in period t (LD_t) is proportional to accumulated pollution through period t:

$$LD_t = b_t AL_t \quad (b_t > 0),$$

where b_t is the increase in pollution damage per unit increase in accumulated pollution in period t, or the marginal cost of pollution damages. Substituting for AL_t in the last equation gives:

$$LD_t = b_t \sum_{\tau=1}^{t} a_\tau Q_\tau^\lambda.$$

This equation states that pollution damage in period t depends on the value of a_τ, b_t and historical production ($Q_1,\ldots,Q_t$).

Consider first the socially efficient rates of production and levels of pollution. The socially efficient production rates over tsime are determined by maximizing the present value of net social benefits. Net social benefit in period t equals net private benefit minus pollution damage in period t. Let B_t be net private benefit in period t. Then, net benefit in period t is:

$$NSB_t = B_t - LD_t.$$

Overall net social benefit is the present value of the NSB_t values, namely:

$$NSB = \sum_{t=1}^{T} NSB_t (1 + r)^{-t},$$

where r is the discount rate. Maximizing NSB gives the socially efficient production rates for all periods, namely:

$$Q^*_{s1},\ldots,Q^*_{sT}.$$

Substituting the efficient production rates into $L_t = a_t Q_t^\lambda$ and $LD_t = b_t \Sigma a_\tau Q_\tau^\lambda$ gives the socially efficient rates of pollution and pollution damages, respectively, in all periods.

If pollution damages increase over time, then the socially efficient production rates decrease over time. Decreasing production rates would raise commodity prices over time, provided there are no offsetting decreases in demand for the commodity. Investments in pollution control technologies could moderate the growth in pollution damages by lowering pollution flows per unit of production and/or the marginal cost of pollution damage (b_t).

When pollution is a pure externality, reducing pollution does not benefit the acting party. Hence, the acting party ignores the pollution damages and simply maximizes net private benefit, which is:

$$NPB = \sum_{t=1}^{T} NPB_t(1 + r)^{-t},$$

where NPB_t is net private benefit in period t. The resulting privately efficient production rates and pollution damages are greater than the socially efficient production rates and pollution damages.

Is it in society's best interest to impose restrictions on acting parties to eliminate excessive production or pollution? If the present value of net social benefits minus the present value of the cost of imposing the restrictions is greater than the present value of net private benefits, then it is in society's best interest to impose the restrictions. The next section discusses alternative public policies for restricting pollution.

Establishing Property Rights for Environmental Resources

Unless polluting firms decide to reduce pollution on their own and/or a regulatory agency, such as the U.S. EPA, taxes or regulates pollution, profit-maximizing firms have little economic incentive to reduce production and/or pollution to the socially efficient level. Eliminating Pareto-relevant pollution externalities warrants some form of public action when it is economically justified (benefits exceed costs). Chapter 5 explained that a principal cause of external diseconomies is the absence of a complete system of property rights. One way to eliminate Pareto-relevant pollution externalities is to establish complete property rights for environmental resources. These rights would establish whether polluters (acting parties) have a right to pollute environmental resources or whether pollution victims (affected parties) have a right to unpolluted environmental resources. Either system of rights provides an economic incentive to negotiate a socially efficient rate of production or pollution abatement.

Several factors impede a property rights solution to pollution externalities. First, even when complete property rights for environmental resources are specified, the cost of settling resource disputes between acting and affected parties is high, particularly when there are several acting and/or affected parties (high transaction costs). Second, a property rights solution is unlikely when pollution damages emanate from a large number of diffuse sources such as with nonpoint source pollution. In contrast, a property rights solution is feasible for point sources of pollution, such as thermal water pollution from a nuclear power plant.

Third, even if pollution damages can be linked to specific residual emissions, as with point source pollution, allocation of damages to specific pollutants is difficult when the damages from two or more pollutants are multiplicative rather than additive. Fourth, a property rights solution is unlikely when there is nonrivalry in the consumption of pollution. Nonrivalry is present when pollution damages to one affected party do not diminish damages to other affected parties. For example, health impairment from inhalation of secondary cigarette smoke is not diminished by the number of parties exposed to the smoke. The externality that occurs when there is nonrivalry in consumption is an *undepletable externality*. One consequence of undepletable externalities is that the efficient level of pollution is not the same for all affected parties and is different than the socially efficient level. While a property rights solution to pollution externalities can be difficult to achieve, support for this type of solution has increased in recent years.[7]

Pollution Abatement Policies

When there is pure competition in product markets and residual emissions are not proportional to production, the privately efficient level of pollution abatement is less than the socially efficient level of abatement (emission-restricted model). This occurs because firms or households ignore the external costs of pollution on other firms or households. External environmental costs arise when firms and households have free access to the residual assimilation capacity of the environment. In other words, the residual assimilation capacity of the environment is unpriced. This section evaluates and compares several policies for alleviating point and nonpoint sources of environmental pollution.

POINT SOURCE POLLUTION. Environmental pollution occurs when the residual assimilation services of the environment are free or unpriced. Because polluters do not have to pay for these services, a large amount of residuals is emitted, which increases the likelihood of environmental pollution, impairment of human health and loss of ecological services. This section discusses how taxes, emission charges, emission standards, tradable emission permits and environmental liability can be used to control point sources of pollution. Elements of these policy instruments can be used in controlling nonpoint sources of pollution. Their application to nonpoint source pollution is more complex because the links between sources of pollution and pollution are not well-defined.

Taxes. Pollution externalities can be reduced to their socially efficient levels by charging polluters to use the residual assimilation capacity of the environment. The efficient price for pollution is a Pigouvian tax on residual emissions that cause pollution. The tax equals the difference between marginal social benefit and marginal private benefit of pollution abatement at the socially efficient level of abatement (ab in Figure 9.7C).

Imposing a Pigouvian tax of ab per unit of pollution causes the marginal private benefit curve to shift upward until it intersects MCA at A_s. The tax shifts MPBA upward by the amount of the tax because decreasing pollution decreases tax payments, which increases the marginal private benefit of pollution abatement. After the tax, the original level of pollution abatement, A_r, is no longer efficient because an additional unit of pollution abatement increases total return more than it increases the total cost of abatement. How much should abatement be increased? There is an economic disincentive for extending pollution abatement beyond A_s. Increasing abatement beyond A_s causes total abatement cost to increase more than total revenue, which reduces total net return. After the tax, the privately and socially efficient levels of pollution abatement are identical. In other words, A_s is the profit-maximizing level of pollution abatement with a Pigouvian tax. Pigouvian taxes have been criticized on the grounds that they might cause marginally profitable firms to leave the industry.[8]

Other types of financial incentives have been used to reduce environmental emissions and pollution, namely: taxing inputs and/or production, subsidizing defensive expenditures made by affected parties, compensating affected parties for damages inflicted by acting parties, and subsidizing emission reduction by acting parties. The proposed carbon tax of $30 per ton proposed by the United States in 1993 was designed to reduce carbon dioxide emission. European countries have high taxes on gasoline. Taxes have been proposed and used to reduce municipal solid waste and congestion on highways and to encourage recycling.

An example of a defensive expenditure is the cost of substituting bottled water for polluted water. Is it efficient to subsidize defensive expenditures, say, by offering government-financed rebates to individuals who purchase bottled water in areas with polluted groundwater? No. Households exposed to pollution will allocate their limited income to consumer goods and defensive activities so as to maximize their utility. When utility is maximized, the added utility per dollar spent on goods and defensive activities is equal. Therefore, utility- and profit-maximizing behaviors will automatically result in an efficient allocation of income or capital to defensive activities. In fact, subsidies to acting parties distort incentives to reduce pollution. Subsidizing bottled water reduces its price and increases the quantity demanded, which reduces the health risk from polluted groundwater. Lower vulnerability to polluted groundwater reduces incentives to protect groundwater from pesticide contamination and increases the likelihood of groundwater contamination.

Compensating affected parties for the damages caused by environmental pollution is also inefficient. Compensating households for each 100 gallons of polluted groundwater they consume creates a disincentive to purchase bottled water and results in an inefficient level of defensive expenditures. In this case, compensation causes the level of defensive expenditures to be too low.

Finally, subsidizing the acting party may also be inefficient. An example of this is the reimbursement given to landowners to cover part of the cost of soil- and water-conservation practices. Consider providing landowners with a unit subsidy equal to t. At first glance, it might appear that a unit subsidy to reduce emissions has the same effect as a unit tax on emission, namely, it achieves the socially efficient level of pollution abatement, A_s. There are important differences. A subsidy, unlike a tax, increases firm profits (it is like an additional source of revenue), which keeps some firms that are inefficient in reducing pollution from leaving the industry and attracts new firms into the industry in the long run. Less exit and greater entry of firms are likely to increase production and emissions—just the opposite of what is needed. In summary, subsidies to acting and/or affected parties, as well as compensation to affected parties, are generally socially inefficient.

While Pigouvian taxes theoretically achieve the socially efficient level of pollution, their implementation requires considerable information about polluting activities. The regulatory authority must know the following: a) the marginal cost of pollution abatement for all parties, b) the marginal private benefit of pollution abatement (MPBA), c) the marginal social benefit of pollution abatement (MSBA), and d) the emission levels for all acting parties. Without this information, it is not possible to determine the Pigouvian tax. Several second-best pollution-control policies have been developed and utilized, including emission charges, emission standards, tradable emission permits and liability rules.

Traditional analysis indicates that the imposition and level of a tax or subsidy creates incentives for firms to enter the industry when a subsidy is imposed, or possibly leave the industry when a tax is levied. Rent-seeking behavior provides an alternative explanation of firm entry and exit relative to the size of the industry that would exist under a system of complete property rights for environmental resources.[9] Rent-seeking behavior consists of actions designed to improve the financial position of a firm, household, industry or special interest group.[10] In the case of a Pigouvian subsidy, rent-seeking behavior stimulates firms to enter the industry to compete for the excess rents generated by the subsidies. Excess rents arise when the value of the subsidy exceeds the cost of pollution control.

While a Pigouvian tax provides no opportunity for rent-seeking behavior by polluting firms, it can stimulate general rent-seeking behavior in which interest groups compete for the revenues generated by the tax. For example, conservation organizations might lobby to have the tax revenues spent on developing less-polluting technologies. Industrial groups might lobby to spend the revenues on retraining workers displaced by pollution control regulations. If the tax revenues are not evenly distributed among communities, those communities receiving a disproportionate share of the revenue would be able to upgrade the provision of public goods such as parks, libraries and educational facilities. This might stimulate a general movement of population to those areas receiving a disproportionate share of the tax revenues.

Emission Charges. The emission charges approach, also called the emission charges and standards approach, requires the regulatory authority to establish an environmental standard and a uniform charge per unit of emission for each source. The charge is adjusted until the standard is achieved. In essence, the charge is the price that the polluter pays for using the assimilative capacity of the environment. The environmental standard can be an *ambient standard* or an *emission* or *effluent standard*. An ambient standard establishes a minimum standard for environmental quality. For example, the current ambient standard for nitrate concentrations in drinking water in the United States is 10 parts per million. An effluent standard establishes the mean or maximum permissible discharge of a residual for a given source. The Clean Air Act Amendments of 1990 established a goal of reducing total emissions of sulfur dioxide from coal-fired power plants by 50 million tons. Ideally, effluent standards are based on the likely benefits to society of avoiding various kinds of pollution.

The efficient level of pollution abatement for a private firm is determined by balancing the costs and benefits of pollution control. The benefits of pollution control equal the reduction in emission charges paid to the environmental authority. Total emission charges equal total emission times the charge per unit of emission. When the per unit emission charge does not depend on the level of emissions, the marginal benefit to the firm of reducing pollution equals the per unit emission charge. A firm minimizes the cost of pollution abatement by equating the marginal benefit to the marginal cost of pollution abatement as shown in Figure 9.9. The efficient abatement level for a per unit emission charge of f_1 is A_1. When abatement is below A_1, it is less costly to increase pollution abatement than it is to pay the emission charge. Conversely, when the abatement level is greater than A_1, reducing abatement is cost-effective because abatement cost decreases more than emission charges increase. Similarly, A_2 is the most efficient level of pollution abatement when the emission charge is f_2.

The regulatory authority achieves the desired level of emission from all sources by increasing or decreasing the uniform emission fee. For a per unit fee of f_1, total emission from all sources is:

$$R(f_1) = \sum_{j=1}^{J} R_j(f_1),$$

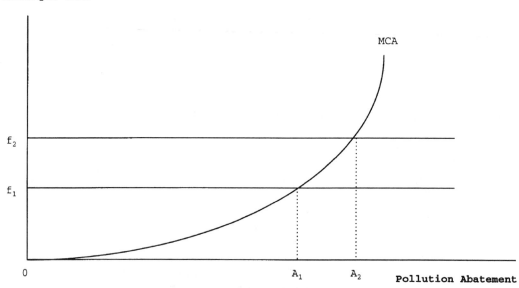

Figure 9.9. Efficient pollution abatement for two different emission charges (f). MCA = marginal cost of abatement.

where $R_j(f_1)$ is the residual emission by firm j when the fee is f_1 and J is the number of firms or emission sources. If $R(f_1)$ is greater than the desired level of emissions, then the effluent standard is not exceeded. In this case, the regulatory authority must increase the emission charge until the desired emission level is achieved. Raising the emission charge from f_1 to f_2 shifts the marginal benefit of abatement upward, which increases the efficient level of abatement and decreases emissions, other things constant. Conversely, if $R(f_1)$ is more than the desired level, then the unit charge must be decreased in order to decrease abatement until the desired level of emissions is achieved. Correct adjustments to the emission charge require the regulatory authority to monitor the level of emissions from all sources.

While emission charges do not necessarily result in the socially efficient level of emission reduction, they are the most cost-effective policy for achieving a given environmental standard. The most cost-effective policy achieves the environmental standard at least cost to society. This is illustrated in Figure 9.10. For simplicity, only two polluting sources are considered, X and Y. Both sources are assumed to maximize the net benefit of emission reduction. Source X has a lower cost of pollution abatement than source Y ($MCA_X < MCA_Y$). For a uniform per unit charge of f_1, the most efficient allocation of emission abatement between sources X and Y is A_X and A_Y, respectively. Hence, the source with the lower cost of abatement (X) has greater abatement ($A_X > A_Y$). For this reason, the average cost of achieving a particular emission reduction is automatically minimized with emission charges. This is true regardless of whether the polluting sources operate in a purely or imperfectly competitive market. One study showed that the total cost of reducing emission of certain halocarbons was $110 million using emission fees versus $230 million using emission standards.[11]

Not only are emission charges more cost effective, but unlike a Pigouvian tax, they do not require knowledge of the marginal cost of pollution abatement or the private and social benefits of emission reduction for each source. Emission charges operate more smoothly, however, when the regulatory agency knows the average cost of controlling different sources of pollution. This information can be used by the regulatory authority to determine the level of emission charges needed to achieve the desired environmental standards. Without some knowledge of average cost, the authority has to use an iterative approach in which emission charges are adjusted up or down until the standard is achieved. Such adjustments can create considerable uncertainty for polluting firms especially when several adjustments are needed to arrive at the desired standard.

In addition to being cost effective, emission charges have four additional advantages. First, they provide an incentive for polluting sources to invest in new abatement technologies. In the long run, the firm can change abatement technology. If the cost of adopting an abatement technology is less than the saving in emission charges, then there is an economic incentive for the firm to adopt that technology. Suppose technological improvement reduces the marginal cost of pollution abatement from MCA_1 to MCA_2 as shown in Figure 9.11. The efficient level of pollution abatement increases from A_1 to A_2. Second, emission charges internalize part of the cost to society of utilizing the residual assimilation capacity of the environment. Imposing emission charges has the same effect as pricing the residual assimilation capacity of the environment. Such charges increase the marginal cost of pollution abatement and the marginal production cost. Part of the increase in marginal pro-

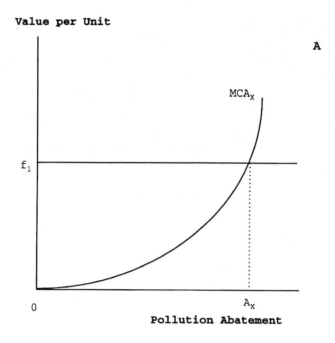

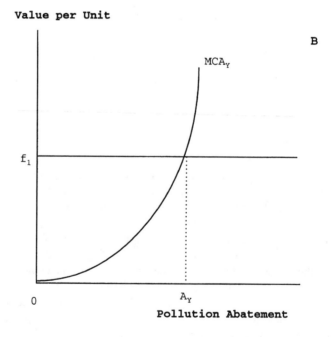

Figure 9.10. Efficient allocation of pollution abatement between sources X (*A*) and Y (*B*) for an emission charge of f_1. MCA = marginal cost of abatement.

duction cost is passed on to households in the form of higher product prices. Higher product prices reduce the quantity demanded of the product, other things equal.

Third, emission charges are a source of public revenue that can be used to develop less-polluting technologies, mitigate the adverse effects of pollution, and/or reduce budget deficits. It has been estimated that charges for sulfur dioxide and particulate matter emitted to the air by stationary sources would generate annual tax revenues of $1.8 billion to $8.7 billion. If the revenues from such charges are substituted for revenues generated by taxes on labor (corporate) income, there would be an efficiency gain of between $630 million ($1 billion) and $3.05 ($4.9) billion in 1982 dollars.[12] Fourth, emission charges are consistent with the polluter-pays principle, which maintains that polluters should pay for the external costs that their pollution imposes on society.

Emission charges have several drawbacks. First, the establishment of emission standards for all pollutants is a monumental task, although less difficult than determining the social damages caused by various emission levels. Furthermore, there can be very different opinions regarding the appropriate ambient or effluent standards. For example, ambient drinking water standards for nitrate–nitrogen and atrazine are continually being debated because of the uncertainty regarding the adverse health effects of these substances. Second, changes in the number of emission sources and pollution abatement technology necessitate periodic adjustments in emission charges, which can be disruptive to polluting industries.

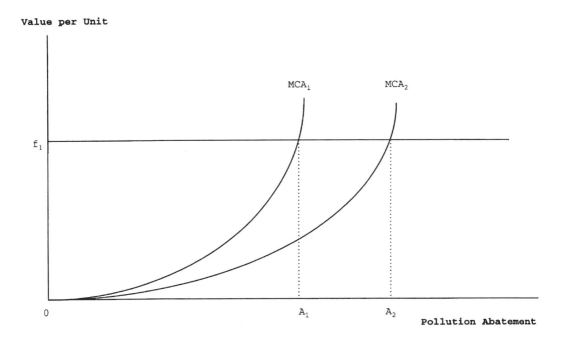

Figure 9.11. Effects of improvement in abatement technology on efficient abatement. MCA = marginal cost of abatement; and f_1 = emission charge.

Third, emission charges are not effective in controlling emissions of toxic residuals because such emissions must be precisely controlled. Fourth, a uniform emission fee is not efficient when emissions from different sources generate unequal unit damages. Consider a uniform charge on surface runoff that has a nitrate–nitrogen concentration in excess of the drinking water standard of 10 parts per million. Suppose one source is upriver from a reservoir that serves as a public drinking water supply and another source is located below the reservoir. A uniform charge on emissions from both sources is not cost effective because emissions from the upriver source have a much greater potential impact on human health than do emissions from the downriver source. While this deficiency can be alleviated by setting higher charges in areas where nitrate pollution of water poses a higher risk to human health, the administrative cost of implementing differential charges is greater than for uniform charges.

Fifth, even though emission charges achieve the desired reduction in emissions at least cost, entirely different approaches might be more efficient. If the objective is to reduce sulfur dioxide emissions from electrical power–generating facilities, then a unit charge on emissions from these facilities is cost effective. Reducing the demand for electricity through conservation would be more cost effective. Conservation includes subsidizing the use of efficient electrical lighting in new facilities, increasing the energy efficiency of lightbulbs and electrical appliances, and expanding space heating with passive solar energy.

Sixth, environmental standards are not likely to be achieved when emission charges are too low and/or the standard is not enforced. No western country has set emission fees at a high enough level to achieve its environmental standards. In Eastern Europe, emission charges are a major part of Poland's air pollution control policy. Poland's air emission charges are generally low, however, even though they have been increased since 1989 and emission charges or fines are not enforced because of the economic difficulties in restructuring the industrial sector.[13] Several countries in Western Europe, notably Germany, France and the Netherlands, have used emission charges to reduce water pollution. Except for the Netherlands, emission charges in this region have been set at too low a level to be effective in reducing water pollution.

Emission Standards. An *emission standard* requires the regulatory authority to set an environmental standard for each emission source and to monitor the emissions for compliance with the standard. This is essentially a command-and-control approach to control of point sources of pollution. In some cases, the regulatory authority requires each source to use the best available emission control technology. Emission standards are the most popular method of controlling point sources of pollution in the United States. A major advantage of emission standards is that unlike incentive-based emission charges, they ensure that emission at each source and total emission do not exceed a certain level when the standards are enforced. Enforcement usually entails penalizing sources that violate the standard.

A major disadvantage of environmental standards is their inefficiency. Because information on the most cost-effective abatement technology for each source is usually not available, standards and/or abatement technologies for each source are typically based on general requirements and conditions in the industry. Because the most cost-effective technology varies across facilities and time, the emission standards approach ends up requiring levels of pollution abatement and, in some cases, abatement technologies for individual sources, that are inefficient.

Studies indicate that the cost to polluters of reducing air and water pollution in the United States is twice to 10 times as costly as the least cost alternative.[14] Furthermore, there is no guarantee that environmental standards established by a body of elected representatives, such as Congress, or by a regulatory authority, such as the Environmental Protection Agency, are efficient from an economic viewpoint (marginal benefit equals marginal cost of abatement). Some economists argue that establishment and enforcement of emission standards by public bodies should be undertaken only when emissions generate high social costs and compliance with the standards significantly reduces social costs.

Emission standards are generally less efficient than emission charges in achieving the same level of pollution reduction as illustrated in Figure 9.12. Consider a *uniform emission standard* that requires sources X and Y to achieve the same level of emission reduction, namely, A. Marginal cost of abatement (MCA) is higher for source X than source Y ($MCA_X > MCA_Y$). The cost of reducing total emission by 2A is the area under MCA_X up to A plus the area under MCA_Y up to A.

With a uniform emission charge, efficient abatement occurs where the emission charge equals the marginal cost of abatement. Hence, the efficient emission reduction is A_X for source X and A_Y for source Y. The total cost of abatement with the uniform emission charge is the area under MCA_X up to A_X plus the area under

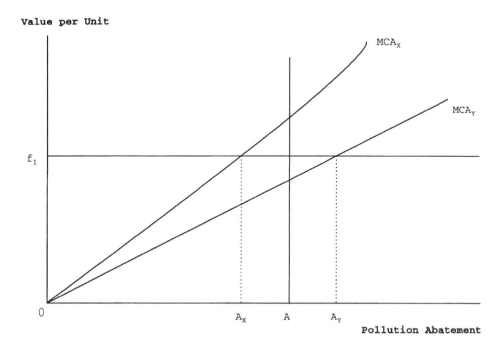

Figure 9.12. Comparison of abatement for sources X and Y with a uniform emission standard (A) and a uniform emission charge (f_1). MCA = marginal cost of abatement.

MCA_Y up to A_Y. Because this area is less than the area below the two marginal cost of abatement curves up to A, the total cost of emission abatement is less with a uniform emission charge of f_1 than with an emission standard of A. This occurs because the uniform standard requires both sources to achieve the same level of abatement regardless of their marginal costs of abatement. In contrast, the uniform emission charge allocates more emission reduction to the source with the lower marginal cost of abatement.

Some of the inefficiency that results when all pollution sources are required to achieve the same emission standard can be reduced by varying the standard according to the type of facility. For example, if older facilities have a higher marginal cost of abatement than newer facilities, then total abatement cost can be reduced by requiring less abatement for older than for newer facilities. Alternatively, the abatement technology for each firm can be based on the size of its facility. Facilities above a certain size can be required to use a more efficient abatement technology than facilities below that size.

The inefficiency of emission standards is especially high when the regulatory authority controls not only the level of emissions at each source but also the technology each source must use to reduce emissions. It is quite common for the regulatory authority to require use of the best available emission control technology. Requiring use of a particular abatement technology has two negative consequences. First, it can increase the social cost of reducing emissions by not allowing the emission sources to select the most cost-effective technologies for achieving the standard. For example, in Figure 9.12, let MCA_X be the marginal cost of abatement for the technology specified by the regulatory authority and let MCA_Y be the marginal cost of abatement with the most cost-effective abatement technology for that particular source. Requiring the source to use a specific technology increases the total cost of achieving A by the area between MCA_X and MCA_Y up to A.

Second, allowing the regulatory authority to determine the abatement technology to be used at each source might decrease the incentive to develop new, more cost-effective technologies. For example, there is a strong disincentive for a firm to develop an abatement technology that results in more efficient control of emissions when the regulatory authority's response to new abatement technologies is to tighten the emission standard and/or to require all emission sources to use that technology. On the positive side, the cost of monitoring the installation and operation of pollution control technologies is much lower than the cost of monitoring specific emissions.

Emission standards have certain advantages relative to emission charges despite their inefficiency. Consider an ambient standard that restricts the maximum permissible concentration of a pollutant. Empirical evidence suggests that an ambient emission standard is likely to result in less total emissions than is an emission charge for the following reason.[15] With an emission charge, the implicit economic value to the firm of reducing pollutant concentrations below the maximum permissible concentration is zero. Hence, it is efficient to increase emissions as long as the pollutant concentration is below the maximum permissible concentration. An emission standard typically has a nondegradation clause that disallows a firm from increasing emissions when the concentration is below the ambient standard. As noted earlier, the inherent inefficiency of emission standards relative to emission charges can be reduced by regulating total emissions at each source and allowing sources the flexibility to choose the most efficient abatement technology for achieving a particular standard.

Determining the maximum permissible concentration of a pollutant for achieving an ambient standard can be much more complicated than setting an emission charge because concentrations are influenced not only by emissions but also by environmental conditions at the time emission occurs. For example, a rainfall that occurs shortly after a particular herbicide has been applied to an agricultural field is likely to cause concentrations of that herbicide in surface runoff to exceed the ambient standard. To account for such events, the ambient standard is typically compared with herbicide concentrations averaged over several time periods. For example, the ambient standard of 3 parts per billion for atrazine in drinking water is violated when the average concentration of atrazine in quarterly samples of drinking water exceeds the standard.

Second, a uniform emission standard guarantees achievement of a specific level of environmental quality. An emission charge achieves that same quality only if it is set at the proper level. This advantage is especially important in reducing the risk of pollution damages during emergencies, and controlling the release of toxic substances. For example, the risk of health-related problems increases dramatically when an air inversion traps hydrocarbon emissions in the Los Angeles, California, or Denver, Colorado, airshed. The increase in risk is caused not by the level of emissions, but by atmospheric conditions. The risk of pollution damages can, however, be decreased by temporarily invoking a restrictive emission standard.

An emission charge cannot achieve a quick reduction in emissions because it influences long-run decisions such as the use of a particular emission control technology. What works in emergency situations is direct intervention by the regulatory authority to reduce the level of emissions and the risk of pollution damages. The damages caused by an air inversion in Los Angeles or Denver can be reduced by restricting nonessential operations and transportation that contribute to hydrocarbon emissions. Such restrictions would be lifted when the emergency passes. In the case of toxic substances, it is vitally important to control emissions. An emission standard is better suited to handling emergency situations and controlling the release of toxic substances than an emission charge.

Third, the greater efficiency of emission charges relative to emission standards is achieved only when the emission charge is set at the proper level. Too high a charge can make an emission charge more costly than an emission standard. Determining the correct charge requires knowledge of the marginal costs of abatement for all firms in the industry. In general, the regulatory authority does not have such information.

Tradable Emission Permits. Tradable emission permits (TEPs) restrict the contribution that different sources make to ambient concentrations of a pollutant *(ambient permit system)*, emissions from a source or area *(emission permit system)* or a combination of the two *(pollution offset system)*. Permits can be bought and sold according to the terms of trade specified in the permit. Tradable emission permits are like emission standards because they require the regulatory authority to set an upper limit on either ambient concentrations or total emissions.

The main goal of TEPs is to minimize the cost of achieving predetermined environmental standards. In the ambient permit system, permits are defined in terms of the pollutant concentration allowed in a particular location. If a source has emissions that contribute 10 units per month to ambient concentrations in a particular location and each permit in that location is for two units of concentration, then the source would need to purchase five permits for that location. If the source is issued only three permits, then it must buy two permits or install a technology to reduce emissions by four units. Source-to-source differences in the ambient concentrations of emissions are usually caused by differences in abatement technology and local environmental conditions. If emissions from a source influence ambient concentrations in three locations, then that source might have to trade permits in the three locations. For this reason, the cost of an ambient permit system can be very costly to sources.

In an emission permit system, the regulatory authority issues permits in each of several zones, which specify a maximum emission level. The number of permits issued in each zone depends on the environmental standard established for that zone. The regulatory authority issues a certain number of emission permits in each zone. Total emissions allowed by all permits equal the environmental standard for that zone. For example, if the regulatory agency wants to limit total sulfur dioxide emissions to 100,000 tons per day and each permit allows a fixed emission of 10,000 tons per day, then emission sources are issued 10 permits.

An emission permit is less costly for sources than ambient permits because each source only has to trade permits in the zone in which it is located. Emission permits are likely to be more costly to administer than will be ambient permits. This occurs because in an emission permit system the regulatory authority has to increase or decrease the number of emission permits until the desired level of environmental quality is achieved. Adjusting the number of emission permits issued for that zone gives rise to the same sort of price uncertainty for sources as adjusting emission charges.

In the pollution offset system, permits are defined in terms of the level of emissions allowed within a zone. Trading of permits is allowed provided it does not violate the environmental standard for that zone. This system results in least-cost attainment of environmental standards regardless of the initial distribution of permits. The only information required to implement the pollution offset system is the contribution that the emissions from each source makes to ambient concentrations. For example, suppose the maximum ambient concentration for a pollutant has already been reached in a zone and source A wants to increase its emissions by purchasing a permit from source B. If one unit of A's emissions contributes twice as much to ambient concentrations as does one unit of B's emissions and each permit allows one unit of emission, then A would have to purchase two permits from B for each unit of increase in its own emissions. In order to sanction this and other possible trades, the regulatory authority must know how much a unit of emissions from each source contributes to ambient concentrations. Because pollution offset permits are issued for individual zones, sources need only be concerned about permit trading in one zone. This makes offset permits less costly to sources than ambient permits.

Tradable emission permits give sources a property right to emit the amounts specified in the permit. In essence, TEPs confer the right to use a certain portion of the environment's residual assimilation capacity. For TEPs to be effective, the regulatory authority must track emissions and permit holdings for all sources.

Unlike emission standards that establish a fixed emission standard for each source, it is possible for TEPs to be traded on an international, national or regional basis by emission sources as well as members of the general public. For example, the TEPs for sulfur dioxide emissions from coal-fired power plants established by the Clean Air Act Amendments of 1990 can be traded on a nationwide basis. Part of the cost of using TEPs is the transaction costs of trading permits. Tradable emission permits are generally more cost effective than are environmental standards because sources with low (high) marginal costs of abatement are allowed to sell (buy) permits to sources with high (low) marginal costs of abatement. Because each source minimizes the cost of abatement by setting marginal cost of abatement equal to the price of the permit, the marginal cost of abatement is equalized across all sources.

A market for TEPs is illustrated in Figure 9.13. Suppose the pollution abatement goal is A and each source reduces pollution by .5A before TEPs are issued. At this level of abatement, the marginal cost of abatement is higher for source X than for source Y, $MCA^{(1)}_X > MCA^{(2)}_Y$. Let TEPs be issued that allow pollution to be reduced by A. Suppose the initial distribution of permits is .5A to each source. For this distribution, there is an incentive for source X to buy permits from source Y provided the price of permits is less than $MCA^{(1)}_X$. Likewise, there is an incentive for source Y to sell permits to source X provided the price of permits is greater than $MCA^{(2)}_Y$. A permit price of m, as shown in Figure 9.13, satisfies both conditions.

Because $m < MCA^{(1)}_X$, it is cost-effective for source X to buy permits from source Y and increase its emissions from A-.5A to A-.25A. Similarly, since $m > MCA^{(2)}_Y$, it is cost effective for source Y to sell permits to source X and reduce its own emissions from A − .5A to A − .75A. In equilibrium, the quantity of permits source X wants to buy equals the quantity of permits source Y wants to sell. Marginal cost of abatement is equalized across sources after trading. Emission trading minimizes the overall cost of achieving an environmental standard. Anything that impedes the free exchange of permits prevents achievement of the full efficiency benefits of TEPs.

In addition to providing more cost-effective control of emissions than a strict emission standard, TEPs internalize the cost of using the residual assimilation capacity of the environment. Requiring sources to pay for the right to emit certain residuals increases the cost of production of commodities that generate those residuals. Higher production costs result in higher commodity prices, which reduce consumption. There is a net gain to society because consumption of commodities that generate potentially polluting residuals is decreased. By allowing TEPs to be bought and sold by the general public, the market price of emission permits reflects society's preferences for environmental quality. For example, suppose an environmental group believes that the total emission standard for a particular residual is too high (environmental quality is not being adequately protected). The group can reduce total emissions by purchasing emission permits and holding them off the market.

Emission charges, emission standards and TEPs internalize some of the cost of using the residual assimilation capacity of the environment. The cost and equity implications for emission sources, the administrative cost to the regulatory authority, and the relative effectiveness of each policy option can be quite different when the benefits and costs of emission control are uncertain. First, the need for the regulatory authority to adjust emission fees in response to economic growth and inflation is costly for the regulatory authority and increases uncertainty for emission sources. For example, inflation lowers the real cost of an emission charge and decreases the

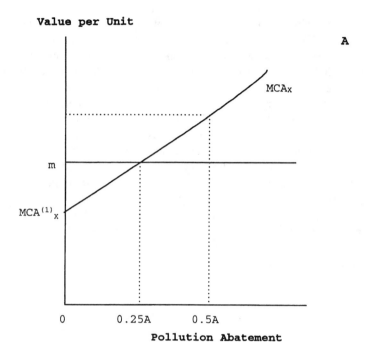

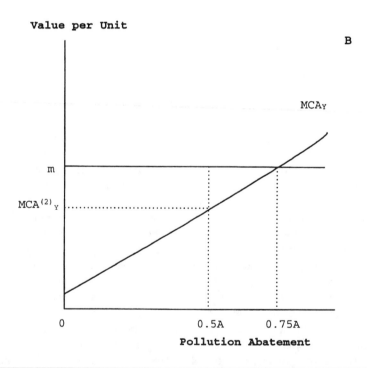

Figure 9.13. Market for tradable emission permits for source X (*A*) and source Y (*B*). Source X is high cost; and source Y is low cost. MCA = marginal cost of abatement; and m = permit price.

incentive to reduce emissions. In order to maintain emission standards during inflationary periods, the regulatory authority would need to periodically increase emission charges. There is no need for the regulatory authority to adjust for inflation or economic growth when TEPs are used. Inflation and economic growth automatically increase the price of permits, but standards are still achieved provided they are enforced.

Second, while the social cost of reducing emissions is lower with TEPs than with emission standards, the cost burden to emission sources is generally higher when sources have to pay for the permits. If the regulatory agency decides to auction the permits to the highest bidder, then sources make an initial outlay of capital to acquire permits. The greater the competition in bidding for permits, the higher the price of permits and the greater the initial cost of acquiring them. After the initial allocation of permits is made, sources can buy and sell permits. One study estimated that air emission permits for certain halocarbons are six times more costly for emission sources than emission standards.[16] For this reason, sources are likely to resist TEPs.

An alternative to auctioning TEPs is to distribute them free of charge to emission sources. Free distribution is less costly to sources, but it raises an equity issue. For example, if the allocation is based on historical emission levels, then sources with low emissions would receive fewer permits than would sources with high emissions. This method of distribution rewards sources that have done a poor job and penalizes sources that have done a good job of controlling emissions. Free distribution of permits does not internalize any of the social costs of emissions. There is some cost internalization, however, when permits are traded. Finally, permit auctioning generates public revenues that can be used to reduce pollution damages; free distribution does not.

Third, regional differences in pollution damages are easier to handle with TEPs than with emission charges. Suppose a ton of sediment from eroding cropland does greater damage to stream ecology in region A than in region B. It is more efficient to place a higher per unit charge on sources in region A than in region B than to have a uniform charge in both regions. Establishing differential charges is difficult, especially when there is limited information on per unit damages in the two regions. In addition, differential charges might be opposed on equity grounds. Because sediment movement in the two regions is driven primarily by natural conditions (topography and rainfall), differential charges for sediment might be considered unfair because they penalize sources based on factors beyond their control. Regional differences in sediment damages can be handled by issuing fewer sediment discharge permits in region A than in region B and disallowing permit trading between regions.

Fourth, preference for emission charges or TEPs is influenced by the uncertainty regarding the costs and benefits of using these two policy options.[17] The preference for either policy option is based primarily on the consequences of selecting an inappropriate level of emission charges or number of emission permits. First, consider a case where the marginal benefit of pollution abatement drops rapidly as abatement increases but the marginal cost of abatement is relatively constant; slope of marginal benefit curve exceeds slope of marginal cost curve for pollution abatement. The marginal benefit curve would be steep when insufficient abatement caused concentrations of a highly toxic substance to exceed some threshold level.

The risk in this situation is that the environmental authority might select too low an emission tax, resulting in insufficient emission reduction and exceedance of the threshold level. Because TEPs would allow the authority to achieve an emission reduction level that does not exceed the threshold, they are preferable to taxes. Therefore, when the slope of the marginal benefit curve exceeds the slope of the marginal cost of abatement curve, TEPs are preferable to emission taxes.

Second, suppose the marginal cost of pollution abatement increases rapidly as pollution abatement rises but the marginal benefit of pollution abatement is relatively constant (slope of marginal cost of abatement curve exceeds slope of marginal benefit curve). The risk here is that the regulatory authority selects too strict an environmental standard and issues too few emission permits. This would require firms to undertake a substantial amount of emission reduction, which is very costly. In this situation, an emission tax would be less costly because it would allow firms to avoid the high cost of pollution abatement by paying the tax. Not surprisingly, a combination of emission taxes and TEPs is superior to either policy taken separately when there is uncertainty regarding marginal benefits and costs.[18]

Finally, environmental standards and TEPs are likely to be preferred to emission charges when a) setting standards or allocating TEPs to sources is more politically acceptable than imposing a large tax burden on emission sources, b) the environmental authority is more concerned about controlling the level of emissions than internalizing the social cost of pollution (although TEPs do both), and c) the environmental authority has more familiarity with emission standards and permits than emission taxes. For example, in the United States there is considerable experience with using site-specific permits to control point sources of water pollution and national emission standards to control air pollution.

Environmental Liability.[19] Another way to control point sources of environmental pollution is to make acting parties financially responsible for pollution damages incurred by affected parties. This approach to pollution control is called *environmental liability*, or EL. A prime example of environmental liability is the highly publicized lawsuit settlement for pollution damages caused by the *Exxon Valdez* oil spill in Prince William Sound in Alaska, which occurred in March 1989. This accident, the largest oil spill in United States history, spilled 40,000 tons of crude oil into a highly productive marine ecosystem. Crude oil spread over about 10,000 mi^2 (25,900 km^2), including 1,200 miles (1,932 km) of coastline. It is estimated that the *Exxon Valdez* oil spill killed 300,000 to 645,000 birds and 4,000 to 6,000 marine mammals. The lawsuit brought against Exxon by federal and state governments was settled in October 1991. It required Exxon to pay a $25 million fine, $100 million in criminal restitution and $900 million over 10 years for civil damages. After legal fees and expenses, about $745 million went toward environmental restoration.[20] Environmental liability allows affected parties to sue acting parties for damages transmitted through the environment. In the case of the *Exxon Valdez* oil spill, the suit only covered damages from job and income losses caused by the spill, not damages to the environment itself.

The assignment of property rights in environmental quality is quite different for EL than for emission charges, environmental standards or TEPs. Emission charges, such as a Pigouvian tax, implicitly assume that property rights to environmental quality belong to the acting party because the environmental authority is usually a government agency. Government agencies can only tax something that they do not own. Environmental standards assume that property rights belong to the environmental authority because the authority is the one who sets and enforces the standards. Affected parties have the right to environmental quality with EL because they can sue for damages related to the environment.

Environmental liability differs from traditional liability laws in several respects. First, the acting party is usually a firm rather than a household. Firms are in a much better position than households to reduce the damages awarded in a liability case because they can incorporate. Second, the affected party does not influence the probability of an accident. Consider a midair collision of two airplanes that occurs when plane A enters plane B's airspace. There is just cause to believe that the collision could have been avoided if both planes had maneuvered to avoid the collision. In other words, the behavior of both parties influences the probability of an accident. This is typically not the case with environmental accidents. The wreck of the *Exxon Valdez* was not influenced by the behavior of affected parties such as fishermen. In some cases, the behavior of affected parties does influence the extent of pollution damages. For example, health problems from eating fish taken from polluted waters can be avoided provided affected parties heed the posted warnings. Finally, EL cases typically involve a large number of affected parties, which makes mass torts appropriate.

The expected marginal benefits and marginal cost of preventative action are illustrated in Figure 9.14. The expected marginal benefit of preventative action taken by an acting party, designated as E(MB), equals the expected reduction in damage claims from an additional unit of preventative action. E(MB) reflects three probabilities: that damages occur, that a legal claim is made against the acting party, and that damages are awarded to the affected parties given that a claim is made. Because E(MB) is likely to become smaller as the level of preventative action increases, the E(MB) curve is negatively sloped. The marginal cost of preventative action (MNC) curve is positively sloped because the cost of additional units of preventative action is expected to rise as preventative action increases. The efficient level of preventative action occurs at C* where E(MB) = MNC. The price corresponding to the efficient level of preventative action, p*, internalizes the cost of polluting activities. In this respect, EL provides an economic incentive to control pollution much the same way as an emission charge.

Environmental liability has certain advantages. First, it does not require collective action as do emission charges, emission standards or TEPs. Only the parties to a liability claim are required to take action. Second, EL substantially reduces the need for information because it does not require collective action. Information is gathered only when a liability claim is made. Third, EL does not require pollution sources to be in compliance with a standard, pay taxes on emissions, or incur the cost of purchasing and trading emission permits. Only those parties to a liability claim incur costs. Fourth, EL does not require emissions to be monitored, at least in cases where environmental damages are obvious. Hence, information and transaction costs are likely to be lower for EL than for other pollution control policies. Sixth, EL is effective in dealing with toxic residuals. For example, during the 1987–1991 period, the release of toxic chemicals to the environment by chemical

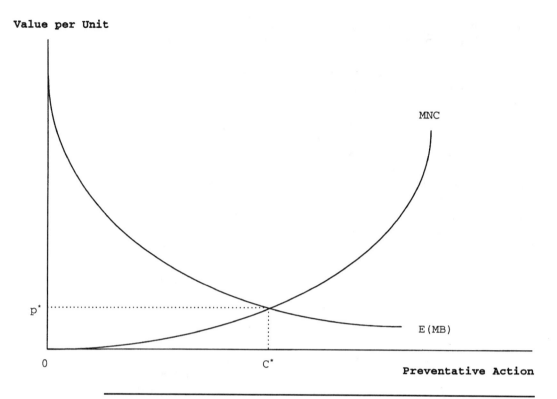

Figure 9.14. Efficient level of preventative action (C) to avoid pollution damages. E(MB) = expected marginal benefit; and MNC = marginal environmental cost.

companies in the United States was reduced by 35 percent, in large part due to the Comprehensive Environmental Response, Compensation and Liability (Superfund) Act of 1980.[21]

The EL approach does have certain drawbacks. First, it does not work well when property rights are incomplete. If the right to environmental quality is not assigned to some party, it is not clear who should pay for pollution damages. If no one has the right to environmental quality, then no one can claim damages from environmental pollution. Second, actual damages may exceed compensation to affected parties due to imperfections in the legal system. Legal roadblocks include a) difficulty in apportioning damages to a large number of affected parties (mass torts), b) inability to identify liable parties due to lags between a pollution event and damages caused by that event, c) statutes of limitations, d) complexity and cost of court cases, and e) ability of acting party to avoid payment of damages by declaring bankruptcy.

Third, the fairness of the damage settlement is colored by the liability rules that apply. For example, under the rule of comparative negligence, the acting party is liable only when the minimum level of preventative action is not taken and the affected party takes due care to avoid damages. Because it is relatively easy to prove that minimum preventative action was taken, the affected party is unlikely to receive compensation for damages. This sort of difficulty can be reduced by imposing a more restrictive form of liability. For example, the Superfund legislation incorporates retroactive, strict and joint-and-several liability for accidental oil spills and leakage of hazardous residuals.[22] Retroactive liability means the law applies to offending activities that occurred prior to its enactment. Strict liability makes the acting party liable regardless of the level of preventative action taken, as long as due care was exercised by the affected party. This law is effective because there are a limited number of qualifying cases (spills and leakage) and conventional monitoring of damaging activities is difficult.[23] With the joint-and-several rule, however, an acting party can be ordered to pay for the entire damages in the event other acting parties cannot be located or are not able to pay damages.

Fourth, the insurance market, which is the basis for limiting an acting party's liability for damages, may distort economic incentives to reduce or prevent pollution. One such distortion is moral hazard, which refers to the decline in the acting party's incentive to prevent pollution after insurance has been purchased. Fifth, EL could reduce the incentive for affected parties to make defensive expenditures. When an EL system incorporates a strict liability rule, it provides an economic incentive for acting parties to reduce pollution much the same way as an emission charge. As shown in Figure 9.14, the expected price for the efficient level of preventative action (p^*) internalizes the cost of polluting activities much the same way as an emission charge. However, there is an important difference. Emission charges are efficient because they achieve pollution reduction at least cost and do not reduce the incentive for affected parties to make defensive expenditures. The latter occurs because emission fees are paid to the environmental authority. An EL system that requires an acting party to compensate affected parties for damages incurred reduces the incentive for affected parties to make defensive expenditures. This aspect of an EL system makes it inefficient relative to emission charges.

Not all EL systems compensate affected parties. Under the Superfund legislation, parties that are responsible for oil spills and leakage from hazardous residuals that pose serious health and environmental risks must pay for the cost of cleaning up the site using remediation measures specified by the U.S. EPA. Superfund does not involve direct compensation to affected parties.

Environmental liability is an effective way to reduce environmental pollution especially when property rights to environmental quality are assigned, frequency of damages is low, and damages are high enough to warrant compensation or environmental remediation. On the negative side, the drawbacks to EL prevent it from dominating other policies for controlling pollution such as emission charges, environmental standards and TEPs.

In summary, differences in the location, sources and consequences of pollution, as well as in the economic and political feasibility of different approaches to pollution control suggest that a variety of pollution control approaches is likely to be superior to any single approach.

NONPOINT SOURCE POLLUTION. Designing effective policies to control nonpoint source pollution is difficult because there is a large number of diffuse pollution sources, the relationships between production, emission and pollution are complex, and data regarding these relationships are limited. These same difficulties inhibit the application of the production-restriction and emission-restriction models to nonpoint source problems. Determining the socially efficient level of pollution with the production-restriction model requires specifying the marginal pollution damage function for each source of pollution. The marginal pollution damage function is derived from the pollution function, $L = L(R)$, which relates pollution (L) to residual emission (R). Residual emission, in turn, is affected by production. Determining the socially efficient level of pollution abatement with the emission-restriction model requires determining the marginal cost of pollution abatement for each source. Marginal cost of pollution abatement depends on the relationship between production and pollution. Application of the two models to nonpoint source pollution requires use of sophisticated biophysical simulation and economic analysis.

This section utilizes the emission-restriction model to evaluate nonpoint source pollution. The emission-restriction model is selected because it affords greater flexibility in evaluating nonpoint source pollution than does the production-restriction model. Recall that in the emission-restriction model, production (Q) is a function of input use (X) and residual emission (R), namely, $Q = Q(X,R)$. This production function implies that inputs and residual emissions are substitutes. Such substitution is not allowed in the production-restriction model. With the emission-restriction model, crop production can be maintained and nonpoint source pollution decreased by increasing input use. For example, herbicide emissions can be decreased by increasing tillage operations, which requires greater use of labor, fuel and equipment.

The following nonpoint source pollution problem is evaluated using the emissions-restriction model. Runoff from agricultural fields in a watershed is polluted by the herbicide litrex (fictitious name), which is used by farmers to reduce weed losses due to weed infestation in corn and sorghum production. Runoff enters the main stream draining the watershed. The stream eventually flows into a reservoir that serves as a drinking water supply for a nearby community. Litrex does not contaminate the drinking water used by residents of the watershed.

Litrex pollution of the reservoir is an external diseconomy that is influenced by economic factors and environmental processes including crop selection, rates and timing of litrex application, timing and intensity of rainfall events, recharge of streams by groundwater, proximity of agricultural fields to the stream, hydrology, topography, vegetative cover in riparian areas, and chemical properties of litrex (potency and persistence). An economic model is used to determine the relationship between input use, crop yield and farm income, and a biophysical model is used to simulate how litrex influences crop yield and contamination of surface runoff and stream water.

Several policies can be used to reduce litrex pollution of the reservoir. These include imposing a tax on litrex, a charge on emissions of litrex, restrictions on the use of litrex, mandating the use of specific production methods, cost sharing (subsidies), and tradable emission permits. Some of these policies were evaluated in the previous section dealing with point source pollution. The remainder of this section evaluates the strengths and weaknesses of these policies in reducing litrex pollution of the reservoir.

Input Tax. An input tax is a tax on an input that contributes to nonpoint source pollution. Agricultural input taxes are more popular in Europe than the United States. Iowa, Wisconsin and Illinois have taxes on fertilizers. Consider a uniform tax on litrex. Farmers are likely to respond to the tax in three ways. First, they might substitute another herbicide for litrex. This response would increase production cost and/or decrease crop yield, both of which reduce net farm income. Second, farmers might reduce their use of litrex and increase tillage operations. Tillage reduces weeds. This response increases production cost because tillage operations require additional labor, fuel and equipment. Third, farmers might switch to crops that do not require litrex. The third response is the least likely because it requires major changes in farming operations.

Maintaining the same production level while reducing herbicide emissions requires that input use be increased as demonstrated by the movement from a to b on isoquant Q_0 in Figure 9.6. Increasing input use with the same level of production causes marginal cost of production to increase. Under pure competition, a higher marginal cost of production reduces the profit-maximizing level of production. Therefore, a tax on litrex has the desired effect of reducing litrex pollution. Unfortunately, crop production and farm income are likely to be reduced by the tax.

At what level should the tax be set? The answer depends on the desired level of pollution abatement. Recall from Figure 9.7 that the socially efficient pollution abatement occurs where marginal cost equals marginal social benefit of pollution abatement. The optimal tax on litrex would reduce litrex pollution by the socially efficient amount. Marginal social benefit of reducing litrex pollution in the reservoir can be estimated with various non–market valuation methods (see Chapter 12). Marginal cost of pollution abatement can be estimated by combining biophysical simulation and economic analysis. Most taxing authorities are not in a position to undertake such sophisticated analysis. Therefore, the tax is expected to be suboptimal.

Taxing an input has several other drawbacks. First, the tax is an inefficient way to reduce pollution because user of the taxed input pays the same tax regardless of their contribution to pollution. For example, a farmer that uses split application of litrex (application on every other crop row) would pay the same tax as farmers who do not use split application. Likewise, farmers who reduce their runoff by installing riparian buffer strips along waterways pay the same tax as farmers who do not install riparian buffer strips.

Second, keeping the amount of pollution abatement at the socially efficient level over time would necessitate periodic adjustments in the tax to reflect changes in the marginal cost and marginal benefit of abatement. For example, if a new technology reduces the marginal cost of abating litrex pollution, then the socially efficient level of abatement increases. The tax would need to be increased in order to achieve this higher level of abatement. Not only would a tax increase be unpopular with farmers, but keeping the tax at the efficient level would require the monitoring of changes in abatement technology.

Third, a tax on litrex would be inappropriate for farmers who operate in an area where drinking water is not polluted by litrex. In this respect, a tax is inefficient because it does not target those areas with the most severe nonpoint source pollution. One advantage of a tax is that it has a relatively low administrative cost because the taxing authority does not have to monitor for pollution.

Emission Charges. An emission charge is a fee imposed on emissions of a polluting substance. The purpose of the charge is to reduce emissions of the substance to a socially acceptable level. In the case of nonpoint source pollution, the socially acceptable level is typically defined in terms of ambient concentrations of the pollutant. Ambient drinking water standards in the United States are currently 10 parts per million for nitrate–nitrogen and 3 parts per billion for atrazine, which is a herbicide. Drinking water standards are applied to quarterly samples drawn from drinking water sources. If the average concentration of the pollutant during the previous four quarters exceeds the standard, the water source is in violation of the standard.

A charge on emissions of litrex can be determined as follows. Suppose the last four quarterly samples taken from the water supply reservoir have the following concentrations of litrex: 8, 8, 10 and 6 parts per billion. Let the drinking water standard for litrex be 5 parts per billion. Because the average concentration of litrex is 8 parts per billion, the water in the reservoir violates the drinking water standard. Bringing the reservoir into compliance with the standard would require reducing the average concentration of litrex by 3 parts per billion.

The following procedure can be used to determine the optimal emission charge for achieving the desired reduction in litrex pollution of the reservoir. First, the watershed is divided into subwatersheds, each containing a group of farms that affects water quality in a particular segment of the stream. Second, a biophysical–economic model is used to estimate the minimum losses in net farm income from reducing litrex concentrations in surface runoff from each subwatershed. Third, the income losses associated with specific reductions in litrex pollution are used to determine the marginal cost of pollution abatement in each subwatershed (MCA_i for $i = 1,...,n$, where n equals the number of subwatersheds). Fourth, the marginal abatement cost curves are summed across subwatersheds to obtain the aggregate marginal cost of abatement in the watershed (MCA_R). This procedure is illustrated for three subwatersheds in Figure 9.15.

The efficient charge on litrex emissions is m and the efficient pollution abatement levels are a_1, a_2 and a_3 for subwatersheds 1, 2 and 3, respectively. Note that the marginal cost of pollution abatement is different in all three subwatersheds and the amount of abatement is higher (lower) in subwatersheds having a lower (higher) marginal cost of abatement. Subwatershed 3, which has the highest marginal cost of abatement, is allocated the lowest abatement and subwatershed 1, which has the lowest marginal cost of abatement, is allocated the highest abatement. Because pollution is expressed in terms of the concentration of litrex, the weighted average of a_1, a_2 and a_3, namely, A_w, equals 3 parts per billion where weights are equal to the flow of water in each subwatershed.

Input Restrictions. Input restrictions limit the rate and/or location of application of an input. For example, Mississippi and Nebraska restrict the use of fertilizers. Input restrictions are effective in areas where soil, climatic and hydrologic conditions favor nonpoint source pollution. For example, claypan soils have much higher runoff rates than sandy soils. If the objective is to reduce litrex concentrations in runoff, then lower rates of application could be required in claypan soils than in sandy soils. Locational restrictions would limit use of litrex in environmentally sensitive areas such as wellheads and along streams, lakes and other water bodies that supply drinking water. Locational restrictions typically take the form of setback restrictions such as no application of litrex within 100 feet (30.5 meters) of a water body.

Input restrictions are more limiting than taxes. A tax simply increases the price

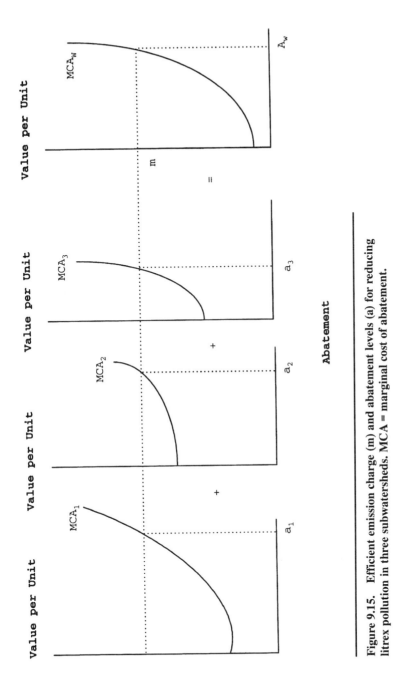

Figure 9.15. Efficient emission charge (m) and abatement levels (a) for reducing litrex pollution in three subwatersheds. MCA = marginal cost of abatement.

of the input, but does not restrict its use. Farmers are likely to respond to restrictions on the use of litrex in much the same way they respond to a tax on litrex, namely, employing substitute herbicides and/or increasing tillage operations, or switching to crops that do not require the use of litrex. Input restrictions increase the marginal cost of production, which decreases production. Unlike a tax, which reduces production in all areas where litrex is used, restrictions on the use of litrex would only affect production in areas where the restrictions are imposed. In this respect, input restrictions are effective in reducing nonpoint source pollution in specific areas. This feature of input restrictions makes them potentially more efficient in reducing nonpoint source pollution than taxes.

Cost Sharing. Nonpoint source pollution can be reduced by cost sharing of best management practices. A best management practice is a production method designed to alleviate a specific type of pollution. With cost sharing, a federal and/or state agency reimburses farmers for part of the cost of installing one or more best management practices. Cost-sharing subsidies are justified on the grounds that pollution is an external diseconomy, which means someone other than the polluter is the primary beneficiary of pollution reduction. With cost sharing, society pays for part of the cost of pollution abatement. Examples of best management practices for reducing nonpoint source pollution include terraces and conservation tillage for reducing sediment in streams, and soil nitrogen testing and split application of pesticides for reducing nitrate and pesticide pollution of water.

Best management practices can be quite effective in reducing nonpoint source pollution. Selecting nitrogen application rates based on crop growth requirements and soil nitrogen levels has allowed nitrogen application rates in some states to be reduced by as much as 30 percent with no loss in crop yields. Adoption of best management practices changes the input mix. For example, integrated pest management substantially reduces pesticide application rates but requires more intensive management. Adoption of a best management practice can increase or decrease marginal cost of production (after cost sharing). Increases (decreases) in marginal production cost cause production to decrease (increase).

Cost sharing of best management practices on a voluntary, first-come, first-serve basis is not an efficient way to reduce nonpoint source pollution. This occurs because cost-sharing funds are not allocated to areas according to the severity of pollution. Consequently, pollution reduction per dollar of cost sharing is likely to be low. Nevertheless, farmers and agencies strongly favor voluntary cost sharing because of its nonregulatory nature. One way to improve the efficiency of cost-sharing subsidies without losing the appeal of a voluntary approach is to allocate cost-sharing funds to areas where pollution damages are high and best management practices are effective in reducing pollution.

Mandated Production Methods. In this approach, a regulatory agency mandates the use of specific production methods for reducing nonpoint source pollution. At one time, the U.S. EPA required point sources of pollution to use the best available technology for reducing emissions. Mandated production methods can be used to reduce litrex pollution of runoff by requiring farmers to use alternative production methods, such as split application or integrated pest management. Mandating the use of specific production methods is likely to be unpopular with farmers because it restricts their freedom. Another drawback of mandatory production methods is the high administrative cost. Not only would the regulatory agency have to keep track of whether the mandated production methods are being used, but it would have to monitor new abatement methods to identify the least-cost abatement method for reducing pollution. Finally, mandatory production methods reduce the incentive for polluters to invest in new abatement technologies.

Tradable Emission Permits. Using a tradable emission permit to control nonpoint source pollution would require the regulatory authority to establish an upper limit on either the ambient concentration or total emission of the litrex. It would then issue enough TEPs to ensure that the limit is not exceeded. The upper limit on litrex emissions from the watershed would be set at a level that provides reasonable assurance that the drinking water standard is not exceeded. Farmers would then buy and sell (trade) TEPs based on the price of the permits and their marginal costs of abatement. Theoretically, TEPs are an efficient way of achieving an emission standard because the regulatory agency does not have to set emissions for each source.

There are several reasons why it can be costly to administer TEPs for nonpoint source pollution control. First, there are a potentially large number of pollution sources. Each source would need to be monitored for compliance with its permits. Second, the agency would need to keep track of the ownership of permits. Third, the number of permits would need to be changed whenever the ambient standard is changed. Such changes generate uncertainty for polluters. Fourth, the agency would need to determine whether to allow permit trading between sources in the same watershed or between sources in different watersheds.

Some types of nonpoint source pollution are especially difficult to control with TEPs. For example, suppose source A in the upper end of a watershed holds five permits and source B in the lower end of a watershed holds eight permits. Each permit allows an emission of 1,000 tons of sediment per year. Because source B is located in the lower end of the watershed, a 1,000-ton reduction at source B achieves a greater reduction in sediment delivery to the stream than a 1,000-ton reduction at source A. Therefore, there is a greater reduction in sediment pollution when source A buys three permits from source B, than when source B buys three permits from source A. This means the value of a permit in reducing sediment pollution depends on how the permits are distributed among pollution sources. In this case, there is no assurance that the desired reduction in emission will be achieved. While it is possible to develop TEPs that are location sensitive, it would be more complex and could reduce the efficiency advantages of TEPs.

Other Considerations. Several other factors need to be considered in designing effective policies for controlling nonpoint source pollution. Care must be taken to avoid pollution policies that reduce one type of pollution but increase another type of pollution. For example, if taxing or restricting a target herbicide causes farmers to substitute another herbicide, then the substitute herbicide could become a source of pollution. Similarly, if taxing or restricting the use of a herbicide causes farmers to increase tillage operations, then soil erosion and sedimentation of streams would increase.

Another consideration in designing nonpoint source pollution policies is the potential tradeoff between surface and groundwater pollution. Allowing farms in sandy soil to use higher rates of herbicide than farms in claypan soil makes sense when the objective is to reduce herbicide pollution of surface runoff. It is likely, however, to increase the risk of groundwater pollution in the sandy soil. Providing cost sharing and other subsidies for conservation tillage in order to reduce erosion and sedimentation would increase the use of herbicides to control weeds. Because conservation tillage reduces runoff, more of the unused herbicide percolates through the soil into groundwater, especially in porous soils. Therefore, while conservation tillage is likely to decrease pollution of surface runoff and increase groundwater pollution, accounting for the complex interactions between different types of nonpoint source pollution is a major challenge in designing effective nonpoint source pollution policies.

Summary

Residual loads from consumption and production are either recycled, which reduces the demand for natural resources, or emitted to the environment. Residual emissions accumulate in or are assimilated by the environment. Environmental pollution is an external diseconomy that results from the accumulation of nondegradable residuals and/or residual emissions in excess of the environment's assimilative capacity. Pollution can reduce the supply of exhaustible resources, the productivity of natural resources, and the amenity and ecological services provided by the environment. Pollution-related damages to humans and ecosystems are influenced by the pollutant's potency and persistence and the vulnerability of human and ecological systems to pollutants. A major goal of environmental economics is to determine socially efficient levels of environmental pollution and to evaluate public policies for pollution abatement.

Static analysis of environmental pollution is appropriate when current pollution damages depend only on current residual emissions. Dynamic analysis is appropriate when current damages are influenced by current and past emission levels. Economically efficient levels of pollution are determined based on the production-restricted and the pollution-restricted models. The production-restricted model requires a fixed, proportional relationship between residual emissions and production rates. This relationship is nonlinear in the emission-restricted model because emission levels can be reduced by varying input use and/or technology.

In the static production-restricted model, household utility or satisfaction is a function of consumption and pollution, production is a function of input use, emission is a function of production, and pollution is a function of emissions. This model indicates that a) the privately efficient production rate exceeds the socially efficient production rate, b) the socially efficient production rate approaches the privately efficient production rate as the production rate beyond which pollution occurs increases, and c) zero pollution is generally not socially efficient.

The most general static emission-restricted model specifies that production is a function of input use, residual emission and pollution. This specification implies that input use and emission levels are substitutes and pollution adversely affects production. Utility is a function of consumption and pollution exposure. Pollution exposure depends on pollution levels and defensive expenditures. The latter are expenditures that reduce the risk of pollution damage. The emission-restricted model indicates that the socially efficient rate of pollution abatement occurs where the marginal social benefit equals the marginal cost of pollution abatement and that the socially efficient abatement rate exceeds the privately efficient abatement rate. Under these conditions, the privately efficient rates of production and levels of pollution under pure competition are greater than under imperfect competition. Similar results hold for the dynamic model of efficient production and pollution abatement.

Several factors reduce the likelihood of a property rights solution to pollution externalities: high transaction costs, inability to trace certain pollution damages to specific residual emission sources, damages from different pollutants being multiplicative rather than additive, and nonrivalry in consumption of pollution. The theory of environmental pollution indicates that pollution occurs because the residual assimilation services provided by the environment are unpriced. The socially efficient price for pollution is a Pigouvian tax on residual emission equal to the difference between marginal social benefit and marginal private benefit at the socially efficient level of abatement. Due to high information requirements, Pigouvian taxes are difficult to implement. Alternative approaches include emission charges, emission standards, tradable emission permits, and environmental liability rules.

Emission charges establish an environmental standard and a uniform charge per unit of emission for each emission source. While emission charges do not generally result in the socially efficient level of emission reduction, they require considerably less information than do Pigouvian taxes, are cost effective in achieving an environmental standard, provide an incentive for polluting sources to invest in new abatement technologies, do not require monitoring of emissions from all sources, internalize part of the cost to society of utilizing the environment's residual assimilation capacity, are a source of public revenue, and are consistent with the polluter-pays principle.

Emission standards set environmental standards for each emission source, sometimes specify the pollution control technology used at each source, and monitor sources for compliance with the standards. Major advantages of emission standards are achievement of the desired level of emission at each source and reduction of emissions to acceptable levels during emergencies.

Tradable emission permits restrict total emissions by issuing permits that can be traded among sources of emissions and other parties. Permits confer the right to use a certain portion of the environment's residual assimilation capacity. Tradable emission permits are cost effective in achieving an environmental standard and automatically adjust for inflation and economic growth.

Environmental liability makes acting parties financially responsible for pollution damages incurred by affected parties. Environmental liability allows affected parties to sue acting parties for damages transmitted through the environment. Advantages of EL are that it does not require collective action; substantially reduces the need for information by a central environmental authority; does not require a large number of pollution sources to be in compliance with a standard, to pay taxes on emissions, or to incur the cost of purchasing and trading emission permits; reduces information and transaction costs by not requiring the monitoring of emissions; and is effective in limiting emissions of toxic residuals. All four policies for reducing pollution have advantages and disadvantages. Typically, a combination of policies is more effective and efficient in reducing pollution than is any single policy.

Control of nonpoint source pollution can be achieved using a tax on the polluting input, a charge on emissions, restrictions on the use of the input, cost sharing of best management practices, mandated production methods and tradable emission permits. Taxes are the easiest policy to administer. Input taxes, however, a) do not account for the extent to which a firm contributes to pollution, b) need to be periodically adjusted to account for changes in the marginal benefit and marginal cost of abatement, and c) penalize firms whose emissions do not cause pollution.

An emission charge is an economically efficient way to reduce nonpoint source pollution. Determining the socially optimal level of nonpoint source pollution requires information on the marginal cost of abatement for all sources. Input restrictions allow the regulatory authority to target areas where use of the input causes pollution. While more limiting than input taxes and emission charges, input restrictions allow better targeting of nonpoint source pollution control efforts. Cost sharing of best management practices is an effective way to reduce nonpoint source pollution. The voluntary nature of cost sharing causes the cost per unit of pollution reduction to be high.

Mandatory production methods are a heavy-handed way of reducing nonpoint source pollution. Not only are they likely to be unpopular with polluting firms, but they entail a high administrative cost because the regulatory agency has to monitor abatement methods to ensure that the most cost-effective pollution control methods are being mandated. Tradable emission permits for nonpoint source pollution would be costly to administer. In addition, some types of nonpoint source pollution are difficult to control with TEPs.

Questions for Discussion

1. Why is it important to distinguish between residual emissions and pollution? Of what value is this distinction in developing efficient policies to reduce environmental damages?

2. What types of residual emissions are especially threatening to humans and the environment?

3. Suppose the demand function (D), marginal production cost (MPC) function and marginal pollution damages (MLD) function are as follows:

D: $p = 40 - 0.5Q$
MPC $= 20 + 0.4Q$
MLD $= 1.6Q$

Calculate a) the privately and socially efficient rates of production, b) net social benefit with socially efficient production, and c) the Pareto-relevant externality.

4. Marginal private benefit of pollution abatement (MPBA), marginal social benefit of pollution abatement (MSBA) and marginal cost of pollution abatement (MCA) are as follows:

MPBA $= 5 - A$
MSBA $= 12 - 2A$
MCA $= 0.75A$

A is the level of pollution abatement. Calculate a) the privately efficient pollution abatement, b) the socially efficient pollution abatement, and c) the Pigouvian tax.

5. Consider two pollution sources (X and Y) that have the following marginal costs of pollution abatement:

$MCA_X = 1.5 A$
$MCA_Y = 0.5 A$

A is the level of pollution abatement. Suppose a uniform emission charge of 2 is levied on both sources. Calculate the efficient level of pollution abatement and the corresponding cost of pollution abatement for each source.

6. Consider an emission standard that requires the two sources in question 5 to reduce pollution by an amount equal to one half of the total amount of abatement achieved with the uniform emission charge. Show that the emission charge results in lower total cost than the emission standard.

7. An environmental authority distributes for free just enough TEPs to sources X and Y (in question 5) to restrict total emission to the socially efficient level attained with a Pigouvian tax (in question 4). What is the efficient distribution of permits between the two sources after trading?

8. Countries such as France, Germany and the Netherlands favor emission charges, whereas the United States has relied on nontradable emission permits to control point sources of water pollution. Why are two different approaches used to deal with the same type of pollution?

9. Why is nonpoint source pollution more difficult to control than point source pollution?

10. Is a regulatory agency interested in reducing nonpoint source pollution more apt to favor a tax on the polluting input or an emission standard? Why?

Further Readings

Anderson, Fredrick R, Allen V. Kneese, Phillip D. Reed, Serge Taylor and Russell B. Stevenson. 1987. *Environmental Improvement Through Economic Incentives.* Washington, D.C.: Resources for the Future.

Baumol, William J. and Wallace E. Oates. 1988. *The Theory of Environmental Policy,* 2nd ed. Cambridge, England: Cambridge University Press.

Cropper, Maureen L. and Wallace E. Oats. 1992. "Environmental Economics: A Survey." *Journal of Economic Literature* 30(2):675–740.

Dales, J.H. 1968. *Pollution, Property and Prices.* Toronto, Canada: University of Toronto Press.

Duttweiler, D.W. and H.P. Nicholson. 1983. "Environmental Problems and Issues of Agricultural Non-Point Source Pollution." In *Agricultural Management and Water Quality,* F.W. Schaller and G.W. Bailey, eds. Ames: Iowa State University Press.

Forsund, Finn R., and Steinar Strom. 1988. *Environmental Economics and Management: Pollution and Natural Resources.* New York: Croom Helm.

Griffin, R.C. and D.W. Bromley. 1984. "Agricultural Runoff as a Non-Point Externality: A Theoretical Development." *American Journal of Agricultural Economics* 66:547–552.

Maler, Karl-Goran. 1974. *Environmental Economics: A Theoretical Inquiry.* Baltimore, Maryland: The Johns Hopkins University Press.

Milon, J. Walter. 1987. "Optimizing Nonpoint Source Controls in Water Quality Regulations." *Water Resources Bulletin* 23:387–396.

Setia, P. and R. Magleby. 1987. "An Economic Analysis of Agricultural Non-Point Pollution Control Alternatives." *Journal of Soil and Water Conservation* 42:427–431.

Taylor, Michael L., Richard M. Adams and Stanley F. Miller. 1992. "Farm-Level Response to Agricultural Effluent Control Strategies: The Case of the Willamette Valley." *American Journal of Agricultural Economics* 17:173–185.

Tietenberg, T. 1985. *Emissions Trading: An Exercise in Reforming Pollution Policy.* Washington, D.C.: Resources for the Future.

Notes

1. Thomas E. Drennen and Harry M. Kaiser, "Global Warming and Agriculture: The Basics," *Choices,* Second Quarter 1994, pp. 38-40.

2. National Wildlife Federation, "26th Annual Environmental Quality Index," *National Wildlife,* Feb.-March 1994, p. 40.

3. U.S. Environmental Protection Agency, *The Quality of Our Nation's Water: 1994,* EPA 841-S-95-004 (Washington, D.C.: EPA 1995).

4. *Ibid.,* p. 41.

5. World Resources Institute, *World Resources Institute, 1992-93: A Guide to the Global Environment* (New York: Oxford University Press, Inc., 1992), p. 194.

6. Dina L. Umali, *Irrigation-Induced Salinity,* World Bank Technical Paper Number 215 (Washington, D.C.: World Bank, 1993).

7. T. L Anderson and D. R. Teal, *Free Market Environmentalism* (San Francisco, California: Pacific Research Institute for Public Policy, 1981).

8. R. Collinge and W. E. Oates, "Efficiency in Pollution Control in the Short and Long Runs: A System of Rental Emission Permits," *Canadian Economics Association* 15(1982):346–354.

9. Jean-Luc Migue and Richard Marceau, "Pollution Taxes, Subsidies, and Rent Seeking," *Canadian Economics Association* 26(1993):356–365.

10. Richard B. McKenzie and Gordon Tullock, "Rent Seeking," chapter 15 of *The New World of Economics: Explorations into the Human Experience,* 3rd ed. (Homewood, Illinois: Richard D. Irwin, 1981).

11. A. Palmer et al., *Economic Implications of Regulating Chlorofluorocarbon Emissions for Nonaerosol Applications* (Santa Monica, California.: The Rand Corporation, 1980).

12. David Terka, "The Efficiency Value of Effluent Tax Revenues," *Journal of Environmental Economics* 11(1984):107–123.

13. Michael A. Toman, "Using Economic Incentives to Reduce Air Pollution Emissions in Central and Eastern Europe: the Case of Poland," *Resources,* Fall 1993, pp. 18-23.

14. Wallace E. Oates, "Taxing Pollution: An Idea Whose Time has Come?" *Resources,* Spring 1988, pp. 5-7.

15. Scott F. Atkinson and T. H. Tietenberg, "The Empirical Properties of Two Classes of Design for Transferable Discharge Permit Systems," *Journal of Environmental Economics and Management* 9(1982):101–121.

16. Palmer et al. (1980).

17. Martin L Weizman, "Prices vs. Quantities," *Review of Economic Studies* 41(1974):477–491.

18. John T. Weitzman, "Optimal Rewards for Economic Regulation," *American Economic Review* 68(1978):683–691.

19. This section draws heavily from Peter Zweifel and Jean-Robert Tyran, "Environmental Impairment Liability as an Instrument of Environmental Policy," *Ecological Economics* 11(1994):43–56.

20. Rick Steiner, "Probing An Oil-Stained Legacy," *National Wildlife,* April–May 1993, pp. 4–11.

21. Karen Schmidt, "Can Superfund Get on Track?," *National Wildlife,* April–May 1994, pp. 10–17.

22. Paul R. Portney and Katherine N. Probst, "Cleaning Up Superfund," *Resources,* Winter 1994, pp. 2–5.

23. James J. Opaluch and Thomas A. Grigalunas, "Controlling Stochastic Pollution Events through Liability Rules: Some Evidence from OCS Leasing," *Rand Journal of Economics* 15(1984):142–151.

CHAPTER 10

Natural and Environmental Resource Accounting

A country could exhaust its mineral resources, cut down its forests, erode its soils, pollute its aquifers, and hunt its wildlife and fisheries to extinction, but measured income would not be affected as these assets disappeared.

—ROBERT REPETTO AND OTHERS, 1989

Material balances and ecological economics indicate that the economy and the environment are linked by a two-way flow. Exhaustible and renewable resources flow from the environment to the economy, and some of the residuals produced by economic activity enter the environment. Natural and environmental resource capacity is degraded when these flows exceed assimilative capacity. High exhaustible resource flows into the economy run the risk that exhaustible resource stocks are depleted more rapidly than substitutes are developed. When the flows of renewable resources into the economy exceed regeneration rates, resource productivity diminishes and the probability of species extinction rises. Residual emissions in excess of the assimilative capacity of the ecosystem raise the likelihood of natural and environmental resource degradation.

Rapid depletion of exhaustible resources, overexploitation of renewable resources, and residual emissions in excess of assimilative capacity generally indicate that the economy has become too large relative to the ecosystem. Because traditional measures of economic performance do not account for changes in natural and environmental resource capacity, they are poor indicators of the economy's long-term sustainability. Natural and environmental resource accounting (NERA) rectifies certain deficiencies in traditional measures of economic performance by taking account of changes in natural and environmental resource capacity.

This chapter discusses NERA. Deficiencies in traditional measures of economic performance such as *gross national product, gross domestic product* and *net national product* are identified. Deficiencies include inappropriate treatment or total disregard of *defensive expenditures, resource capacity* and *residual pollution damages*. Three alternative accounting methods are examined: *physical, monetary* and *satellite accounts*. Resource-specific accounting is discussed for soil erosion. The relationship between NERA and *sustainable development* is discussed in terms of sustainable income.

Accounting Deficiencies

Gross national product (GNP), gross domestic product (GDP), net national product (NNP) and other national income accounts were introduced to the United States in 1942. These accounts are designed to monitor temporal changes in aggregate economic performance. The theoretical basis for national income accounts is the Keynesian macroeconomic model. This model explains economic activity in terms of consumption, investment, government expenditures and exports. Many countries use the United Nations System of National Accounts (SNAs) to track aggregate economic activity.[1] The SNAs consist of a standard set of procedures for constructing national income accounts.

When national income accounts indicate that economic conditions are unfavorable, appropriate fiscal and/or monetary policies are implemented. For example, the economy is said to be in a recession when growth in inflation-adjusted GNP falls for two consecutive quarters. Policy responses to recession include stimulating consumption by lowering taxes (fiscal policy) and/or stimulating investment by reducing interest rates on federal reserve bank loans to member banks (monetary policy). Conversely, if inflation is too high, then opposing fiscal and monetary policies (higher taxes and higher interest rates) are used to dampen economic activity.

National income accounts were developed during a period when the link between natural and environmental resource capacity and the economy was subordinate to devising Keynesian-based macroeconomic policies for reducing economic stagnation. The latter issue arose out of the economic disruption caused by the Great Depression of the mid-1930s. As a result, national income accounts do not address the effects of economic growth on natural and environmental resource capacity.[2] This is in stark contrast to material balances and ecological economics, which address the adverse effects of unbridled economic growth on natural and environmental resource capacity. While GNP and NNP are good indicators of short-term economic welfare, they are poor indicators of long-term economic welfare. As Devarajan and Weiner[3] point out, overexploitation (underexploitation) of natural resources causes GNP to overestimate (underestimate) long-term economic welfare.

NERA adjusts national income to obtain a measure of *sustainable income*. Daly and El Serafy[4] define sustainable income as the maximum amount that can be consumed in the current period without reducing consumption in future periods or without being less well-off in the future. Hicks's concept of income is very compatible with the notion of sustainable income. He recommends that "a man's income [be defined as] the maximum value which he can consume during a week, and still be as well off at the end of the week as he was at the beginning."[5] Sustainable income implies that consumption is at a level that does not deplete the stock of manufactured and natural capital.

National income accounts are not good measures of sustainable income. For example, NNP accounts for depreciation in manufactured capital, such as structures and equipment (NNP = GNP – depreciation on manufactured capital), but not depreciation of natural capital. Many of the adjusted NNP measures developed in NERA explicitly consider depreciation of both manufactured and natural capital. Therefore, NERA is more compatible with the concept of sustainable income than is conventional national income accounting.

Whether sustainable income requires that total manufactured and natural capital be kept constant or each form of capital be held constant has been addressed by

10. Natural and Environmental Resource Accounting

Daly and Cobb.[6] They distinguish between weak and strong sustainability. *Weak sustainability* of income requires keeping the total capital stock intact. It is based on the notion that manufactured capital and natural capital are substitutes for one another. *Strong sustainability* requires holding constant both manufactured and natural capital. The rationale for strong sustainability is that manufactured capital and natural capital are complements rather than substitutes in production. Daly and Cobb support strong sustainability because they believe that manufactured capital and natural capital are complements rather than substitutes. They admit, however, that weak sustainability is an improvement over the current practice of ignoring natural capital in national income accounts. The remainder of this section discusses the three major deficiencies in national income measures, namely: treatment of *defensive expenditures*, accounting for changes in *resource capacity*, and handling *residual pollution damages*.

DEFENSIVE EXPENDITURES. Defensive expenditures are expenditures made to reduce the adverse welfare effects of resource depletion and environmental degradation. They are different from national defense expenditures. Examples of defensive expenditures include expenditures to clean up an oil spill and to decontaminate soil polluted by a hazardous waste. A specific defensive expenditure is the $750 billion that will be needed to clean up the nation's 1,500 hazardous waste sites currently on the Superfund National Priorities List. Because defensive expenditures made by firms are treated as intermediate expenditures in national income accounts, they are excluded from GNP. Expenditures made by government, as well as medical and relocation expenses incurred by households to reduce the adverse effects of environmental pollution, are included in GNP. For example, government expenditures made to clean up hazardous wastes at military installations are defensive expenditures. Because defensive expenditures do not increase economic welfare, they should be subtracted from, not added to, aggregate measures of economic welfare.

Defensive expenditures are not necessarily linked to resource and environmental degradation. In developed countries, there is generally a close relationship between the two. As incomes increase, the demand for services that reduce the harmful effects of resource degradation and environmental pollution also increases. In contrast, low-income developing countries typically have much smaller defensive expenditures because most of their income is spent on the basic necessities of life. Developing countries whose economies are heavily dependent on the exploitation of natural resources typically experience significant resource or environmental degradation.

There are two problems with defensive expenditures.[7] First, there is some ambiguity as to what constitutes a defensive expenditure. Suppose a household moves into a neighborhood that is located downwind of a coal-fired power plant and the household has full knowledge that the neighborhood has poor air quality. Should expenditure on an air purification system for the household be considered a defensive expenditure? Clearly, the expenditure could be avoided by purchasing a house in an unpolluted neighborhood. Second, defensive expenditures can be double counted. If a household constructs a privacy fence in the back of their property to reduce noise pollution from a nearby highway and the fence increases the value of the property, then the fence might be counted twice, first as a defensive expenditure and second in the value of the property.

RESOURCE CAPACITY. Changes in natural and environmental resource capacity are not considered in national income accounts. The term *capacity* is more appropriate than degradation because it allows for both appreciation and depreciation of natural and environmental resources. Examples of reduced capacity include high soil erosion, impairment of the carrying capacity of rangeland due to overgrazing, depletion of fossil fuels, deforestation, air pollution and water pollution. Increased capacity results from new discoveries of oil and gas resources, higher petroleum recovery rates due to technological progress, recycling and capacity-increasing investments.

Many resource costs are excluded from national income accounts because they are unpriced. For example, user fees for grazing on public rangeland (most of which are located in the western United States) are determined by Congress rather than by supply and demand conditions. If the grazing fee for public rangeland is set below the full resource cost of grazing, as many believe, then resource users shift livestock grazing from private to public rangeland, which results in overgrazing and reduced capacity of public rangeland. As a result, the retail price of meat products is lower and consumption of meat products is higher than they would be if the grazing fee reflected the full resource cost of grazing. Because the cost of reduced capacity on public rangeland due to overgrazing is not reflected in retail meat prices and consumption, national income accounts are overstated. Similarly, other resource costs, such as off-site damages from soil erosion and air or water pollution, are not reflected in commodity prices and national income accounts because they are unpriced.

Ignoring changes in resource capacity in national income accounts has negative consequences. First, not adjusting NNP for decreases (increases) in capacity inflates (deflates) NNP and overstates (understates) economic progress. In cases where capacity is decreasing, failure to adjust NNP inflates the growth rate, which overstates economic progress.

Second, failure to adjust NNP for changes in resource or environmental capacity, combined with the goal of unlimited growth, accelerates exploitation of natural and environmental resources. During the period when frontier areas were being developed, competition between the environment and the economy was limited because natural resources were abundant relative to the scale of the economy and technology was not natural resource intensive. Rapid growth in population and resource-intensive technologies increased per capita use of natural or environmental resources and caused the capacity of several natural systems to be exceeded.

Third, because national income accounts do not reflect depletion in natural and environmental resource capacity, there is little incentive to consider ecological carrying capacity in developing and evaluating economic policies. As a result, policies designed to increase NNP run a higher risk of ecological degradation. This is not the case for manufactured capital, such as structures and equipment. Depreciation and investment in manufactured capital are considered in national income accounts. Specifically, NNP is derived from GNP by subtracting depreciation on manufactured capital. As long as changes in natural capital are excluded from national income accounts, there will continue to be underinvestment in technologies and products that protect natural resources and the environment. In many developed countries and some developing countries, laws have been passed that provide incentives to develop technologies and products that protect natural and environmental resources.

Does technological change and the development of substitutes obviate the need to maintain or enhance natural capacity? When a new computer chip is developed, it does not negate the need to depreciate computers that use the old computer chip. In general, equipment that embodies old technology is depreciated even though it is eventually replaced with equipment that embodies new technology. Likewise for natural or environmental resources. The actual or potential substitution of energy from fission or fusion for fossil energy does not justify ignoring the reduction in an economy's productive capacity caused by the depletion of fossil fuels. Investment in the development of safe and effective fission or fusion energy (manufactured capital) enhances and depletion of fossil fuels (natural capital) diminishes an economy's capacity to produce energy. Changes in both types of capital need to be reflected in national income.

Changes in natural capacity are measured by changes in the stock of natural capital, which includes natural and environmental resources. An increase (decrease) in natural capital results in an increase (decrease) in productive capacity. This is similar to the depreciation adjustment for manufactured capital (structures and equipment) in the national income accounts. The rationale for the latter adjustment is that wear and obsolescence reduce the capacity of manufactured capital to generate future income. Capital depreciation is subtracted from GNP to obtain NNP. Investment, which contributes to GNP, offsets capital depreciation. If new investment in manufactured capital exceeds depreciation, then there is capital formation that enhances the productive capacity of the economy. Likewise, if investment falls short of depreciation, productive capacity is diminished.

A similar rationale justifies depreciation or depletion of natural capital.[8] Depletion of exhaustible resources and overexploitation of renewable resources decrease the stocks of these resources, which diminishes the capacity of natural systems to sustain economic activity. Gross national product can be adjusted for changes in natural capital by subtracting from GNP the decrease in the value of natural capital or adding to GNP the increase in the value of natural capital. Making this adjustment requires that changes in the stock of natural capital be expressed in monetary terms. Monetization can be based on *replacement cost, willingness to pay, current market prices* or *user cost*.

Because a renewable resource has the capacity to regenerate itself, replacement cost equals the expenditure required to restore the services lost by exceeding the rate of regeneration. For example, if soil erosion in a particular location is below the rate of regeneration, replacement cost is not applicable because there is no loss in soil productivity. In fact, topsoil depth in that location would increase over time. However, if soil loss exceeds the rate of regeneration by 2 tons per acre per year, then the replacement cost equals the cost of installing the least expensive best management practice for reducing the erosion rate by 2 tons per acre per year.

Depletion of an exhaustible resource decreases the stock of the resource and permanently reduces the services that that resource provides. In other words, there is a permanent reduction in the capacity of the ecosystem to sustain economic activities. It is appropriate to reduce GNP for the economic losses associated with the reduction in exhaustible resource stocks. Losses in capacity can be estimated several ways. First, they can be estimated by the expenditures required to develop another resource that would provide the same services as those lost by depletion of the original resource. This is pseudo-replacement cost. Second, losses could be equated to the amount that users of the depleted resource would be willing to pay to avoid depletion. Third, losses could be estimated based on current market prices. This approach is unacceptable because it reduces GNP by the entire value of the resource. This implies that ownership of the resource provides no net gain in income.

Fourth, capacity losses resulting from the extraction of exhaustible resources can be estimated with the user cost method. This method avoids having to place monetary values on changes in the stock of the resource.[9] In the user cost approach, net revenue from the sale of an exhaustible resource (gross revenue minus extraction cost) is divided into a capital or user cost element and a value added element that represents sustainable or true income: namely,[10]

Net revenue (R) = User Cost (U) + True Income (X).

User cost is equivalent to a depletion allowance. El Serafy[11] argues that true income represents the portion of net revenue that can be consumed in perpetuity provided the user cost component is invested in a renewable resource. Ideally, the annual return from investment in the renewable resource would yield the same level of true income as provided by the exhaustible resource before it was exhausted.

The division of net revenue between user cost and true income is based on the following ratio of true income to net revenue:

$X/R = 1 - [1/(1 + r)^{n+1}]$,

where X is true income, R is net revenue, r is the discount rate and n is the reserves-to-use ratio for the current period. n equals the number of time periods it takes to exhaust the resource at current rates of use. The ratio does not determine the optimal rate of resource depletion. Rather the optimal rate of depletion determines the ratio. For a given level of reserves, n is determined by the optimal rate of depletion, which in turn depends on r and other factors.

The proportion of net revenue claimed as true income (X/R) increases with respect to n and r. In other words, true income becomes a larger fraction of net revenue as the discount rate and/or reserves-to-use ratio increase. The relationships among X/R and n and r are shown in Table 10.1. Holding life expectancy constant, a lower (higher) discount rate implies a slower (faster) optimal rate of depletion, which requires that a smaller (larger) fraction of net revenue be set aside for depletion. Likewise, holding the discount rate constant, a larger (smaller) reserves-to-use ratio or higher (lower) life expectancy results in a longer (shorter) period to exhaustion, which allows a greater (smaller) proportion of net revenue to be claimed as income. User cost or depletion allowance (U), which equals R − X, is determined as follows:

$U = R - R(X/R)$.

The other side of the coin from depletion of natural capital is investment in natural capital. Investment in natural capital involves developing technology to improve the efficiency of extraction and use of exhaustible resources and/or developing renewable resource substitutes for exhaustible resources. For example, investment in electricity generated from solar energy is a way to reduce reliance on electricity generated from fossil fuels. Investment in natural capital would be included in GNP much the same way as investment in manufactured capital. If in-

Table 10.1. X/R for different life expectancies and discount rates

Life Expectancy (years)	Discount Rate (percent)		
	2	5	10
2	6	14	25
5	11	25	44
10	20	42	65
20	34	64	86
50	64	92	99
100	86	99	100

Source: Salah El Serafy and Ernst Lutz, Environmental and Natural Resource Accounting, *Environmental Management and Economic Development*, Gunter Schramm and Jeremy J. Warford (eds.) (Baltimore, Maryland: The John Hopkins University Press, 1989), p. 31.
R = net revenue; and X = true income.

vestment in natural capital is greater (less) than depletion, then the productive capacity of the economy increases (decreases).

Combining manufactured and natural capital allows examination of net changes in capital formation. There are direct linkages between the two accounts. Certain technologies increase the productivity and decrease the use of renewable natural capital but require greater use of both manufactured capital and exhaustible natural capital. Rapid increases in the productivity of agricultural land (renewable natural capital) have allowed countries like the United States to produce more food and fiber with less land. Land productivity gains have come at the expense of massive increases in the use of farm machinery and equipment and inorganic fertilizers. While these agricultural inputs are usually considered to be manufactured capital, they make extensive use of fossil fuel, which is exhaustible natural capital. Such linkages are more the norm than the exception and suggest that manufactured and natural capital are complements rather than substitutes.

RESIDUAL POLLUTION DAMAGES. Residual pollution damages are human-related damages from environmental degradation that are not alleviated by defensive expenditures. Recall that individuals make defensive expenditures in order to reduce the adverse effects of environmental degradation. Such expenditures do not necessarily eliminate all adverse effects of pollution. For example, expenditures made by an individual for medical treatments to reduce the debilitating effects of emphysema are considered defensive expenditures. Expenditures on medical treatment would improve the quality of the individual's life compared with what it would be without treatment. Medical treatment, however, does not necessarily restore the individual's health to what it was before emphysema. The disease is still likely to result in some health impairment. The remaining health damages are a residual pollution damage.

Another example of a residual pollution damage is an uncompensated, environmentally related property damage. Residents of Times Beach, Missouri, were evacuated from their community in 1983 after it was discovered that a highly toxic substance (dioxin) was used in making roads. While the Superfund compensated the residents for the loss of their homes and the cost of relocation, the evacuation undoubtedly caused damages in the form of social and psychological hardships for which compensation was not received. Uncompensated damages are a residual pollution damage.

Unlike defensive expenditures, residual pollution damages are not reflected in market transactions. Damages awarded in court cases could be used to approximate residual damages. Alternatively, the monetary value of residual pollution damages could be estimated using nonmarket valuation methods (see Chapter 12). In the emphysema and the Times Beach examples, the affected parties could be asked their willingness to pay to avoid uncompensated damages. Total willingness to pay by affected parties is an estimate of residual pollution damages. Estimation of residual pollution damages is hampered by limited *a priori* knowledge of when, where and how such damages occur and measurement difficulties. This situation does not diminish the need to account for residual pollution damages.

Resource Accounting Methods

While deficiencies in traditional national income accounts are widely recognized, there is not general agreement about how best to resolve them. Three methods have been used or proposed: physical accounts, monetary accounts and satellite accounts. Each method is discussed below.

PHYSICAL ACCOUNTS. Physical accounts of natural and environmental resources enumerate changes in resource quantity and quality in physical units such as barrels of oil or tons of carbon dioxide emissions. Physical accounts should represent resource stocks and flows in a consistent manner. Specifically, the stock at the end of a period should equal the stock at the beginning of a period plus net additions (gross additions minus use). Physical accounting systems were first introduced by the Norwegian government in 1974 and adopted by the French government in 1978 and the Canadian government in 1986.

The Norwegian system classifies resources into two broad categories: material resources and environmental resources.[12] Material resources include minerals (oil and natural gas), hydrocarbons (coal and forests), stone, gravel, sand, biological resources (occurring in air, water and land) and inflowing resources (solar radiation, hydrologic cycle, wind and ocean currents). Material resources are reported in stock and flow (use) accounts. Stock accounts for minerals include developed reserves, undeveloped reserves, new fields and revaluation (revised estimates of reserves). Flow accounts monitor extraction and imports by households, industry and government. Biological stocks include reserves, recruitment (new additions), revaluation and natural mortality.

Norway's environmental resource accounts keep track of the status of air, water and soil resources and include an emissions account and a state account. The emissions account records total emissions of waste products into the air, water and land by emission source. A typical Norwegian emissions account is given in Table 10.2. Changes in specific emissions over time and space (region) are recorded in the state account. Norway has found it more difficult to track environmental resources than material resources because it is more difficult to define environmental quality than stocks of material resources. Norwegian resource and environmental accounts have been used as a basis for forecasting future use of resources and associated changes in environmental quality. In terms of usefulness, energy mineral and air emission accounts have been most successful, land accounts have been moderately successful, and fisheries, forestry and other mineral resource accounts have been least successful.

The French system of resource accounts uses a somewhat different classification of resources than the Norwegian system. Four categories of resources are considered: nonrenewable, physical environment, living organisms and ecosystems. Accounting for these resources is done in three accounts: central, peripheral and agent. Central accounts track changes in the state of the resource between time periods. Peripheral accounts link resources to one another and to human activities. Agent accounts track the flows between resources and economic activities.

Consider the French water accounts. The central account for water tabulates changes in the stock of water from various supply sources, including snow, glaciers, ground water, lakes, ponds, reservoirs and rivers. The peripheral account for water is essentially a water balance that identifies where water originates (evapotranspiration sources) and its destination (runoff, percolation and interception), plus consumptive use of water in human activities (urban, industry, agriculture and hydropower). The agent account for water shows expenditures made and income earned on water-related activities such as water treatment, drainage, development of drinking water supplies, irrigation and flood control.[13] In addition, there is a water-quality balance sheet that shows the actual and desired volume of water in four water-quality categories by river basin. In addition to being more comprehensive than the Norwegian accounts, the French system reports expenditures needed to support different economic activities. Hence, the French resource accounts are in physical and monetary terms.

Worldwide physical accounts have been developed for specific resources. The World Resources Institute monitors global changes in several resources including land (agricultural, forest and rangeland), wildlife and habitat, energy, freshwater, oceans and coasts, atmosphere and climate. A physical account for worldwide soil depletion from human activities is illustrated in Table 10.3. Resource flows (production, extraction and consumption) are periodically reported by organizations such as the Worldwatch Institute.[14]

While physical accounts avoid the difficulties of placing monetary values on flows, lack of a common unit of measurement makes it difficult to compare resource or environmental accounts with one another and with national income accounts. The latter comparison is especially important for investment and public policy analyses. Suppose a proposed 1,000 megawatt hydroelectric power facility increases electricity sales by $5 million per year but reduces wildlife habitat by 1,000 acres (405 hectares) and eliminates 15 miles (24.2 km) of free flowing river. How does one compare the energy benefits to the environmental losses? Direct comparison is not

Table 10.2. Norwegian emissions account (metric tons)

Source	Sulfur Dioxide		Nitrogen Oxide		Carbon Monoxide	
	1980	1982	1980	1982	1980	1982
Agriculture	2	2	2	3	19	21
Manufacturing	108	84	30	20	34	68
Transportation	11	11	38	37	35	40
Households	6	5	28	29	390	407
Total	127	102	98	89	478	536

Source: David Pearce, Anid Markandya and Edward B. Barbier, Accounting for the Environment, *Blueprint for a Green Economy* (London: Earthscan Publications, Ltd, 1989), p. 97.

Table 10.3. Worldwide human-induced soil depletion, 1945–1990

Region	Total Degraded Area (million acres)	Degraded Area as a Percentage of Vegetative Area
World	4,911	17.0
Europe	547	23.1
Africa	1,236	22.1
Asia	1,868	19.8
Oceania	257	13.1
North America	239	5.3
Central America and Mexico	157	24.8
South America	609	14.0

Source: World Resources, 1992–93, A Guide to the Global Environment, Report by the World Resources Institute (New York: Oxford University Press, 1992), p. 112.

possible because the economic benefits are in monetary terms and the environmental losses are in physical terms. Placing a monetary value on environmental losses makes it possible to calculate and compare the economic benefits, net of environmental losses, for several hydroelectric development alternatives.

MONETARY ACCOUNTS. The monetary approach to natural and environmental resource accounting is founded on the premise that income aggregates, such as GNP and NNP, give a distorted view of economic progress. Distortion results from inappropriate or lack of treatment of defensive expenditures, depletion in resource capacity and residual pollution damages. Incorporating these elements into national income accounts yields a resource-inclusive measure that can be used to determine whether or not the economy is developing in a sustainable manner. The construction of resource-inclusive monetary measures of economic activity is especially important for countries whose economies are highly dependent on the use of natural resources.

Several monetary indicators of economic welfare have been developed. One of the earliest indicators is Net National Welfare (NNW) which was proposed by Nordhaus and Tobin.[15] Net National Welfare is GNP minus the cost of pollution and other factors that degrade the quality of life, plus the value of uncompensated household services such as cleaning, cooking and repairing. Only the cost of pollution pertains to resource or environmental degradation. Household services are included in NNW because they increase the quality of life. Since 1940, inflation-adjusted (real) NNW in the United States increased half as much as real GNP on a per capita basis.

Uno[16] estimated NNW for Japan by subtracting from GNP the cost of reducing environmental damages from water contamination, air pollution and waste disposal. Damages were estimated as the cost of achieving established environmental standards. Because the choice of a standard is somewhat arbitrary, damage estimates obtained with this method are somewhat subjective. Furthermore, NNW does not consider defensive expenditures and residual pollution damages. Keeping these limitations in mind, NNP behaves quite differently than GNP. Real NNW and GNP for the 1955–1985 period in five-year increments are compared in Figure 10.1. Net National Welfare averaged 23 percent less than GNP and NNW increased by a factor of 5.8 compared with 8.3 for GNP.

Adjusted GNP. A more comprehensive monetary measure of economic welfare is *adjusted GNP*, which is defined as follows:

Adjusted GNP = GNP − Depreciation of Manufactured Capital
− Defensive Expenditures
− Net Changes in Resource/Environmental Capacity
− Residual Pollution Damages

Gross national product minus depreciation in manufactured capital is simply NNP. Therefore, adjusted GNP is NNP minus the sum of the three adjustments discussed above. Pearce et al.[17] refer to this measure as sustainable income.

There are several ways to calculate defensive expenditures, net changes in resource or environmental capacity, and residual pollution damages. Repetto et al.[18] examined how net changes in resource capacity affect national accounts. They identified three methods of accounting for changes in resource capacity. These methods are illustrated using the hypothetical data given in Table 10.4. The first method is the *balance sheet method*. It gives a national balance sheet presentation of national resource accounts. In this method, net changes in the value of stocks are the difference between the ending and beginning values of the stock. In Table 10.4, the beginning stock is 100 units and the unit resource value (price minus input cost) is $1.00, resulting in a beginning stock value of $100. Twenty units of the resource are discovered during the period, the previous estimate of the resource is revised downward by 30 units, and extensions of existing deposits are 15 units. Current production (extraction) of the resource is 20 units. Net change in stocks is −15 (stock declined) giving an ending stock of 85 units (100 − 15). Unit resource value at the end of the period is $3, resulting in an ending stock value of $255 ($3 × 85). Hence, the

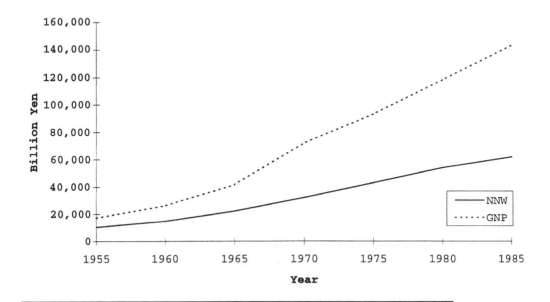

Figure 10.1. Net national welfare (NNW) versus gross national product (GNP) for Japan, 1955–1985.

SOURCE: K. Uno, *Economic Growth and Environmental Change in Japan–Net National Welfare and Beyond* (Mimeo), Tsukuba, Japan: Institute of Socioeconomic Planning, University of Tsukuba, 1988.

Table 10.4. Accounting for net changes in natural resource capacity

	Physical Units	Unit Value ($)	Value ($)
Beginning stock	100	1.00	100
Additions			
Discoveries	20	1.60	32
Net revisions	−30	1.60	−48
Extensions	15	1.60	24
Net	5	—	8
Reductions			
Production	−20	1.60	−32
Net changes	−15	1.60	−24
Reevaluations			
Beginning stock			200
Transactions			−21
Ending stock	85	3.00	255

Source: Robert Repetto et al., The Need for Natural Resource Accounting, *Wasting Assets, Natural Resources in National Income Accounts* (Washington, DC: World Resources Institute, 1989), p. 23.

net change in the value of the stock is $155 ($255 − $100).

The second method is based on the *value of depletion*. In this method, the value of extraction or production in the current period is subtracted from NNP. Value of depletion equals resource depletion times average per unit value of the resource. For the data in Table 10.4, value of depletion is −$32 (−20 × $1.60).

The third method of adjusting GNP for changes in resource capacity is the *net price method*. This method subtracts from NNP the value of total net change in the resource, not just the value of depletion. Because total net change in the stock of the resource is −15 units and the average unit value of the resource is $1.60, value of the total net change in stock is −$24 (−15 × $1.60). The net price method is different from the balance sheet method because it does not consider changes in the value of the beginning stock due to either inflation or deflation of resource prices. Procedures for depreciating manufactured capital (structures and equipment) are based on the book value of the asset not on its replacement cost. Book value does not reflect changes in asset values due to inflation or deflation. Therefore, the balance sheet method is inconsistent, whereas the net price method is consistent with current methods for depreciating manufactured capital in national income accounts. This example illustrates that there is more than one way to account for changes in resource capacity. The same is true for environmental capacity.

All three methods of accounting for changes in resource capacity are based on current prices. If national income accounts are expressed in constant (inflation-adjusted) prices, then changes in resource capacity should likewise be expressed in constant prices. Changes in capacity are converted from current to constant prices by dividing amounts in current prices by an appropriate price index. For example, if the selected price index increased from 210 to 220 between the beginning and the end of the period, then general prices increased by 4.76 percent [(220 −210)/210 × 100]. The value of the total net change in resource capacity is −$22.91 in constant prices (−$24/1.0476).

Repetto et al.[19] compared Indonesia's gross domestic product (GDP) to net domestic product (NDP) for the period 1971 to 1984. Indonesia is a developing country whose economy is heavily dependent on extraction of exhaustible natural resources. NDP was calculated by subtracting the value of net depletion in petroleum, timber and soil resources from GDP. Net depletion is positive (negative) when gross investment in natural resources is less (greater) than depletion. A positive net depletion indicates that the country's endowment of natural capital has been drawn down to finance consumption and/or debt repayment. NDP is less (greater) than GDP when net depletion is positive (negative).

Gross domestic product and NDP for Indonesia are compared for the 1971–1984 period (in constant 1973 prices) in Figure 10.2. Two findings are noteworthy. First, NDP exceeds GDP in the early 1970s because net additions to petroleum reserves are positive; however, NDP is consistently below GDP after 1975 due to depletion of natural capital. From 1975 to 1984, NDP is below GDP by an average of 17.7 percent. Second, annual average growth in NDP is 3.1 percent below annual average growth in GDP (7.1 − 4.0 percent) for the entire 1971–1984 period. Therefore, ignoring the depletion of natural capital in Indonesia results in a significant overstatement in both the level of and growth in economic activity.

Index of Sustainable Economic Welfare. The index of sustainable economic welfare (ISEW) was developed by Daly and Cobb.[20] This index adjusts GNP for:

1. Improvements or worsening in the distribution of income.
2. Changes in net capital stock that include the stock of fixed reproducible capital but exclude the value of land and human capital.
3. Capital accumulation financed from borrowed foreign sources is excluded from net capital formation.

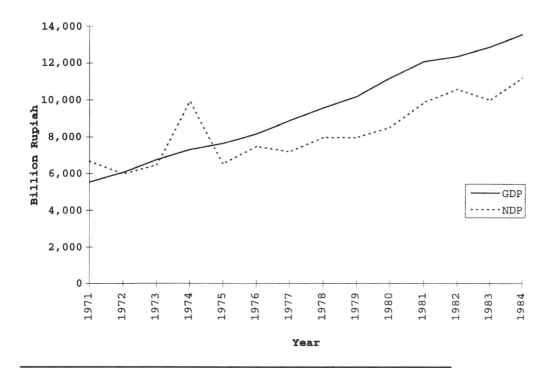

Figure 10.2. Gross domestic product (GDP) versus net domestic product (NDP) for Indonesia, 1971–1984.

SOURCE: Robert Repetto et al., The Need for Natural Resource Accounting. *Wasting Assets, Natural Resources in National Income Accounts* (Washington, DC: World Resources Institute, 1989).

4. Depletion of natural capital, namely, fuels, minerals, wetlands and farmland, is subtracted.

5. Environmental damages from air, water, noise pollution and climate modification are subtracted.

6. Imputed value of leisure is omitted.

7. Value of unpaid household labor is included.

Daly and Cobb calculate two versions of ISEW, both on a per capita basis. The full index (ISEW$_f$) includes all seven adjustments listed above and the partial index (ISEW$_p$) excludes the adjustments for depletion of natural capital (item 4) and environmental damages (item 5). For the entire 1950–1986 period, annual average growth is 0.53 percent for ISEW$_f$ and 0.84 percent for ISEW$_p$, both of which are considerably below the 1.90 percent growth in GNP. For the 1970–1980 and 1980–1986 periods, growth in ISEW$_f$ is −0.14 and −1.26 percent, respectively (negative growth), whereas GNP rises by 2.04 and 1.84 percent, respectively. In the 1980–1986 period, growth in ISEW$_p$ is −0.84 percent (negative growth).

With the exception of the fourth and fifth items, the adjustments in GNP listed above are for factors other than natural and environmental resource capacity and residual pollution damages. The difference between ISEW$_f$ and ISEW$_p$ indicates how many percentage points GNP is overstated by ignoring depletion in natural capital and environmental damages. Subtracting this difference from growth in GNP approximates the growth in GNP adjusted for natural and environmental resource capacity and residual pollution damages. Unadjusted and adjusted growth in GNP are reported in Table 10.5.

Except for the 1960–1970 period, annual average growth in adjusted GNP is 0.31 percentage points below growth in unadjusted GNP. The difference between the two growth rates is greatest in the 1970–1980 period, namely, 0.80 percentage points. While the differences in growth rates seem small, they have major implications for increases in GNP. For example, if GNP grows at an annual unadjusted rate of 1.84 percent, which is the rate for the 1980–1988 period, then GNP doubles in 38 years. If, however, GNP grows at 1.42 percent, which is the adjusted growth rate for this period, then GNP doubles in 49 years, a 29 percent longer doubling time.

Differences between adjusted and unadjusted growth rates are likely to be greater in developing countries that have economies that are highly dependent on use of natural resources. In a study by Repetto et al., adjusting for natural resource depletion in Indonesia reduced the annual growth rate in GDP from 7.1 percent to 4.0 percent in the 1971–1984 period, a difference of 3.1 percentage points.[21] In-

Table 10.5. Growth in unadjusted and adjusted GNP in the United States for various time periods[a]

Period	Unadjusted GNP	Adjusted GNP	Unadjusted − Adjusted GNP
1950-1986	1.90	1.59	0.31
1960-1970	2.64	2.68	−0.04
1970-1980	2.04	1.24	0.80
1980-1986	1.84	1.42	0.42

[a]Based on the index of sustainable economic welfare (ISEW) from Herman E. Daly and John B. Cobb, Jr., *For the Common Good: Redirecting the Economy Toward Community, The Environment, and a Sustainable Future* (Boston, Massachusetts: Benton Press, 1989).
GNP = gross national product.

donesia's GDP would double in 10 years based on the unadjusted rate and 17.5 years based on the adjusted rate. Hence, the doubling time increases by 75 percent for Indonesia compared with only 29 percent for the United States by shifting from the unadjusted to the adjusted rates of growth in GNP or GDP.

SATELLITE ACCOUNTS. A third method of dealing with deficiencies in national income accounts is to utilize satellite accounts. Satellite accounts are natural resource accounts that are kept separate from the core national income accounts. They can be reported in either physical or monetary units. Satellite accounts are supported by the United Nations Statistical Office and are part of the United Nation's SNAs. The SNAs include satellite or reconciliation accounts that track changes in the value of reproducible tangible assets such as forests and nonreproducible tangible assets such as oil, natural gas and agricultural land. Only privately owned natural assets used in commercial production are included in the United Nations (U.N.) satellite accounts. Publicly owned natural resources such as national parks are excluded from the satellite accounts because the SNAs only cover transactions involving assets in the private market economy.

Changes in the value of natural resources in the U.N. satellite accounts result from annual variability in stock levels and prices. Changes in stocks result from growth, discoveries, depletion, extraction and natural losses. The U.N. Statistical Office prefers to value natural resources based on their current market prices. In the absence of market values, the U.N. recommends using the present value of the asset determined by discounting future net income from the asset. The latter equals current price times expected sales minus extraction and management costs. While a system of satellite accounts is endorsed by the U.N. Statistical Office, most countries employ only the core national income accounts. Combined with conventional measures of economic performance, such as GNP, satellite accounts portray how economic growth affects the natural resource base. For this reason, the Bureau of Economic Analysis in the U. S. Department of Commerce has developed a set of satellite accounts as well as a set of integrated accounts for oil and minerals.

Resource-Specific Accounting

NERA can be applied to all resources or to specific resources. Public policies to alleviate specific forms of natural resource depletion, such as soil erosion or soil salinization, have stimulated interest in applying resource accounting methods to specific resources. Consider the soil conservation policy that established the Soil Erosion Service in 1933 (renamed the Soil Conservation Service and the Natural Resources Conservation Service) within the United States Department of Agriculture. The goal of the Soil Erosion Service was to reduce soil erosion in the Great Plains and avert another Dust Bowl. In the mid-1980s, the Soil Conservation Service expanded its programs to include agricultural nonpoint source water pollution. The impetus for this expansion was the mounting evidence that offsite damages from cropland erosion in the United States are two to three times greater than onsite damages.[22] Annual offsite erosion damages in the United States are given in Table 10.6. Damages result primarily from sediments, fertilizers, pesticides and salts carried in runoff. Waterborne sediment damages attributed to agriculture ($3.6 billion) are equivalent to 10 percent of net farm income in 1986.[23]

Natural resource accounting of onsite and offsite erosion damages is justified because the damages constitute a real economic cost to society. Accounting for erosion damages can significantly alter the economic value of agricultural production. Faeth et al.[24] compared conventional net farm income to net economic value for a conventional corn-soybean rotation in Pennsylvania. Net farm income ignores and net economic value includes onsite and offsite erosion damages. Results of their analysis are summarized in Table 10.7.

Gross operating margin is average revenue minus average variable cost. Onsite erosion damage is calculated using the Erosion Productivity Impact Calculator and offsite erosion damages are based on a per unit damage of $8.16 per ton for the northeastern United States.[25] Net operating income with onsite damage is gross operating income minus onsite damage ($45 per acre) or $20 per acre. The per acre subsidy received by farmers from the government-sponsored commodity program

Table 10.6. Annual offsite erosion damages in the United States in 1990 ($ million)

Damage Category	Best Estimate	Range of Estimates
Freshwater recreation	2,404	995–7,580
Marine recreation	692	497–2,772
Water storage	1,260	756–1,761
Navigation	866	616–1,078
Flooding	1,130	755–1,787
Roadside ditches	618	310–929
Irrigation ditches	136	68–184
Freshwater commercial fishing	69	61–96
Marine commercial fishing	451	443–612
Municipal waste treatment	1,114	573–1,655
Municipal and industrial use	1,382	768–1,848
Steam power cooling	28	24–39
Total	10,150	5,826–20,341

Source: Marc O. Ribaudo, *Water Quality Benefits from the Conservation Reserve Program*, Agricultural Economic Report 606 (Washington, DC: U.S. Department of Agriculture, Economic Research Service, 1989).

Table 10.7. Net farm income and net economic value for corn-soybean rotation in Pennsylvania (1990 dollars per acre per year)

	Net Farm Income		
Economic Measures	Excluding Onsite Damage	Including Onsite Damage	Net Social Benefit
Gross operating margin	45	45	45
− Onsite erosion damage		25	25
= Net operating income	45	20	20
+ Government subsidy	35	35	
= Net farm income	80	55	
− Offsite erosion damage			47
= Net economic value			−27

Source: Paul Faeth, et al., *Paying the Farm Bill: U.S. Agricultural Policy and the Transition to Sustainable Agriculture* (Washington, DC: World Resources Institute, March 1991), p. 33.

for corn adds another $35 per acre to net farm income. Net farm income is $80 per acre when ignoring onsite erosion damage, and $55 per acre when considering onsite erosion damage. Therefore, a corn-soybean rotation is profitable from the farmer's viewpoint.

Government subsidies for crop production are a transfer payment from taxpayers to farmers. Because transfer payments do not contribute to economic welfare, they are excluded from net social benefit. Accordingly, there is no entry for net farm income in the column labeled net social benefit. While farmers generally ignore offsite erosion damage in making their production decisions, it is a real social cost of agricultural production. Net social benefit equals net operating income minus offsite damage, namely, −$27 per acre ($20 − $47). This example shows that corn and soybean production is socially inefficient in Pennsylvania because it reduces net social benefit by $27 per acre.

Implications for Sustainable Development

There is a close link between sustainable income and sustainable development. Sustainable income accounts for depreciation in both manufactured and natural capital, defensive expenditures and residual pollution damages. Sustainable development implies constant sustainable income when the economy is at its optimal scale and increasing sustainable income when the economy is below its optimal scale. Sustainable income equals NNP or NDP minus defensive expenditures, minus the net change in natural and environmental resource capacity minus residual pollution damages. What are the consequences of evaluating economic progress based on sustainable income versus unadjusted national income? When the two income measures are similar, there is no need to make a distinction. However, when the two income measures are dissimilar, the differences in economic progress can be substantial. Net National Welfare for Japan (Figure 10.1), NDP for Indonesia (Figure 10.2) and adjusted GNP for the United States (Table 10.5) are proxies for sustainable income. All three measures are significantly below the corresponding unadjusted national income measure, and growth in adjusted income is generally less than growth in unadjusted income.

Basing economic and resource management decisions on national income rather than sustainable income can be detrimental to the achievement of sustainable development, particularly when there are significant differences between the two income measures. Consider a hypothetical example involving the countries of Angora and Bangora. Angora's economy is highly dependent on extraction of natural resources and environmental protection is a low priority. Bangora's economy is tourist based, extraction of natural resources is minor, and environmental protection is a high priority. Angora's historical growth rates are 8 percent in GNP and 4 percent in resource depletion, implying a 4 percent growth rate in sustainable income. Bangora's historical growth rates are 6 percent in GNP and 1 percent in resource depletion, implying a 5 percent growth rate in sustainable income. The International Bank Corporation (IBC) is evaluating loan applications from Angora or Bangora. There are only enough funds to make a loan to one of the countries, and IBC prefers to make loans to countries that have a high growth in GNP. Under these conditions, the loan is made to Angora.

Does this decision contribute to sustainable development? The decision is consistent with IBC's loan policy because the historical growth rate in GNP is 2 percentage points higher in Angora than in Bangora. It is not the best decision for sustainable development. Bangora has a higher rate of growth in sustainable income than Angora (5 percent versus 4 percent), and the growth in resource depletion is over three times greater in Angora than in Bangora (4 percent versus 1 percent). The IBC's loan decision favors a country that is depleting its natural and environmental resources more rapidly. Making a loan to Angora results in a greater relative increase in economic activity but at the expense of a greater relative depletion in natural capital.

Summary

Gross national product and GDP do not measure sustainable income that is the maximum amount than can be consumed in the current period without reducing consumption in future periods or without being less well off in the future. Natural and environmental resource accounting adjusts GNP or GDP for defensive expenditures, changes in natural and environmental resource capacity and residual pollution damages. Defensive expenditures are made by households, firms and governments to reduce the adverse effects of natural resource depletion and environmental degradation. Medical expenses for treatment of pollution-related diseases such as emphysema are defensive expenditures. Under the current United Nations System of National Accounts, defensive expenditures made by households and governments are added to, rather than subtracted from, GNP.

Depreciation in manufactured capital (structures and equipment) are subtracted from GNP to obtain NNP because depreciation reduces the productive capacity of these assets. Depletion in natural capital, however, which reduces natural and environmental resource capacity, is not subtracted from GNP even though it decreases the economy's long-run productive capacity. Examples of depletion in natural capital include excessive soil erosion, impairment of rangeland carrying capacity from overgrazing, depletion of fossil fuels, deforestation, air pollution and water pollution. Appreciation of natural capital results from oil and natural gas discoveries, higher petroleum recovery rates due to technological progress, net gains in wetlands and investments that increase natural and environmental resource capacity. Residual pollution damages are human-related damages from environmental degradation that are not alleviated by defensive expenditures.

Physical, monetary and satellite accounts are three principal ways to handle deficiencies in national income accounts. Physical accounts enumerate changes in the stocks of natural and environmental resources in physical units such as barrels of oil or tons of carbon dioxide emissions. Norway, France and Canada employ physical accounts. Monetary accounts adjust GNP for defensive expenditures, changes in natural resource or environmental capacity and/or residual pollution damages. Adjusted measures for GNP or GDP have been developed such as net national welfare, sustainable income and the index of sustainable economic welfare. Changes in the value of natural resource assets can be measured using the balance sheet method, value of depletion method and net price method. The net price method is the most reasonable method.

Satellite accounts are physical or monetary accounts that keep track of depletion of natural capital. Such accounts are kept separate from core national income accounts. The United Nations Statistical Office recommends that satellite accounts be used to trace changes in the value of privately owned reproducible tangible assets including forests and nonreproducible tangible assets such as oil, natural gas and agricultural land.

Adjusting conventional economic measures for defensive expenditures, resource capacity and residual environmental damages generally reduces the level of income and rate of economic growth. Subtracting depletion of petroleum, timber and soil resources from Indonesia's GDP reduces annual economic growth by 3.1 percent. Growth in per capita GNP for the United States declined from 1.90 percent to 1.59 percent after adjusting for depletion of fuels, minerals, wetlands, and farmland, and environmental damages from air, water, noise pollution and climate modification. Measuring the benefits of economic activity in terms of sustainable income is more consistent with sustainable development.

Natural and environmental resource accounting is applicable to individual resources. A study of corn and soybean production in Pennsylvania showed that accounting for onsite and offsite erosion damages makes the net social benefit of production negative even though production is profitable to farmers.

Questions for Discussion

1. Conventional national income accounts were developed over 50 years ago. Natural and environmental resource accounting is a relatively new approach to correcting certain deficiencies in conventional national income accounts. Some analysts argue that natural resource and environmental accounts should be kept separate from national income accounts to protect the integrity of the latter. Do you agree or disagree? What are the advantages of combining conventional and natural resource or environmental accounts? What factors might prevent widespread adoption of natural and environmental resource accounting?

2. The stated objective of the World Resources Institute is to address the question: How can societies meet basic human needs and nurture economic growth without undermining the natural resources and environmental integrity on which life, economic vitality, and international security depend? Do you think this issue is more applicable to developing than to developed countries? Why?

3. A family buys a house located near a cement factory that generates lots of dust. The family knew about the dusty conditions before purchasing the house. Prior to moving into the house, the family installs an air purification system. Is the cost of the air purification system a legitimate defensive expenditure?

4. The following table shows that the difference between the growth rates for unadjusted GNP and adjusted GNP in the United States decreased from the 1970–1980 period to the 1980–1986 period. During the late 1960s and 1970s, many laws were passed to protect natural resources and the environment. Might there be a connection between the decline in the difference between growth in unadjusted and adjusted GNP and the proliferation of environmental laws? Discuss.

Period	Unadjusted GNP	Adjusted GNP	Difference
1970–1980	2.04	1.24	0.80
1980–1986	1.84	1.42	0.42

5. Comment on the following statement. The NNP accounts for depreciation of manufactured capital but ignores depletion of natural capital. Therefore, NNP is an inappropriate measure of sustainable income.

Further Readings

Alfsen, Knut H., Torstein Bye and Lorents Lorentsen. 1987. *Natural Resource Accounting and Analysis: The Norwegian Experience, 1977-1986,* Oslo, Norway: Central Bureau of Statistics of Norway.

Bartelmus, Peter, Carsten Stahmer and Jan van Tongeren. 1989. "SNA Framework for Integrated Environmental and Economic Accounting." Paper presented at the 21st Conference of the International Association for Research in Income and Wealth, Lahnstein, Germany, August 21.

Daly, Herman E. 1986. "On Sustainable Development and National Accounts." In *Economics and Sustainable Environments: Essays in Honor of Richard Lecomber,* D. Collard, D. Pearce and D. Ulph, eds. New York: Macmillan & Co.

El Serafy, Salah. 1991. "The Environment as Capital." In *Ecological Economics: The Science and Management of Sustainability*, Robert Costanza, ed. New York: Columbia University Press. pp. 168–175.

Peskin, Henry M. 1976. "A National Accounting Framework for Environmental Assets," *Journal of Environmental Economics and Management* 2:255-262.

Theys, J. 1989. "Environmental Accounting in Development Policy: The French Experience." In *Environmental Accounting for Sustainable Development,* Yusurf J. Ahmad, Salah El Serafy and Ernst Lutz, eds. Washington, D.C.: The World Bank.

Notes

1. United Nations, Department of Economic and Social Affairs, *A System of National Accounts, Statistical Papers,* Series F, No. 2. Rev. 3 (New York: United Nations, 1968).

2. Repetto et al. point out, "as Keynesian analysis largely ignored the productive role of natural resources, so does the current system of national accounts." Robert Repetto et al., "The Need for Natural Resource Accounting," *Wasting Assets, Natural Resources in National Income Accounts* (Washington, D.C.: World Resources Institute, 1989), p. 1.

3. S. Devarajan and R. J. Weiner, *Natural Resource Depletion and National Income Accounts,* Memo, J. F. Kennedy School of Government, Harvard University, 1988.

4. Herman Daly, "Toward a Measure of Sustainable Social Net Product," in Yusuf J. Ahmad, Salah El Serafy and Ernst Lutz, eds. *Environmental Accounting and Sustainable Income* (Washington, D.C.: World Bank, 1989); Salah El Serafy, "Absorptive Capacity, the Demand for Revenue, and the Supply of Petroleum," *Journal of Energy and Development* 7(1981):73–88; Salah El Serafy, "The Proper Calculation of Income from Depletable Natural Resources," in Yusuf J. Ahmad, Salah El Serafy and Ernst Lutz, *Environmental Accounting and Sustainable Income* (Washington, D.C.: World Bank, 1989).

5. John Hicks, *Value and Capital,* 2nd ed. (Oxford: Oxford University Press, 1946), p. 172.

6. Herman E. Daly and John B. Cobb, Jr., *For the Common Good: Redirecting the Economy Toward Community, the Environment, and a Sustainable Future* (Boston, Massachusetts: Beacon Press, 1989).

7. David Pearce, Anid Markandya and Edward B. Barbier, "Accounting for the Environment," *Blueprint for a Green Economy* (London: Earthscan Publications, Ltd, 1989), p. 111.

8. Anne Harrison, "A Possible Conceptual Approach to Introducing Natural Capital into the SNA;" and Henry Peskin, "Environmental and Non-Market Accounting with Some Reference to Indonesia," in Yusuf J. Ahmad, Salah El Serafy and Ernst Lutz, *Environmental Accounting and Sustainable Income* (Washington, D.C.: World Bank, 1989).

9. Salah El Serafy and Ernst Lutz, "Environmental and Natural Resource Accounting," in *Environmental Management and Economic Development.* Gunter Schramm and Jeremy J. Warford, eds. (Baltimore, Maryland: The John Hopkins University Press, 1989), pp. 22-38.

10. *Ibid.* (1989).

11. Salah El Serafy, "The Proper Calculation of Income from Depletable Natural Resources," *Environmental and Resource Accounting and Their Relevance to the Measurement of Sustainable Income,* Ernst Lutz and Salah El Serafy, eds. (Washington, D.C.: World Bank, 1988).

12. *Organization for Economic Cooperation and Development, Natural Resource Accounting: The Norwegian Experience,* prepared by A. Lone, Environment Committee, Group on the State of the Environment, Paris, France, 1988.

13. P. Corniere, "Natural Resource (1) Accounts in France: An Example: Inland Waters," *Information and Natural Resources* (Paris, France: Organization for Economic Cooperation and Development, 1986).

14. Lester R. Brown, Hal Kane and Ed Ayers, *Vital Signs, 1993*, Worldwatch Institute (New York: W. W. Norton & Co., 1993).

15. W. D. Nordhaus and J. Tobin, "Is Growth Obsolete?" M. Moss, ed., *The Measurement of Economic and Social Performance: Studies in Income and Wealth*, No. 38 (New York, New York: National Bureau of Economic Research, 1973).

16. K. Uno, *Economic Growth and Environmental Change in Japan-Net National Welfare and Beyond,* Mimeo, Institute of Socioeconomic Planning, University of Tsukuba, Japan, 1988.

17. Pearce, Markandya and Barbier (1989), p. 108.

18. Repetto et al. (1989).

19. Robert Repetto et al., "The Need for Natural Resource Accounting," *Wasting Assets, Natural Resources in National Income Accounts* (Washington, D.C.: World Resources Institute, 1989).

20. Daly and Cobb (1989).

21. Repetto et al. (1989).

22. E. H. Clark, J. A. Haverkamp and W. Chapman, *Eroding Soils: The Off-Farm Impacts* (Washington, D.C.: The Conservation Foundation, 1985).

23. Paul Faeth et al., *Paying the Farm Bill: US. Agricultural Policy and the Transition to Sustainable Agriculture* (Washington, D.C.: World Resources Institute, March 1991).

24. *Ibid.* (1991).

25. J. R. Williams, P. T. Dyke and C. A. Jones, "EPIC–A Model for Assessing the Effects of Erosion on Soil Productivity," *Proceedings of the Third International Conference on State-of-the-Art in Ecological Modeling,* Colorado State University, Fort Collins, Colorado, 1982.

CHAPTER 11

Benefit–Cost Analysis of Resource Investments

> *... we are spending $9 billion more per year to comply with the Clean Water Act than we are benefiting from that compliance.*
>
> —*U.S. Water News,* March 1994

Private and public investments are made to develop, preserve, enhance, restore and protect natural and environmental resources. Whether to increase oil and gas exploration and development on private land is a private investment decision. Charging higher fees for livestock grazing on public rangeland is designed to protect the quality of rangeland and to increase the rate of return on publicly owned resources. Expanding the boundaries of a national park or wilderness area to reduce encroachment by private development is a public resource investment in natural area protection. Private resource investments are usually guided by the goal of maximizing profit subject to financial and technical constraints. Public resource investments are typically undertaken with the goal of advancing social, economic, cultural and environmental conditions.

The primary objective of this chapter is to develop and apply economic efficiency criteria for evaluating public resource investments. Such evaluation is referred to as *public investment* or *benefit–cost analysis*. Benefit–cost analysis is a monetary assessment of economic impacts of a resource investment. It does not consider nonmonetary social, cultural and environmental impacts. Yet, a comprehensive evaluation of a public resource investment should consider how the investment affects social, economic, cultural and environmental values. One way to conduct a comprehensive evaluation of a public resource investment is to utilize multiple criterion decision analysis, which allows integration of impacts expressed in monetary and nonmonetary terms. From this perspective, benefit–cost analysis is an important but by no means the only element in a comprehensive assessment of public resource investments.

The topics covered in this chapter deal with various aspects of benefit–cost analysis including socially efficient resource investment, continuous versus discrete time, investment evaluation period, efficiency and equity implications of the discount rate, selection of the discount rate, treatment of capital and operating costs, economic versus financial feasibility, local versus global efficiency, independent versus interdependent investments, capital rationing, primary and secondary benefits and costs, risk and uncertainty, alternative investment evaluation criteria and evaluation of multiple resource investments.

Socially Efficient Investment

An investment in natural and/or environmental resources is *socially efficient* if it has a positive net social benefit and socially inefficient if it has a negative net social benefit. Social efficiency is maximized by selecting the resource investment with the highest net social benefit. This rule ensures that society achieves the maximum social gain per unit of throughput of material and energy resources. Private individuals evaluate resource investments in terms of profit or net private benefit rather than net social benefit. A resource investment is *privately efficient* when it has a positive net private benefit. Private efficiency is maximized by selecting the investment with the highest net private benefit.

Privately efficient investments are not necessarily socially efficient, and vice versa, for several reasons. First, net private benefit typically excludes the environmental costs associated with an investment. Second, a private investment decision is unlikely to account for potential adverse consequences on human health and welfare. For example, the decision to increase coal mining on private land is likely to ignore whether land disturbances caused by mining are within the assimilative capacity of the environment or whether the health and welfare of nearby residents are adversely affected. Ignoring potential environmental and human costs of resource investments causes net private benefit to exceed net social benefit, other things equal. Third, private firms commonly use a higher discount rate than public agencies to calculate the net present value of an investment. A higher discount rate lowers the net present value of future resource extraction, which accelerates depletion of exhaustible resources.

Divergence between private and social net benefits has significant implications for renewable resources. Ponder the consequences of rapid expansion in tiger shrimp production in southern Thailand. Shrimp production, processing and exportation are very profitable. Rapid expansion in shrimp production has boosted the value of agricultural income and exports and accelerated economic development in rural areas of Thailand. Shrimp production involves the pumping of sea water into shrimp ponds. Ponds are intensively managed (very high stocking densities and feeding rates) and highly concentrated along the southeastern coast of Thailand. Wastewater from shrimp ponds has very high concentrations of organic matter and other contaminants. Untreated pond wastewater is drained to the sea.

Shrimp production in southern Thailand is adversely affecting social and environmental conditions. Large areas have been converted from rice production to shrimp production, salt water intrusion has reduced the productivity of remaining riceland, mangrove forests have been destroyed to make way for shrimp ponds and marine ecosystems have been degraded by pond wastewater. Accounting for these externalities causes the net social benefit to be less than the net private benefit of shrimp production. Finally, rapid expansion in shrimp production has adversely affected the incomes, culture and livelihood of fishermen and rural residents not involved in shrimp production.

NET SOCIAL BENEFIT. Basing resource investment decisions on social rather than private efficiency criteria is more consistent with achieving sustainable resource development and use. This section addresses the application of social efficiency criteria to resource investments decisions. Net social benefit (NSB) of a nat-

11. Benefit–Cost Analysis of Resource Investments

ural resource investment is defined in a similar way as the NSB used to derive the efficient intertemporal allocation of an exhaustible resource (Chapter 7) and a renewable resource (Chapter 8). For simplicity, NSB is evaluated for discrete time periods.

The undiscounted net social benefit of q_t is:

$$B(q_t) = V(q_t) - C(q_t),$$

where q_t is the quantity of the resource used in time period t, $V(q_t)$ is the total value of q_t and $C(q_t)$ is the total cost of producing q_t with the investment. $V(q_t)$ equals the area under the demand curve up to q_t. The demand curve includes all the benefits of the resource. $C(q_t)$ equals the area under the supply curve up to q_t. The supply curve accounts for the full social cost of production. The area under the demand curve minus the area under the supply curve equals consumer surplus plus producer surplus. Therefore, $B(q_t)$ equals consumer surplus plus producer surplus for q_t.

NSB of a specific pattern of intertemporal resource use is:

$$NSB(q_1, q_2,...,q_T) = \sum_{t=0}^{T} B(q_t)(1 + r)^{-t},$$

where $(q_1, q_2,...,q_T)$ are the quantities of the resource used in all time periods, T is the length of the evaluation period and $B(q_t)(1 + r)^{-t}$ is the present value of net social benefit of q_t. The summation operator (Σ) sums the present values of net social benefits over the entire evaluation period. A more detailed explanation of present value is given in Chapter 2.

When several investments are being evaluated, NSB is calculated for each investment. Investments for which NSB > 0 are economically feasible. The investment with the highest positive NSB is judged to be economically superior to all other investments. If a choice is being made between two investments (as in the timber harvesting example presented below), then the decision rule can be expressed in terms of the difference between the net social benefits for the two investments. For example, if NSB_A is the net social benefit for investment A, NSB_B is the net social benefit for investment B and both investments have positive net social benefits, then investment A is selected when $NSB_A > NSB_B$ and investment B is selected when $NSB_A < NSB_B$.

TIMBER HARVESTING EXAMPLE. The expression for net social benefit given above is likely to overestimate or underestimate the true net social benefit of a resource investment when changes in the price of the resource influence the prices of substitutes or complements for that resource. Consider the decision of whether to invest in a new or conventional technology for removing timber from mountainous areas. The new technology utilizes helicopters and the conventional technology uses trucks and roads to remove felled timber. Because logging roads are a major source of erosion and water contamination in conventional logging operations, the new technology is expected to have a lower marginal environmental cost of timber harvesting. A lower marginal environmental cost results in a lower marginal social cost of timber harvesting, which reduces the price of timber.

When the price of timber declines, the demand for wood substitutes such as plastic is reduced. As demand for plastic decreases, the price of plastic falls and that reduces consumer plus producer surpluses for plastic. The net social benefit of investment in the new technology includes changes in consumer and producer surpluses for timber as well as changes in consumer and producer surpluses for plastic triggered by the change in timber price.

The change in net social benefit between the conventional (c) and new (n) technology is illustrated in Figure 11.1 in terms of the timber (m) and plastic (p) markets. Marginal social cost of timber production in period t is MSC_{ct} with the conventional technology (trucks and roads) and MSC_{nt} with the new technology (helicopters). The decrease in marginal social cost between the conventional technology and new technology reduces timber price from pm_{ct} to pm_{nt} and increases timber production from qm_{ct} to qm_{nt}. Because wood and plastic are substitutes in certain uses, the reduction in the price of timber causes the demand for plastic to decrease from D_{ct} to D_{nt}. As a result, plastic production declines from qs_{ct} to qs_{nt} and the price of plastic decreases from ps_{ct} to ps_{nt}.

Changes in technology induce second-order price effects. A lower price for plastic causes the demand for timber to shift downward because timber and plastic are substitutes. The downward movement in timber demand reduces net social benefit from timber. Eventually, new equilibrium prices are established for timber and plastic. For simplicity, second-order price effects are not shown in Figure 11.1 and not considered in the evaluation of NSB.

Consumer surplus, producer surplus and total surplus (consumer plus producer surpluses) for the demand and supply curves in Figure 11.1 are reported in Table 11.1. The "net change" column in Table 11.1 indicates whether the price change "increases" or "decreases" surplus. The direction of change in surplus values due to shifts in the demand and/or supply curves depends on the slopes of the two curves. For this reason, the net changes stated in Table 11.1 are only valid for the demand and supply curves illustrated in Figure 11.1.

The economic feasibility of the two technologies is based on the following net social benefit:

$$NSB = (GSB_n - GSB_c) - (K_n - K_c),$$

where GSB_n is the present value of gross social benefit with the new technology, GSB_c is the present value of gross social benefit with the conventional technology,

Table 11.1. Consumer surplus, producer surplus and total surplus resulting from a decrease in timber price

	Before Price Decrease[a]	After Price Decrease[a]	Net Change
Timber			
Consumer surplus	abc	ade	Increases
Producer surplus	bcf	deg	Increases
Total	acf	aeg	Increases
Plastic			
Consumer surplus	hij	klm	Decreases
Producer surplus	jin	mln	Decreases
Total	hin	kln	Decreases

[a]Refer to Figure 11.1 for a graphical explanation of these magnitudes.

11. Benefit–Cost Analysis of Resource Investments   269

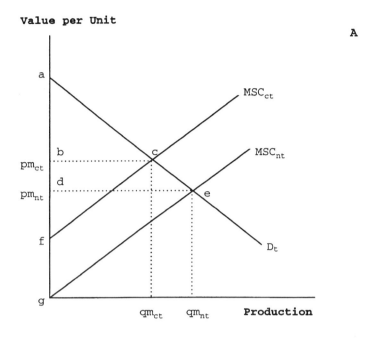

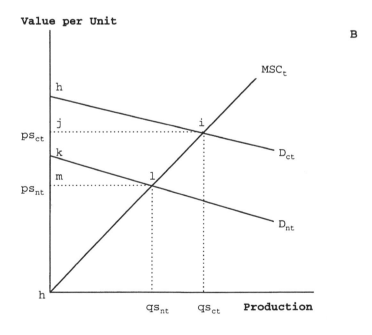

Figure 11.1. Change in net social benefit between a conventional (c) and new (n) technology for timber harvesting in timber (m) markets (A) and plastic (p) markets (B) in period t. D = demand; and MSC = marginal social cost.

K_n is the present value of fixed cost with the new technology and K_c is the present value of fixed cost with the old technology. If $GSB_n - GSB_c$ is positive (negative), then present value of gross social benefit is higher (lower) with the new technology than with the conventional technology. Likewise, if $K_n - K_c$ is positive (negative), then present value of fixed cost is higher (lower) with the new technology than with the conventional technology. Social efficiency increases by selecting the new technology when NSB is positive and the conventional technology when NSB is zero or negative.

The first term $(GSB_n - GSB_c)$ in NSB is calculated as follows. First, the difference in gross social benefits is determined for each time period. For the demand and supply curves given in Figure 11.1, the difference in gross social benefits in period t is:

$$GSB_{nt} - GSB_{ct} = (aeg - acf) - (hin - kln),$$

where (aeg − acf) is the increase in total surplus for timber and (hin − kln) is the decrease in total surplus for plastic in period t. This difference is:

positive if (aeg − acf) > (hin − kln),

negative if (aeg − acf) < (hin − kln), or

zero if (aeg − acf) = (hin − kln).

The second term in NSB $(K_n - K_c)$ is the difference in the present values of fixed costs for the two technologies. Only the difference in fixed costs of harvesting appears in the second term because differences in variable costs are already accounted for in the calculation of $GSB_n - GSB_c$.

NSB can be used to evaluate the social efficiency of a variety of natural and environmental resource investments. In addition to comparing resource extraction technologies, NSB can also be used to evaluate the social efficiency of employing an emission charge versus tradable emission permits to reduce air pollution from coal-fired power plants (environmental policy analysis) or of expanding a national park or wilderness area (natural resource management). While NSB can be used to evaluate any number of investment options, the calculations become more complex and time consuming as the number of investment alternatives increases.

Special Topics in Investment Analysis

Application of resource investment criteria requires consideration of several factors, including continuous versus discrete time, length of investment evaluation period (T), selection of discount rate (r), treatment of capital and operating costs, local versus global efficiency, independent versus interdependent investments, capital rationing, primary and secondary benefits and costs, and risk and uncertainty. This section examines each of these topics.

CONTINUOUS VERSUS DISCRETE TIME. The formula for calculating net social benefit depends on whether the benefits and cost occur continuously over time or at discrete points in time. As indicated earlier in the chapter, NSB for *discrete time* is:

$$NSB(q_0, q_1,...,q_T) = \sum_{t=0}^{T} B(q_t)(1 + r)^{-t},$$

where q_t is the amount of the resource extracted or the services provided by the resource at the end of time period t, $B(q_t)$ is the undiscounted net social benefit derived from q_t, r is the discount rate and $(1 + r)^{-t}$ is the discrete discount factor in period t.

NSB in *continuous time* is:

$$NSB[q(1), q(2),...,q(T)] = \int_{0}^{T} B[q(t)]e^{-rt}dt,$$

where the integral operator ($\int$) accumulates the present values of $B[q(t)]$ over the entire evaluation period. It is the continuous counterpart of the summation operator (Σ). e^{-rt} is the continuous discount factor in period t.

Money flows in discrete time are calculated as of the end of each time period. For example, if t stands for quarters, then all benefits and costs are evaluated as of the end of each quarter and r is the quarterly discount rate. If the annual discount rate is 0.08, then the quarterly discount rate is .02 (0.08/4). Fixed cost is usually incurred in the very beginning of the evaluation period, designated as time period zero. Let $V(q_t)$ and $C(q_t)$ take on the following values at the end of each quarter:

	_0_	Quarter _1_	_2_	_3_	_4_
$V(q_t)$	0	50	75	75	50
$C(q_t)$	100	0	0	0	0

For an annual discount rate of r = 0.08:

$$\begin{aligned}NSB &= (0 - 100)(1 + .02)^{-0} + (50 - 0)(1 + .02)^{-1} + (75 - 0)(1 + .02)^{-2} \\ &\quad + (75 - 0)(1 + .02)^{-3} + (\$50 - 0)(1 + .02)^{-4} \\ &= \$138.01.\end{aligned}$$

INVESTMENT EVALUATION PERIOD. The investment evaluation period (T) is the period of time during which the investment yields benefits and costs. This period can be less than, but not greater than, the economic life of the investment when a single investment is being evaluated. For example, suppose the timber company wants to decide which of the two technologies (trucks and roads versus helicopters) is superior for harvesting a timber stand over a 10-year period. The evaluation period for this investment is 10 years even if the physical life of one or both investments exceeds 10 years. Any benefits and costs that occur after 10 years are not relevant to this investment decision.

Suppose the timber company wants to compare the two technologies over a 15-year period by considering the benefits and costs of using the roads for other purposes such as timber management and recreation and of using the helicopters to harvest timber stands in other areas. Because the physical life of the helicopters is 12 years and that of the trucks and roads is 15 years, the life of the two investments is different. The two investments need to be evaluated for the same period. One way to handle the unequal life of the investments is to include the annual benefits and costs for three years (13, 14 and 15) from reinvesting in the helicopters in year 13. Annual benefits and costs for helicopters in the reinvestment period are based on the expected life of the helicopters purchased in year 13. This procedure ensures that the length of the evaluation period is the same (15 years) for both technologies.

EFFICIENCY AND EQUITY IMPLICATIONS OF THE DISCOUNT RATE.

The discount rate is important because it influences the economic efficiency of resource investments and intergenerational equity. First, consider how the discount rate influences economic feasibility of renewable resource investments. A low discount rate increases the economic feasibility of renewable resource investments, such as hydroelectric power facilities, which have high initial capital costs and revenues (benefits) that are spread out over many years. For such investments, the present value of benefits increases as the discount rate decreases. Historically, the United States government has used low discount rates to justify water resource development projects such as hydroelectric power generation. This policy has accelerated the development and use of water resources, particularly in the western United States. While these projects have been instrumental in economic development of the western United States, they have taken a major toll on natural and environmental resources. A prime example of this is the dramatic reductions in salmon populations caused by hydroelectric dams on the Columbia River.[1]

Second, consider how the discount rate influences the extraction of exhaustible resources. A low discount rate makes it more profitable to shift resource extraction from the present to the future, which causes the stock of the resource to be extracted more slowly. This intertemporal shift in resource extraction benefits the future generation relative to the current generation. Conversely, a high discount rate makes it more profitable to shift resource extraction from the future to the present, which causes the stock of the resource to be extracted more quickly. This intertemporal shift in resource extraction benefits the current generation relative to the future generation.

Changes in the discount rate can affect the total amount of an exhaustible resource that is extracted. This is called a stock effect of changes in the discount rate. Specifically, a low discount rate makes it more profitable to extract marginal deposits of the resource, which increases total resource extraction. If the stimulating effect of a low discount rate on total resource extraction offsets the shifting of extraction from the present to the future, current extraction rates can be higher. Conversely, a high discount rate makes it less profitable to extract marginal deposits of the resource, which decreases total resource extraction. If the depressing effect of a high discount rate on total resource extraction is offset by the shifting of extraction from the future to the present, current extraction rates can be lower. Hence, the effects of changes in the discount rate on intertemporal extraction of an exhaustible resource and intergenerational equity are ambiguous.

Third, changes in intertemporal use of natural and environmental resources have implications for intergenerational equity. The preceding discussion indicates that the effect of the discount rate on intergenerational equity is likely to be different depending on whether the resource is exhaustible or renewable. In the case of renewable resources, a lower discount rate favors current development and raises the risk of overexploitation, which can adversely affect future generations. A case in point is the extinction of plants and animals due to human activities such as deforestation. In comparison, a low discount rate can delay the extraction of an exhaustible resource, which benefits future generations.

Because changes in the discount rate can have opposite effects on the allocation of exhaustible and renewable resource between current and future generations, it is not clear whether future generations gain or lose by higher or lower discount rates. Perhaps, something as important as the intergenerational allocation of exhaustible resources and use of renewable resources should not be subject to the whim of changes in the discount rate. A more proactive approach to intergenerational allocation of natural resources was proposed by Howarth and Norgaard.[2] They suggest that an equitable intergenerational allocation of resources can be achieved by assigning future generations specific rights to exhaustible and renewable resources. This approach is based on ethical considerations because it implies that future generations should have the same rights of access to natural resources as the current generation.

Much of the literature on sustainable development and preservation of biodiversity discuss various mechanisms for achieving sustainable resource use. El Serafy[3] argues that only a portion of the net revenue generated by an exhaustible resource is true income that can be consumed in perpetuity. The difference between net revenue and true income, so-called user cost, should be invested in a renewable resource. In this manner, when the exhaustible resource is used up, a renewable resource will be available as a substitute.

Fourth, the discount rate influences the weight given to social benefits and costs in efficiency calculations, which has a direct bearing on intergenerational equity. Consider two resource investments. The first investment has relatively high future social cost and the second investment has relatively high future social benefit. The decommissioning cost for a nuclear reactor is an example of high future social cost. The long-term reduction in carbon dioxide emissions and global warming resulting from investments in conservation practices and more efficient combustion technologies are examples of high future social benefits. A high discount rate favors acceptance of the first investment and rejection of the second investment. Acceptance of the first investment causes future generations to bear a disproportionate share of the social cost of the investment. Rejection of the second investment deprives future generations of important benefits. In both cases, a high discount rate benefits the current generation at the expense of future generations. Conversely, a low discount rate favors rejection of the first investment and acceptance of the second investment, which benefits the future generation at the expense of the current generation.

SELECTION OF DISCOUNT RATE. The preceding discussion indicates that the discount rate influences not only the economic feasibility of resource investments but also the intertemporal use of natural and environmental resources. For this reason, considerable attention has been given to the selection of the discount rate in benefit–cost analysis. This section discusses the theoretical basis for selecting the discount rate. There are two primary contenders for the discount rate, the *marginal rate of time preference* and *marginal opportunity cost of capital*.

The argument for using the marginal rate of time preference is based on Figure 11.2, which depicts efficient intertemporal consumption for a household. The household has an indifference map composed of individual indifference curves (I_1, I_2 and I_3) for current consumption (C_1) and future consumption (C_2). Each indifference curve contains all combinations of C_1 and C_2 that provide the same level of household satisfaction. Higher indifference curves contain more preferred combinations of current and future consumption.

The marginal rate of commodity substitution (MRCS) measures how much future consumption the household requires to compensate for a decrease in current consumption without altering total satisfaction. The MRCS equals the slope of a line tangent to the indifference curve at a given point. When current consumption is high relative to future consumption, like at point a, only a relatively small amount of future consumption is needed to offset a one unit loss in current consumption. The MRCS is low at a. When current consumption is low relative to future consumption, like at point b, a relatively large amount of future consumption is needed to offset a one unit loss in current consumption. MRCS is high at b. Because MRCS increases from a to b, the indifference curve is convex (∪).

MRCS is defined as follows:

$$\text{MRCS} = dC_2/dC_1 = -(1 + v),$$

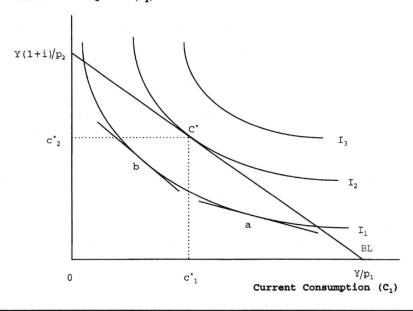

Figure 11.2. Efficient intertemporal consumption for a household. I = indifference curve; and BL = budget line.

where dC_2/dC_1 is the ratio of the change in future consumption to the change in current consumption, v is the marginal rate of time preference between current and future consumption. v increases (decreases) as the ratio of future consumption to current consumption increases (decreases). The expression for MRCS indicates that the household is willing to give up one unit of current consumption as long as it receives $(1 + v)$ units of future consumption. Alternatively, the household is indifferent between one unit of current consumption and $(1 + v)$ units of future consumption. Hence, v is the household's personal discount rate. For $v = 0.07$, the household is indifferent between $1.07 worth of consumption in the future period and $1 worth of consumption in the current period. As v increases (decreases), the absolute value of MRCS increases (decreases).

The budget line for the household, labeled BL in Figure 11.2, contains all combinations of C_1 and C_2 that can be purchased with a household income of Y. It is assumed that all of the household's income is received at the beginning of the current period. If the entire income is spent in the current period, the household can purchase Y/p_1 units in the current period, where p_1 is the price of goods in the current period. If income not spent in the current period earns interest at an annual rate of i, then the maximum consumption in the future period is $Y(1 + i)^t/p_2$ units, where $(1 + i)^t$ is the compound interest factor, p_2 is the price of goods in the future period and t is the time interval between the current and future periods, for example, three years.

In the simple case where there is only one time period separating the current period from the future period ($t = 1$), the budget constraint for the household is BL in Figure 11.2. BL is a straight line that contains all affordable combinations of consumption in the current period and consumption in the future period. The C_1 intercept of BL is the maximum amount of current consumption and the C_2 intercept of BL is the maximum amount of future consumption. The slope of the BL is:

$$dC_2/dC_1 = -[Y(1 + i)/p_2]/(Y/p_1).$$

The numerator is the intercept of BL on the vertical axis and the denominator is the intercept of BL on the horizontal axis of Figure 11.2.

When the prices are the same in both periods ($p_1 = p_2$):

$$dC_2/dC_1 = -Y(1 + i)/Y = -(1 + i).$$

This implies that increasing current consumption by one unit requires future consumption to be reduced by $(1 + i)$ units. Conversely, reducing current consumption by one unit allows future consumption to be increased by $(1 + i)$ units. The slope of the BL gives the rate at which current consumption can be exchanged for future consumption based on the market-determined interest rate (i).

Household equilibrium occurs at C* (C^*_1 in the current period and C^*_2 in the future period) where the BL is tangent to the highest possible indifference curve, namely I_2. At C*, the MRCS equals the slope of the BL. Because MRCS equals $-(1 + v)$ and the slope of the budget line equals $-(1 + i)$, household equilibrium requires:

$$-(1 + v) = -(1 + i) \text{ or } v = i.$$

When a household maximizes satisfaction subject to the budget constraint, the marginal rate of time preference (v) equals the market interest rate (i). Therefore, if the marginal rate of time preference is selected as the discount rate, then the market interest rate is the appropriate discount rate.

Typical firm-level product transformation curves for current (Q_1) versus future (Q_2) production are given in Figure 11.3. Each transformation curve (T_1, T_2 and T_3) represents the maximum amounts of current and future production that can be achieved with a given amount of manufactured and natural capital and a given technology. Higher transformation curves require greater capital investment and indicate greater production. Consider transformation curve T_1. When most of the capital is invested in current production, diverting capital from current production to future production increases future production more than it decreases current production, as indicated by the slope of the line tangent to T_1 at c.

Conversely, when most of the capital is invested in future production, diverting additional capital from current to future production decreases current production more than it increases future production, as indicated by the slope of the line tangent to T_1 at d. The slope of the product transformation curve equals the marginal rate of technical substitution (MRTS):

$$\text{MRTS} = dQ_2/dQ_1 = -(1 + m),$$

where dQ_2/dQ_1 equals the slope of a line tangent to the product transformation curve at a particular point and m equals the marginal opportunity cost of capital. m decreases (increases) as the ratio of future production to current production increases

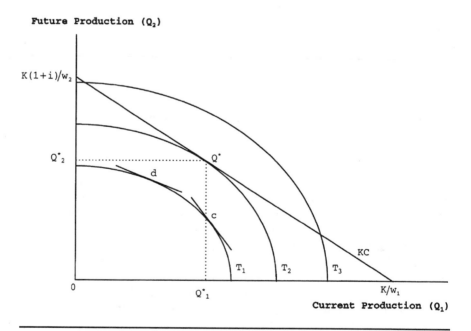

Figure 11.3. Efficient intertemporal production for a firm. T = transformation curve; and KC = capital constraint.

(decreases). Because the increase in production per unit increase in capital (marginal productivity of capital) decreases as the level of production increases, the absolute value of MRTS declines from c to d. Hence, the product transformation curve is concave (⌒).

The firm's most efficient combination of current and future production is determined by maximizing the amount of current and future production that can be achieved with a given amount of capital. Suppose there is a fixed amount of capital (K) available to the firm at the beginning of the current period. All possible allocations of capital between current and future production periods lie on the capital constraint line labeled KC in Figure 11.3. This constraint is derived as follows. The maximum level of current production equals K divided by the per unit cost of current production (w_1). If all capital is allocated to future production, then the original amount of capital (K) plus the interest on capital (Ki^t) are available in the future. This assumes that capital can be invested at the market interest rate of i. Hence, he maximum amount of capital available in the future period is $K(1 + i)^t$ and maximum future production is $K(1 + i)^t$ divided by the per unit cost of future production (w_2).

When there is only one time period separating the current period from the future period t = 1, the maximum capital available in the future period is K + Ki = K(1 + i), where Ki is the interest earned on K in one time period. For t = 1, KC is a straight line connecting current maximum production to future maximum production as shown in Figure 11.3. The slope of KC is:

$$dQ_2/dQ_1 = -K(1 + i)/K = -(1 + i).$$

This slope indicates that increasing capital use by one dollar in the current period reduces the amount of capital available for use in the future period by (1 + i) dollars.

Production equilibrium occurs at Q* (Q^*_1 in the current period and Q^*_2 in the future period) where KC is tangent to the highest product transformation curve, namely T_2. At the tangency point, MRTS equals the slope of KC. Because MRTS = −(1 + m) and the slope of KC equals −(1 + i), production equilibrium implies:

$$-(1 + m) = -(1 + i), \text{ or } m = i.$$

Therefore, in production equilibrium, the marginal opportunity cost of capital (m) equals the market interest rate (i). Hence, if the marginal opportunity cost of capital is selected as the discount rate, then the market interest rate is the appropriate discount rate.

When the household and firm are in equilibrium, the marginal rate of time preference and the marginal opportunity cost of capital both equal the market interest rate. This joint equilibrium is illustrated in Figure 11.4. Households maximize satisfaction by equating their marginal rates of time preference to the market interest rate. Firms maximize profits by equating their marginal opportunity cost of capital to the market interest rate. In practice, the marginal opportunity cost of capital is likely to exceed the marginal rate of time preference because of taxes, subsidies, externalities, uncertainties and imperfections in capital markets. Moreover, there are numerous interest rates in an economy that can be used as a basis for selecting the discount rate. One way to handle the ambiguity regarding the appropriate discount rate is to evaluate the economic feasibility of resource investments for a range of discount rates.

Discount rates used in the evaluation of public resource investments are typically less than discount rates used in evaluation of private resource investments. This occurs because public agencies often require that a certain discount rate be used in evaluating the social benefits of investments. In addition, the private discount rate tends to be higher than the public discount rate because the private rate is often adjusted for risk (to be discussed later) and taxes. Public agencies can spread the risks of different investments over a larger number of investments, and they do not pay taxes. For the same discount rate and stream of net benefits, the net present value of an investment is higher, with zero taxes. Thus, public agencies do not have to use as high a discount rate as do private companies to achieve the same net present value.

The preceding example involves two time periods, current and future. Hence, only one discount rate is needed. Benefits and costs of resource investments generally occur over several time periods during which interest rates, and hence discount rates, are likely to vary. Most benefit–cost analyses do not vary the discount rate over time because there is generally little economic basis for selecting a particular intertemporal pattern for the discount rate. A common procedure is to hold the discount rate constant over time, but to repeat the benefit–cost analysis for different discount rates. This procedure is known as *sensitivity analysis*.

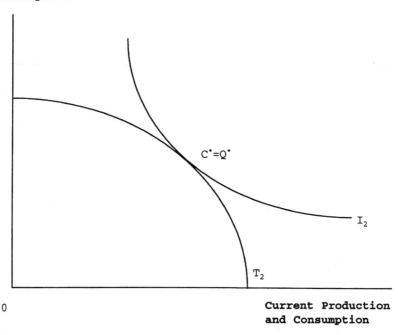

Figure 11.4. Equality of marginal rate of time preference and marginal opportunity cost of capital. I = indifference curve; and T = transformation curve.

CAPITAL AND OPERATING COSTS. Resource investments normally involve a capital cost and an operating cost. *Capital cost* is the lump sum payments made to acquire, develop and produce the resource. In the timber harvesting example, the conventional technology requires capital to construct roads and purchase trucks. The new technology requires capital to purchase helicopters. Other capital costs, such as machinery and equipment for felling trees, salaries for management staff, housing for workers, and property taxes or stumpage fees, are common to both technologies. *Operating cost* for both technologies includes the cost of fuel and oil, labor, replacement parts, insurance and other items.

There is no need to distinguish between capital and operating costs in present value calculations. Both are treated as outlays in the time period in which they occur. One offset to capital cost is *salvage value* which is the expected market value of capital assets at the end of the investment evaluation period. In the timber harvesting example, the expected value of the land (which includes the roads), trucks, helicopters and other capital assets are included in salvage value. Salvage value is treated as a benefit. Some capital assets might appreciate in value (land) and others might depreciate in value (machinery and equipment) due to wear and tear and/or obsolescence.

While normal accounting procedures treat depreciation or amortization of capital and interest payments on borrowed capital as expenses, such expenses are not considered in benefit–cost analysis. Including depreciation or amortization expenses on capital in net social benefit results is double counting because the initial cost of capital assets and their salvage value are already incorporated in net social benefit. Interest payments, which represent the cost of borrowed capital, are important in determining the financial feasibility but not the economic feasibility of a resource investment. Suppose company A can finance the new timber harvesting investment out of equity, whereas company B must use borrowed capital. The zero interest payments for company A and positive interest payments for company B have no bearing on whether investment in the technology is socially efficient or inefficient. However, interest payments do influence the financial feasibility of the investment. If company B projects a negative net return (inclusive of interest payments) and company A projects a positive net return on the investment, then the investment is unprofitable for B and profitable for A.

ECONOMIC VERSUS FINANCIAL FEASIBILITY. The last example leads to the distinction between *economic feasibility* and *financial feasibility*. An investment is economically feasible when net social benefit is positive where net social benefit includes all relevant benefits and costs generated by the investment and the discount rate is based on social rather than private considerations. An investment is financially feasible when gross monetary returns received by the investor are sufficient to cover all monetary costs incurred by the investor where costs include interest on borrowed funds and the discount rate is the interest rate paid on borrowed funds. It is possible for a resource investment to be economically feasible but not financially feasible and vice versa because a) gross benefits and/or costs for society differ from monetary returns and/or costs for the private investor and b) the social discount rate differs from the private discount rate. For example, suppose that an investment generates an external diseconomy. While the social cost of the investment includes the external diseconomy, the private cost excludes it.

LOCAL VERSUS GLOBAL EFFICIENCY. When resource managers are determining how best to allocate their capital budgets among alternative resource investments, they often limit their attention to investment opportunities in their particular region. For example, Grand Teton National Park and Yellowstone National Park, both located in northwestern Wyoming, each has an annual budget for resource investment within the park. If resource investments in each park are independently chosen based on efficiency criteria, then the net social benefit for each park is maximized. Local efficiency is achieved. Selecting investments in each park based on its local efficiency does not necessarily achieve the greatest overall gain in regional net social benefit, which is the sum of the net benefits for the two parks. Investments selected on the basis of local efficiency are likely to differ from investments selected based on regional efficiency.

Global efficiency takes a broader geographic view which entails looking at efficiency for substate, state, multistate or national areas. Simultaneously evaluating the economic feasibility of resource investments for all publicly managed natural resource areas including national parks, forests, wildlife refuges, wild and scenic rivers, wilderness areas, roadless areas and recreation areas in the United States is an example of resource investment analysis from a global perspective. If efficiency is increased by moving from the local to the global level, what is the rationale for making resource investment decisions at the local level?

First, it is less costly in terms of resource planning and economic analysis and more politically acceptable for local managers to identify, evaluate and select resource investments within their specific management area. Asking resource managers from different regions and/or agencies to evaluate and select resource investments in areas managed for different objectives is likely to result in conflict. While such conflict can be resolved, the transaction costs would be high. Second, the regional distribution of resource investments is likely to be inequitable when resource investment decisions are based on a global evaluation. For example, if public investment in soil conservation is allocated only to areas having the most serious erosion problems, investment would be concentrated in just a handful of states or regions.

INDEPENDENT VERSUS INTERDEPENDENT INVESTMENTS. Application of investment criteria is different for *independent* and *interdependent* investments. Independent investments satisfy two conditions. First, selection of one investment does not alter the net social benefit of the other investment. Second, selection of one investment does not prevent selection of the other investment except for financial reasons (capital rationing). Consider two investments: expanding a national park and creating a new wildlife refuge. These investments are independent, provided that a) expansion of the park does not alter the net social benefit of creating the wildlife refuge, b) creating the wildlife refuge does not affect the net social benefit of the park, c) selection of the park does not prevent creation of the wildlife refuge, and d) creation of the wildlife refuge does not prevent selection of the park. Conditions c and d can be violated, however, when investment funds are insufficient to undertake both investments.

Investments are interdependent when one or both of the conditions for independence are violated. Expanding a national park and expanding a wildlife refuge are interdependent investments when a) expanding the park (wildlife refuge) changes the net social benefit of the wildlife refuge (park) or b) expanding the park (wildlife refuge) interferes with the expansion of the wildlife refuge (park) for other than financial reasons. Condition "a" implies the two investments are competing. If the park and wildlife refuge are in the same general area, then net social benefit from park expansion is likely to be greater without expansion of the wildlife refuge because both resource investments provide wildlife viewing and outdoor recreation.

Condition "b" is relevant when the proposed expansions occur in the same area. In this case, the investments are *mutually exclusive*. Selection of the scale of an investment is also a mutually exclusive decision. Suppose three national park expansions are under consideration: 5,000, 10,500 and 30,000 acres (2,025, 4,253 and 12,150 hectares). Selection of any one scale precludes selection of the remaining two scales.

CAPITAL RATIONING. Capital rationing[4] occurs when the decision maker has a fixed capital budget that limits resource investment. The budget constraint is assumed to be effective only in the current period. In future periods, the decision maker is assumed to have the flexibility to acquire additional capital. There are two forms of capital rationing: maximum and specific. *Maximum rationing* requires that the decision maker not exceed the current budget. This form of rationing allows the decision maker to allocate surplus (uninvested) funds to noninvestment activities such as consumption in the case of households and dividends in the case of firms. Surplus government funds can be returned to the treasury. *Specific rationing* is more restrictive than maximum rationing. It requires the decision maker to spend all of the current budget on resource investments. Maximum rationing is more efficient than specific rationing when the return from investing additional funds is less than the value of those funds in noninvestment activities.

Capital rationing is a problem when the amount of current funds available to the decision maker is less than the amount of funds that can be invested at a reasonable return. This problem does not occur when the decision maker has the ability to request and/or borrow additional funds for worthwhile investments. A fixed and inadequate capital budget implies that the decision maker cannot borrow capital to increase current funds even when the projected future earnings from investment are sufficient to pay back the principal and interest on borrowed capital.

PRIMARY AND SECONDARY BENEFITS AND COSTS. Resource investments generate primary benefits and secondary benefits or costs. A primary benefit is the change in consumer plus producer surpluses in the primary market, which is the market most directly affected by the investment. A secondary benefit or cost is the change in consumer plus producer surpluses in secondary markets caused by changes in demand and/or supply in the primary market.[5] In the example of investing in a new timber harvesting technology (Figure 11.1), the primary benefit is the present value of the increase in consumer plus producer surpluses in the timber market. The secondary cost is the present value of the decrease in consumer plus producer surpluses in the plastic market. Net social benefit of investing in the new timber harvesting technology is the primary benefit minus the secondary cost.

When primary and secondary markets are complementary rather than competitive, net social benefit is the sum of primary and secondary benefits. There are no secondary costs. An example of a secondary benefit is the flood protection benefit (decrease in potential flood damages) associated with the construction of a dam and reservoir to supply drinking water for a rural community. Flood protection is a secondary benefit as long as the dam and reservoir are constructed for the primary purpose of supplying drinking water.

There are two types of secondary economic benefits or costs: *technological spillovers* and *pecuniary spillovers*. Technological spillovers are real increases or decreases in productivity caused by resource investments. Improvements in water quality resulting from the new timber harvesting technology and the reduction in flood damages from the water supply reservoir are technological spillovers. Pecuniary spillovers are changes in capital values and/or incomes induced by resource investments. Increases in the recreational value of mountainous forested areas resulting from the new timber harvesting technology or increases in local labor wages caused by the construction of the dam and reservoir are pecuniary spillovers. Technological spillovers are included and pecuniary spillovers are excluded from net social benefit.[6]

Labor income generated by a resource investment is sometimes treated as technological spillover when the workers hired to support the resource investment would be otherwise unemployed. This procedure has two potential drawbacks. First, it is inappropriate to treat the wages paid to previously unemployed workers as a national benefit when the proposed investment simply displaces another investment that might have employed the same workers. Second, it overstates benefits when the employment of surplus labor contributes to inflation. The latter condition is very common when resource investment occurs in areas having a relatively small economic base.

RISK AND UNCERTAINTY. When there is complete knowledge about how a resource investment affects demand and supply for the resource, the net social benefit of the investment is known with complete certainty (as assumed in Figure 11.1). Complete certainty is very rare. Normally, changes in demand and/or supply are subject to risk or uncertainty. A resource investment decision is risky when there are several possible outcomes for changes in demand and/or supply and probabilities can be attached to these outcomes. Uncertainty prevails when probabilistic statements cannot be made regarding demand and/or supply outcomes. A common way to handle risk and uncertainty in resource investment analysis is to increase the discount rate. This approach has several disadvantages. First, the amount of the increase is arbitrary. While the difference in the borrowing rate between a riskless and risky investment might be used for this purpose, the uniqueness of most resource investments tends to invalidate this procedure. Second, when the interest rate is incremented for risk, the increment gets compounded over time. Such compounding overcompensates or undercompensates for risk when there is year-to-year variability in risk. Finally, not all benefits and costs are subject to the same degree of risk or uncertainty. Incrementing the discount rate for risk assumes that all benefits and costs exhibit the same degree of risk and uncertainty.

The preferred way to handle risk is to adjust benefits and costs separately for risk and to discount them using a riskless discount rate. Introducing risk in the timber harvesting example requires the decision maker to assign a probability to each and every possible reduction in marginal social cost (MSC) of harvesting with the new technology. Table 11.2 illustrates how to determine the expected decrease in MSC for three hypothetical reductions in MSC and their respective probabilities of occurrence. The first column gives the possible decreases in MSC, and the second column contains the respective probabilities of a decrease. The expected decrease in MSC (third column) is the decrease in MSC multiplied by its respective probability.

Summing the figures in the third column of Table 11.2 gives the overall expected decrease in MSC, namely, $19.50 per unit. The latter is used in calculating the expected change in consumer and producer surpluses in the timber market due to the new technology. A similar procedure is used to estimate the expected decrease in net social benefit due to a fall in the demand for plastic. In this case, risk arises in terms of the extent to which the decrease in timber price reduces the demand for plastic and, hence, consumer plus producer surpluses for plastic.

Alternative Investment Evaluation Criteria

This section addresses the strengths and weaknesses of the following financial or economic criteria for evaluating the feasibility of resource investments:

Payback period
Average rate of return
Net present value
Annual net benefit
Benefit–cost ratio
Internal rate of return

PAYBACK PERIOD. The *payback period* is a financial rather than an economic criteria. The payback period is the number of years it takes to recoup an initial investment, namely:

$P = V/APR,$

Table 11.2. Expected decrease in marginal social cost (MSC) for new timber harvesting technology

Decrease in MSC ($)	Probability of Decrease	Expected Decrease in MSC[a] ($)
10	0.20	2.00
15	0.25	3.75
25	0.55	13.75
Sum	1.00	19.50[b]

[a]Decrease in MSC times probability.
[b]Overall expected reduction in MSC.

where P is the payback period in years, V is the initial investment cost, and APR is the average annual private return, which equals the average monetary return on the investment minus average monetary costs of the investment excluding investment cost. Investments with a shorter payback period are judged to be superior to investments with a longer payback period.

The payback period is a very crude financial criterion because it a) does not consider the economic life of the investment and b) ignores the time value of money. Regarding the first criticism, consider two projects that cost $25,000 and yield an APR of $5,000 per year. The payback period is five years ($25,000/$5,000). Based on the payback period, both investments are equally attractive. Suppose the first investment has an economic life of 8 years and the second has an economic life of 15 years. The second investment is more profitable because it yields a return for 10 years beyond the payback period compared with only three years for the second investment. Total net return is $50,000 (10 × $5,000) for the first investment versus $15,000 (3 × $5,000) for the second investment.

Turning to the second criticism, suppose both investments have a 10-year economic life, the annual maintenance cost is $3,000 higher with the first investment and the salvage value is $30,000 lower with the second investment. When the time value of money is ignored, the $30,000 higher maintenance cost over the 10-year economic life of the first investment just equals the $30,000 (10 × $3,000) lower salvage value for the second investment. The two projects have the same cash return; however, for an 8 percent discount rate, the present value of the higher maintenance cost ($20,130) exceeds the present value of the lower salvage value ($13,896), which means the first investment has a lower present value than the second investment. After accounting for the time value of money, the second investment is preferred to the first investment despite the fact both investments have the same payback period.

AVERAGE RATE OF RETURN. The *average annual rate of return* is also a financial criterion. It equals the average annual net private return as a percentage of the initial investment cost, namely:

$$R = (APR/V)100.$$

The average annual rate of return has a similar deficiency as the payback period. These deficiencies are illustrated for three investments in Table 11.3. Investments A, B and C have an average annual rate of return of 20 percent. The first deficiency is that the average annual rate of return is positive despite the fact that each investment generates a cash loss of $4,000 ($6,000 − $10,000). A cash loss implies a negative rate of return. This inconsistency arises because the average annual rate of return does not consider that part of the cash return must be used to pay back the initial investment.

The second deficiency of the average annual rate of return is that it ignores the time value of money. For any positive discount rate, investment A has the highest, investment B has the second highest and investment C has the lowest present value of annual net cash return. While the average annual rate of return shows that all

Table 11.3. Average rates of return for three investments

Year (t)	Investment A ($)	Investment B ($)	Investment C ($)
0	−10,000	−10,000	−10,000
1	3,000	2,000	1,000
2	2,000	2,000	2,000
3	1,000	2,000	3,000
Rate of Return (%)	20	20	20

three investments are equally preferred, accounting for the time value of money by discounting the cash flows indicates that A is more profitable than B and C and that B is more profitable than C. The average annual rate of return is a valid financial criterion (the two deficiencies are not present) in very restrictive cases where the investments being compared have uniform cash flows, like investment B, and there is either a 100 percent salvage value or a perpetual cash flow.

NET PRESENT VALUE. The net present value (NPV) of a resource investment in discrete time is:

$$NPV = \sum_{t=0}^{T}(V_t - C_t)(1 + r)^{-t},$$

where V_t equals the primary and secondary benefits and C_t equals the secondary cost generated by the investment at the end of time period t, r is the discount rate or interest rate corresponding to the length of each period (if t is a year, then r is an annual rate) and T is the length of the evaluation period. $V_t - C_t$ is called the net benefit in time period t. Resource investments are economically feasible when their NPV is positive and economically infeasible when their NPV is zero or negative.

The primary benefit of a resource investment is the change in consumer plus producer surpluses in the primary market. A secondary benefit (cost) is the increase (decrease) in consumer plus producer surpluses in secondary markets. When there is insufficient information to estimate consumer and producer surpluses, benefits and costs are defined differently. It is quite common to define V_t as the total economic value of production and C_t as the total cost of production in primary and secondary markets with the investment. An example of how net present value is calculated based on this definition is given in Table 11.4. Net benefit in each period equals (V − C) and discounted net benefit in each period equals (V − C) times the discount factor (f). The NPV of the investment is the sum of the discounted net benefits over the four-year evaluation period, namely $923. Because NPV is positive, the investment is economically feasible.

An example of benefit-cost analysis is Freeman's evaluation of the Clean Water Act (CWA). He estimated CWA's annual benefit to be between $5.7 billion and $27.7 billion, with a most likely estimate of $14 billion (as of 1985). Annual benefit included enhancement in recreation, aesthetic and property values and commercial fishing as well as the reduced cost of treating wastes and supplying drinking water. Cost of complying with the CWA was estimated to be $17.4 billion in 1979 and $33.4 billion in 1988, with an annual average cost for the decade of $23.2 billion.

Table 11.4. Calculation of net present value of an investment

Year (t)	Benefit (V) ($)	Cost (C) ($)	Net Benefit (V – C) ($)	Discount Factor[a] (f)	Discounted Net Benefit (V – C)(f) ($)
0	0	10,000	–10,000	1.000	–10,000
1	3,000	500	2,500	0.952	2,380
2	3,000	500	2,500	0.907	2,268
3	3,000	500	2,500	0.864	2,160
4	3,000 +2,000[b]	0	5,000	0.823	4,115
Sum	14,000	11,500	2,500	—	923

[a] $(1 + 0.05)^{-t}$.
[b] Salvage value.

Estimated net benefit of the CWA is –$9.2 billion ($14 billion – $23.2 billion).[7] Based on Freeman's analysis, the cost exceeds the benefits of the CWA.

ANNUAL NET BENEFIT. A variant of net present value is the *annual net benefit* (ANB). The ANB equals the uniform annual payment over the life of the investment, which has a net present value just equal to the net present value of the investment. ANB is the solution to the following equation:

$$\sum_{t=0}^{T} ANB(1 + r)^{-t} = NPV,$$

namely:

$$ANB = NPVr(1 + r)^T / [(1 + r)^T - 1].$$

An investment is deemed economically feasible when ANB is positive and economically infeasible when ANB is zero or negative. Because the expression $r(1 + r)^T / [(1 + r)^T - 1]$ is positive, ANB has the same algebraic sign as NPV provided r is positive. Therefore, any investment that is economically feasible based on the NPV criterion is economically feasible with the ANB criterion. The two criteria are equivalent. For the benefits and costs given in Table 11.4, ANB = $260.30 ($923 × 0.282), which implies that the investment is economically feasible.

BENEFIT–COST RATIO. The benefit–cost ratio (BCR) for a resource investment is the ratio of the present value of benefits to the present value of costs. Specifically:

$$BCR = \frac{\sum_{t=0}^{T} V_t(1 + r)^{-t}}{\sum_{t=0}^{T} C_t(1 + r)^{-t}}.$$

An investment is economically feasible if its BCR is greater than 1 and economically infeasible if its BCR is less than or equal to 1. For the benefits and costs given in Table 11.4, BCR = 1.08. Because BCR exceeds 1, the investment is economically feasible.

To show the equivalence of the BCR, NPV and ANB criteria, NPV is rewritten as follows:

$$\text{NPV} = \sum_{t=0}^{T} V_t(1+r)^{-t} - \sum_{t=0}^{T} C_t(1+r)^{-t},$$

where the first term on the right of the equality is the present value of benefits (PVB) and the second term is the present value of costs (PVC). When NPV is positive (negative), PVB is greater (less) than PVC, which implies that BCR is positive (negative). In addition, when ANB is positive (negative), NPV is positive (negative), which makes BCR positive (negative). Therefore, the NPV, ANB and BCR criteria lead to the same conclusions regarding the economic feasibility of resource investments provided the discount rate is positive.

INTERNAL RATE OF RETURN. The internal rate of return (IRR) is defined as the discount rate that makes the NPV of an investment just equal to zero. The IRR is found by solving the following equation for ρ:

$$\sum_{t=0}^{T}(V_t - C_t)(1+\rho)^{-t} = 0.$$

ρ is also the discount rate that makes ANB equal to zero and BCR equal to 1. An investment is economically feasible (infeasible) when ρ is greater than (less than or equal to) the discount rate. For the benefits and costs given in Table 11.4, $\rho = 0.0847$.

For the 5 percent discount rate used in Table 11.4, the investment is economically feasible because IRR exceeds the discount rate (0.0847 > 0.05). In this example, the IRR criterion leads to the same conclusion (economic feasibility) as the NPV, ANB and BCR criteria. When capital costs occur early in the evaluation period and benefits are spread out over the entire evaluation period (such as in Table 11.4), the IRR is usually unique and leads to the same investment decision as the NPV, ANB and BCR criteria. This is not always the case. The present value of the net benefits −100, +210, −110 is zero for $\rho_1 = 0$ and $\rho_2 = 0.10$. There are two solutions for ρ. If the discount rate is greater than zero or less than 10 percent, then the investment is economically infeasible based on ρ_1 but economically feasible based on ρ_2. Some net benefits do not have a nonimaginary solution for ρ.[8] For these reasons, the IRR criterion is not a reliable criterion for evaluating the economic feasibility of resource investments.

EXAMPLE. Resource investment analysis involves calculating present values for a wide variety of net benefits. This section employs the financial factors given in Table 11.5 to calculate net present value. The financial factors are for an 8 percent discount rate (r = 0.08) and a 15-year evaluation period (T = 15).

For ease of calculation, cash values are in relatively small denominations. Suppose the conventional timber harvesting investment generates the cash flows given in Table 11.6.

Net present value of the investment is the present value of cash benefits minus the present value of cash costs. Because the $50,000 in investment cost occurs in the beginning of the evaluation period, it does not need to be discounted. Present value

Table 11.5. Financial factors for calculating net present values, r = 0.08 and T = 15

Present Value of 1	Amortization Factor	PV of Annuity of 1 per year	PV of Increasing Annuity of 1 per year	PV of Decreasing Annuity of 1 per year
0.31525	0.11683	8.55948	56.44514	80.50652

PV = present value.

Table 11.6. Cash flows for conventional timber harvesting investment ($)

Item	Amount
Investment cost	50,000
Annual benefit	8,000
Annual O&M cost	3,000
Salvage value	10,000

O&M = operating and maintenance.

of the annual benefit of $8,000 equals the benefit times the present value of an annuity of 1 per year (column 3 in Table 11.5). Therefore:

PV($8,000) = ($8,000)(8.55948) = $68,475.84.

Likewise, the present value of the annual operating and maintenance (O & M) cost of $3,000 is:

PV($3,000) = ($3,000)(8.55948) = $25,678.44.

The present value of the salvage value is $10,000 times the present value of 1 (column 1 in Table 11.5):

PV($10,000) = ($10,000)(0.31525) = $3,152.50.

NPV of the investment is:

NPV = $68,475.84 + $3,152.50 − $25,678.44 − $50,000 = −$4,050.10.

ANB of the investment equals NPV times the amortization factor (column 2 in Table 11.5):

ANB = (−$4,050.10)(0.11683) = −$473.17.

BCR is present value of benefits divided by present value of costs:

BCR = ($68,475.84 + $3,152.50)/($25,678.44 + $50,000) = 0.946.

Finally, IRR = 0.03 for this investment.

In conclusion, the conventional timber harvesting investment is not economically feasible from a private viewpoint because NPV and ANB are negative, BCR is less than 1 and IRR is less than the discount rate. In this example, the IRR criterion is consistent with the other three criteria.

Changes in the benefits and costs, discount rate and evaluation period can change the investment decision. For example, instead of being constant at $3,000 per year, suppose O & M cost increases at a constant rate from $500 in the first year to $4,000 in the 15th year as indicated by line C in Figure 11.5. Present value of the increasing annual O & M cost is the sum of the present value of $500 per year for 15 years as indicated by line A in Figure 11.5 plus the present value of a cost that increases from $0 in the first year to $3,500 in the 15th year as indicated by line B in Figure 11.5. The present value of $500 per year is $500 times the present value of an annuity of 1 per year (column 3 in Table 11.5):

PV($500) = ($500)(8.55948) = $4,279.74.

Present value of the increasing portion of O & M costs is the average annual increase in O & M cost times the present value of an increasing annuity of 1 per year (column 4 of Table 11.5):

PV(increasing O & M cost) = [($3,500)/15][56.44514] = $13,170.53.
Present value of the increasing O & M cost (line C in Figure 11.5) is $17,450.27 ($4,279.74 + $13,170.53).

Net present value of the timber harvesting investment with increasing O & M cost is:

NPV = $68,475.84 + $3,152.50 − $17,450.27 − $50,000 = $4,178.07.

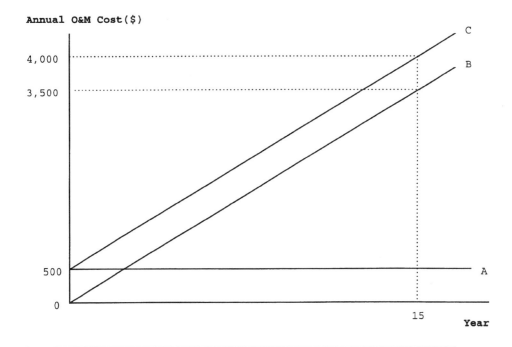

Figure 11.5. Increasing operating and maintenance (O&M) cost for conventional timber harvesting investment. C = O&M cost.

The investment has a positive NPV when O & M cost increases from $500 to $4,000 per year but a negative NPV when O & M cost is constant at $3,000 per year. Therefore, the investment is economically justified from a private viewpoint with this increasing pattern of O & M cost.

Consider the net social benefit of the conventional timber harvesting investment. Suppose the conventional timber harvesting investment causes sedimentation of streams that adversely affects fish reproduction and growth. Let the annual environmental damage (ED) from sediment decrease over time from $1,000 in the first year to $0 in the 15th year. Present value of environmental damage (PED) equals the annual average decrease in damages times the present value of a decreasing annuity of one dollar per year (column 5 in Table 11.5):

PED = [($1,000 − $0)/15][80.50652] = $5,367.10.

Net social benefit of the investment with increasing O & M cost is:

NSB = NPV − PED = $4,178.07 − $5,367.10 = −$1,189.03.

The conventional timber harvesting technology is not socially efficient because it has a negative NSB. Therefore, the investment in the conventional timber harvesting technology is privately efficient but socially inefficient.

Evaluation of Independent and Interdependent Investments

Caution must be used when applying the NPV, ANB, IRR and BCR criteria. One caution has already been mentioned in connection with the IRR. The IRR can be ambiguous regarding economic feasibility when it is applied to investments that do not have a normal cash flow. The latter has a capital cost that occurs early in the evaluation period and benefits that are spread throughout the evaluation period. This section examines in detail the application of the NPV, IRR and BCR criteria to independent and interdependent investments.

INDEPENDENT INVESTMENTS. An independent investment is economically feasible when NPV is positive, IRR is greater than the discount rate or the BCR is greater than 1. In other words, all the evaluation criteria are appropriate for independent investments.

When there is no capital rationing, independent investments that are economically feasible are justified. With specific or maximum capital rationing, there is unlikely to be sufficient funds in the current period to undertake all economically feasible investments. Suppose a capital budget of $40,000 is to be allocated among six independent investments. The evaluation period is five years (T = 5) and the salvage value is 100 percent. Table 11.7 summarizes the investment cost, annual benefit, net present value and internal rate of return for the six independent investments.

11. Benefit–Cost Analysis of Resource Investments

Table 11.7. Investment cost, annual benefit, net present value and internal rate of return for six independent investments

Investment Option	Investment Cost ($)	Annual Benefit ($)	Net Present Value ($) for a discount rate			IRR (%)
			of 10%	of 12%	of 15%	
A	10,000	2,800	6,824	5,768	4,358	28
B	8,000	2,000	4,549	3,749	2,682	25
C	15,000	3,000	5,585	4,326	2,514	20
D	4,000	720	1,213	793	402	18
E	22,000	3,300	4,170	2,379	0	15
F	9,000	900	0	−649	−1,508	10

IRR = internal rate of returns.

The NPV criterion is applied by raising the discount rate until the cost of the economically feasible investments just exhausts the capital budget. For a 10 percent and 12 percent discount rate, investments A through E are economically justified, but the total investment cost exceeds the capital budget ($59,000 > $40,000). When the discount rate is increased to 15 percent, investments A through D are justified and the total investment cost is less than the capital budget ($37,000 < $40,000). There is $3,000 of unused capital budget. With maximum capital rationing, the $3,000 surplus can be dispensed as a dividend in the case of private investment, or returned to the treasury in the case of public investment. If there is specific capital rationing, then a $3,000 investment must be found in order to exhaust the current budget. The opportunity cost of capital or the loss in net return from capital rationing is the IRR on the next investment that would be selected without a capital constraint, namely, 15 percent for investment E.

The IRR criterion is applied by first ranking all investments in descending order of IRR. Investments are then selected in the order of their IRR until the capital budget is just exhausted (specific capital rationing) or nearly exhausted (maximum capital rationing). Investments A through D satisfy this decision rule. While not shown here, the same five investments are selected with the ANB and BCR criteria.

Suppose a seventh investment (G) is identified that costs $7,000 and has an IRR of 19 percent. By not selecting investment D, $4,000 is freed up, which, together with the $3,000 surplus, is just sufficient to undertake investment G. Investment G is superior to investment D because it has a higher IRR (19 percent > 18 percent).

INTERDEPENDENT INVESTMENTS. Interdependent investments are categorized as type I or type II. For a type I investment, choosing A influences the net benefits for B, or vice versa. For a type II investment, choosing A prevents selection of B, or vice versa, which means the projects are mutually exclusive.

Type I Investments. Five cases of a type I interdependent investment are examined using the NPV criterion in Table 11.8. The NPV of B is increased by $50 in Case 1 and reduced by $75 in Case 2 when A is undertaken. Because the NPV of B is still positive after A is undertaken ($150 in Case 1 and $25 in Case 2), B is still economically justified. The $50 increase in B's NPV in Case 1 is credited to A's NPV and the $75 loss in B's NPV in Case 2 is charged against A's NPV. In Case 3,

A causes the NPV of B to become negative, which means B is no longer economically justified when A is undertaken. The $100 present value loss from abandoning B is charged against the NPV of A. In Case 4, B is not economically justified before or after A, so there is no need to adjust the NPV of A. In Case 5, B is not economically justified before A but it is economically justified after A. The $35 NPV of B, which occurs when A is undertaken, is credited to the NPV of A.

Type II Investments. Application of the economic criteria need to be modified when investments are mutually exclusive. Specifically, the investment criteria have to be applied to incremental benefits and costs rather than to the absolute level of benefits and costs. Economic feasibility of two mutually exclusive investments (A and B) are evaluated for T = 5 and 100 percent salvage value in Table 11.9. Comparing the overall NPVs for the two investments leads to the same decision as comparing the NPVs for incremental net benefits. First, compare the overall NPVs. For $r < 18$ percent, both investments have a positive NPV and $NPV_B > NPV_A$, which implies B is superior to A. When r is between 18 percent and 24 percent, both investments have a positive NPV and $NPV_A > NPV_B$, which implies A is superior to B. For r between 24 and 30 percent, $NPV_A > 0$, but $NPV_B < 0$, which makes B economically infeasible. Hence, when r is between 18 percent and 30 percent, investment A is superior to investment B.

Second, suppose the investment decision is based on incremental NPVs. Incremental NPV equals the present value of the incremental benefit minus the present value of the incremental cost. The first possible increment of investment is $1,000 on A. Investment A results in an incremental NPV equal to the NPV of $300 for five years plus the NPV of $1,000 of salvage value in year 5 minus $1,000 of investment cost. The second possible increment of investment, designated as B-A, is to invest an additional $1,000 to achieve B. The NPV of the second increment, designated NPV_{B-A}, equals the present value of $180 ($480 − $300) for five years, plus the present value of $1,000 of additional salvage value in year 5 minus $1,000 of incremental investment cost.

For $r < 18$ percent, $NPV_{B-A} > 0$ and increment B-A is economically feasible.

Table 11.8. Type I interdependence between investments A and B

Case	Present Value of B ($) Before A	After A	Adjustment to Present Value of A ($)
1	100	150	50
2	100	25	−75
3	100	−10	−100
4	−50	−100	None
5	−50	35	35

Table 11.9. Investment cost and annual benefit for two mutually exclusive investments

Investment	Investment Cost ($)[a]	Annual Benefit ($)[b]
A	1,000	300
B	2,000	480

[a] 100 percent salvage value.
[b] For 5 years.

11. Benefit–Cost Analysis of Resource Investments

Selecting increment B-A means choosing investment B. For r between 18 percent and 30 percent, $NPV_{B-A} \leq 0$ and increment B-A is not economically feasible. Only investment A is selected. Therefore, the same investment decisions are reached regardless of whether overall NPV or incremental NPV is utilized. This is not the case for the IRR and BCR criteria.

Straightforward application of the overall IRR criterion leads to incorrect investment choices. When the annual benefit is constant and salvage value is 100 percent of initial investment cost, as in Table 11.9, overall IRR is the ratio of the annual benefit to the investment cost times 100. The overall IRR for investments A and B are 30 percent [($300/$1,000) × 100] and 24 percent [($480/$2,000) × 100], respectively, which implies A is superior to B for a discount rate less than 24 percent. As the NPV criterion indicates, A is superior to B only when r is between 18 and 30 percent.

To obtain correct decisions, the IRR criterion must be applied to investment increments A and B-A. Increment A yields an incremental annual return of $300 for five years and entails an incremental investment cost of $1,000. Increment B-A provides an incremental annual benefit of $180 ($480 − $300) for five years and requires an incremental investment cost of $1,000 ($2,000 − $1,000). The incremental IRRs (IIRRs) for A and B-A are:

$IIRR_A = (\$300/\$1,000)100 = 30$ percent, and

$IIRR_{B-A} = (\$180/\$1,000)100 = 18$ percent.

The IIRR-based decision rule is:

Accept (B-A) when r < 18 percent.
Accept A when 18 percent ≤ r ≤ 30 percent, and
Reject both A and (B-A) when r > 30 percent.

This is the same decision rule used with the NPV criterion.

The overall benefit–cost ratio (BCR) leads to incorrect decisions regarding the selection of mutually exclusive investments. The overall BCR is the ratio of the present value of annual benefit plus salvage value to investment cost. For r = 0.16, the BCRs of investments A and B are:

$BCR_A = \$1,639/\$1,000 = 1.64$, and
$BCR_B = \$2,813/\$2,000 = 1.41$.

For r = 0.20, the BCRs for investments A and B are:

$BCR_A = \$1,299/\$1,000 = 1.29$, and
$BCR_B = \$2,239/\$2,000 = 1.12$.

These BCRs indicate that both investments are economically feasible because $BCR_A > 1$ and $BCR_B > 1$; however, investment A is superior to investment B because $BCR_A > BCR_B$. This result is inconsistent with the NPV criterion, which shows that B is superior to A when r < 18 percent and A is superior to B when r is between 18 percent and 30 percent. Therefore, the overall BCR is not valid for mutually exclusive investments.

Valid results for economic feasibility can be achieved by using the incremental BCR criterion. When r = 0.16, which is less than 0.18, BCR_A = 1.69 and BCR_{B-A} = 1.24. It is efficient to invest in B because BCR_{B-A} > 1. When r = 0.20, which is between 0.18 and 0.30, BCR_A = 1.3 and BCR_{B-A} = 0.94, which implies that investment in A is efficient but investment in (B-A) is inefficient. For r > 0.30, BCR_A and BCR_{B-A} are both less than one and neither A nor B-A are efficient. The incremental BCR criterion leads to the same investment decision as the incremental IRR and NPV criteria. The decision rule based on the incremental BCR is:

Accept A when BCR_A > 1,
Accept (B-A) when BCR_{B-A} > 1, and
Reject both A and (B-A) when BCR_A < 1 and BCR_{B-A} < 1.

In summary, applying the investment criteria to mutually exclusive investments for which incremental IRR > r or incremental BCR > 1 ensures that the increment adds more to benefits than to costs, which increases NPV.

Evaluation of Multiple Resource Investments

The incremental investment criteria for mutually exclusive investments are easily generalized to more than two investments. An example of the incremental IRR criterion is given in Table 11.10. The example assumes there is no capital constraint.

Multiple investments are listed from lowest to highest investment cost. The overall IRR is highest for A and lowest for D. Results of the incremental IRR analysis are given in the last two columns of the table. Because investment cost increases from A through D, incremental IRRs are calculated between no investment and A (A-0), between A and B (B-A), between B and C (C-B) and between B and D (D-B). Because the incremental IRR for C-B is less than any reasonable discount rate, it is not economically feasible to invest in C-B, which means this increment is bypassed. In this case, the increment D-C is irrelevant.

Table 11.10. Incremental IRRs for four mutually exclusive investments[a]

Investment Option	Investment Cost ($)	Annual Benefit ($)	Overall IRR ($)	Investment Increment	Incremental IRR (%)
A	981,800	400,000	40.7	A-0	40.7
B	1,178,160	432,000	36.6	B-A	16.3
C	1,374,520	434,000	31.4	C-B	1.0
D	1,570,880	480,000	30.4	D-B	12.2

[a]Twenty-year evaluation period (T = 20) and 100 percent salvage value.
IRR = internal rate of return.

Incremental IRRs are calculated based on the increment in investment cost and the increment in annual benefit. For example, the incremental IRR for B-A is determined using the increment in investment cost of $196,360 ($1,178,160 − $981,800) and the corresponding increment in annual benefit of $32,000 ($432,000 − $400,000) between A and B. The incremental IRR for B-A is 16.3 percent.

For an 8 percent discount rate, A-0 is efficient because the incremental IRR of 40.7 exceeds 8, B-A is efficient because 16.3 > 8, C-B is inefficient because 1.0 < 8, and D-B is efficient because 12.2 > 8. Because B-A and D-B are efficient, investment D is selected. In fact, investment D is economically feasible when the discount rate is less than 12.2 percent, which is the incremental IRR on the last increment selected, namely, D-A.

Incremental BCRs for the same four mutually exclusive investments are given in Table 11.11. The calculations are based on an 8 percent discount rate and a 20-year evaluation period. The incremental BCR equals the present value of the incremental annual benefit plus the present value of the incremental salvage value divided by the present value of the incremental investment cost. As long as the BCR for an increment is greater than 1, it is efficient to invest in that increment. A-0 is efficient because $BCR_{A-0} = 4.22 > 1$, B-A is efficient because $BCR_{B-A} = 1.85 > 1$, C-B is not efficient because $BCR_{C-B} = 0.32 < 1$, and D-B is efficient because the $BCR_{D-B} = 1.42 > 1$. Selecting increments A-0, B-A and D-B results in investment D. This is the same investment selected using the incremental IRR criterion.

Capital rationing can prevent the selection of some investments. For the four mutually exclusive investments given in Tables 10.10 and 10.11, a current capital budget of $1.2 million only allows investments A and B to be selected. Investment D is not financially feasible, even though it is economically feasible, because the cost of D exceeds the capital budget ($1,570,880 > $1,200,000). However, if there exists an investment E that costs less than $1.2 million and for which $IIRR_{E-A} > 16.3$ percent or $BCR_{E-A} > 1.85$, then investing in E may be economically superior to investing in B.

Because the NPV, incremental IRR and incremental BCR criteria lead to selection of the same mutually exclusive investment, is there any basis for preferring one of these criteria? Provided the decision maker can specify a rate of discount or a range of discount rates, the NPV criterion has the advantage of being direct and easy to apply, especially when multiple investments are being evaluated. To apply the NPV criterion, the NPVs are calculated for the alternative investments and the investment with the highest NPV is selected.

Table 11.11. Incremental BCRs for four mutually exclusive investments[a]

Investment Increment	Incremental Investment Cost ($)	Incremental Annual Benefit ($)	PV of Incremental Benefit ($)	Incremental BCR[b]
A–0	981,800	400,000	210,645[c] + 3,927,200[d]	4.22
B–A	196,360	32,000	42,129 + 314,176	1.85
C–B	196,360	2,000	42,129 + 19,636	0.32
D–B	392,720	18,000	84,258 + 471,264	1.42

[a]Twenty-year evaluation period, 8 percent discount rate and 100 percent salvage value.
[b]Column 3 divided by column 1.
[c]Present value of incremental salvage value (column 1).
[d]Present value of incremental annual benefit (column 2).
BCR = benefit–cost ratio; and PV = present value.

Summary

Public agencies and private firms make investments to develop, preserve, enhance, restore and protect natural and environmental resources. The primary objective of private resource investments is to maximize net private benefit (usually profit) subject to a capital constraint. Public resource investments are selected so as to maximize net social benefit subject to budgetary, ecological, equity and other constraints. Private and public resource investments can be evaluated in terms of their economic efficiency. Net social benefit, which is a measure of social economic efficiency, equals the present value of the changes in consumer plus producer surpluses in all markets affected by the investment minus the present value of the incremental fixed cost of the investment. Net social benefit is expressed in either discrete or continuous time. Calculations made in discrete (continuous) time utilize a discrete (continuous) discount factor. The evaluation period for an investment is the period of time during which the investment yields benefits and costs. An investment is deemed socially efficient when it has a positive net social benefit, and, socially inefficient when it has a zero or negative net social benefit.

The discount rate influences the economic efficiency of a resource investment. Lowering (raising) the discount rate increases (decreases) the net present value and economic feasibility of investments that provide benefits over an extended period of time. Economic theory provides justification for selecting a discount rate that equals the marginal rate of time preference or marginal opportunity cost of capital. Both rates are equal to each other and to the market interest rate when households maximize their satisfaction subject to a budget constraint and firms maximize their returns subject to a capital constraint in the absence of taxes, subsidies, externalities, uncertainties and imperfections in capital markets. Because there are many interest rates in the economy, benefit–cost analysis is often based on a range of discount rates. Discount rates used to evaluate public investments are generally lower than discount rates used to evaluate private investments because public agencies are often mandated to use a specific discount rate that is generally lower than the private rate, they can spread risk over many investments, and they are not required to pay taxes.

In determining the economic efficiency of an investment, there is no need to distinguish between capital and operating costs or to consider depreciation or amortization of capital and interest payments on borrowed capital. An investment is economically feasible when net social benefit is positive, but it is financially infeasible when gross monetary returns are less than capital and operating costs. Investments may be efficient at a local level but inefficient at a global level. Resource investments can be either independent or interdependent. Investments are independent when selection of one investment neither alters the net social benefit of any other investment nor prevents selection of other investments except for financial reasons. Investments are interdependent when selection of one investment alters the net social benefit of other investments (type I) or prevents other investments from being selected for nonfinancial reasons (type II). Specific capital rationing implies that all of the current budget must be spent on resource investments. Maximum capital rationing only requires that the current budget not be exceeded.

Resource investments generate primary benefits and secondary benefits or costs. A primary benefit is the increase in consumer plus producer surpluses in the primary market, which is the market most directly affected by the investment. A secondary benefit (cost) is the increase (decrease) in consumer plus producer surpluses in secondary markets. The preferred way to handle risk and uncertainty is to adjust individual benefits and costs for risk or uncertainty and to discount risk-adjusted benefits and costs using a riskless discount rate.

Four economic-based criteria are commonly used to evaluate resource investments: net present value, annual net benefit, benefit–cost ratio and internal rate of return. Net present value is the present value of primary and secondary benefits minus the present value of secondary costs. Annual net benefit is the uniform annual payment over the evaluation period, which has a net present value just equal to the net present value of the investment. Internal rate of return is the discount rate that makes the NPV of an investment just equal to zero. The benefit–cost ratio is the ratio of the net present value of benefits to the net present value of costs.

With the possible exception of the internal rate of return, economic-based investment criteria generally lead to the same investment decision when applied to independent investments. Applying investment criteria to an investment with type I interdependence requires adjusting the present value of that investment for changes in the present value of related investments. Correct assessment of type II (mutually exclusive) investments requires applying the criteria to incremental rather than absolute benefits and costs.

Questions for Discussion

1. What are some of the major differences between private and public economic evaluation of resource investments?

2. The United States Environmental Protection Agency (U.S. EPA) has the authority to regulate the use of pesticides in areas that are critical habitat for endangered species. The U.S. EPA proposes a regulation that would ban the use of insecticide X in region Z. Cotton farmers in region Z use insecticide X to control insect damage. EPA recommends that cotton farmers switch from insecticide X to insecticide Y. Insecticide Y is less effective and more expensive than insecticide X, but it is not regulated. Synthetic fibers, whose production does not involve insecticides, are a substitute for cotton. How would you evaluate the social efficiency of this regulation?

3. Explain the theoretical basis for selecting the market interest rate as the discount rate.

4. Consider the exploration and development of a 300-million-barrel oil deposit in two offshore areas. The first deposit is located in the Gulf of Mexico, which has a mild climate and is close to workers and facilities. It would take about one year to complete exploration and 10 years to develop and produce the Gulf of Mexico deposit. The second deposit is located in the Bering Sea west of Alaska, which has a severe climate and is far from workers and facilities. It would take about three years to complete exploration and 15 years to develop and produce the Bering Sea deposit. The likelihood of environmental damages from oil development is much higher in the Bering Sea than in the Gulf of Mexico.

a. Does a high discount rate favor oil exploration and development in the Gulf of Mexico or the Bering Sea? Why?

b. Which of the following two procedures is the better way to handle the differences in potential environmental damages of exploration and development in the Bering Sea and the Gulf of Mexico: Use a higher discount rate in the Bering Sea evaluation than in the Gulf of Mexico evaluation? Include expected environmental damages in the benefit–cost analysis? Explain.

5. What are the drawbacks of using the payback period and average rate of return to evaluate the feasibility of resource investments?

6. Investment A has an initial cost of $200 and investment B has an initial cost of $300. The investments are independent. Net benefits in dollars for both investments over nine years are:

	Year								
	1	2	3	4	5	6	7	8	9
A	100	110	120	130	130	130	125	120	115
B	225	230	235	240	230	220	210	200	190

Determine the net present value of each investment when the discount rate is 8 percent. Which investment would you select when there is no capital rationing? Why?

7. Consider three mutually exclusive investments:

	Investment Cost ($)	Annual Net Return ($)
A	1,000	300
B	700	450
C	1,300	600

a. Use the incremental IRR criterion to select the most efficient investment when the discount rate is 15 percent.

b. Would you select the same investment with the incremental BCR and the NPV criteria? Explain.

Further Readings

Brennan, Timothy J. "Discounting the Future: Economics and Ethics." *Resources*, Summer 1995, pp. 3–6.

Mishan, E. J. 1971. *Cost-Benefit Analysis: An Introduction*. New York: Praeger.

Pearce, David W., Edward Barbier, and Anil Markandya. 1990. "Discounting the Future." Chapter 2 in *Sustainable Development: Economics and Environment in the Third World*. London: Earthscan Publications Ltd., pp. 23—56.

Pearce, David W., and R. Kerry Turner. 1990. "Discounting the Future." Chapter 14 in *Economics of Natural Resources and the Environment*. Baltimore: The Johns Hopkins University Press, pp. 211–225.

Randall, Allan. 1987. "Benefit Cost Analysis." Chapter 13 in *Resource Economics: An Economic Approach to Natural Resource and Environmental Policy*. New York: John Wiley & Sons, pp. 233–259.

Notes

1. Kris Wernstedt, Jeffrey B. Hyman and Charles M. Paulsen, "Evaluating Alternatives for Increasing Fish Stocks in the Columbia River Basin," *Resources*, No. 109, Fall 1992, pp. 10–16.

2. Richard B. Howarth and Richard B. Norgaard, "Intergenerational Resource Rights, Efficiency, and Social Optimality," *Land Economics* 66(1990):1–11.

3. Salah El Serafy, "The Proper Calculation of Income from Depletable Natural Resources," in *Environmental and Resource Accounting and Their Relevance to the Measurement of Sustainable Income*, Ernst Lutz and Salah El Serafy, ed. (Washington, D.C.: World Bank, 1988).

4. Jack Hirschleifer, James C. DeHaven, and Jerome W. Milliam, "Investment in Additional Water Supplies" and, "The Practical Logic of Investment Efficiency Calculations," in *Water Supply: Technology and Policy* (Chicago, Illinois: The University of Chicago Press, 1969).

5. Roland N. McKean, *Efficiency in Government through Systems Analysis* (New York: John Wiley & Sons, 1958), pp. 151–167; and Otto Eckstein, *Multiple Purpose River Development* (Baltimore, Maryland: John Hopkins Press, 1958), pp. 202–214.

6. Hirschleifer et al., Chapter VII (1969).

7. Paul R. Portney, *Public Policies for Environmental Protection* (Baltimore, Maryland: Johns Hopkins University Press for Resources for the Future, 1990).

8. Hirschleifer et al. (1969) indicate that net benefits of $-1, 3$ and -2.5 do not have a non-imaginary solution for ρ.

CHAPTER 12

Nonmarket Valuation of Natural and Environmental Resources

The idea of putting a money value on damage done to the environment strikes many as illicit, even immoral.

—PEARCE AND TURNER, 1990

Efficient use of natural and environmental resources requires knowledge of the value of these resources in various uses. The epigraph suggests that some individuals (not Pearce and Turner) object to placing monetary values on environmental damages (as well as benefits) associated with alternative uses of natural and environmental resources. Except for the fact that market prices do not reflect the full social cost of resource use, there appears to be little objection to using market prices as a measure of the scarcity value of resources. Many uses of natural and environmental resources, however, cannot be valued in the marketplace because of incomplete or nonexistent markets. Considering the value of natural and environmental resources for which markets exist and ignoring benefits from and/or damages to resources that do not have markets result in a socially inefficient use of resources. This chapter examines the importance of *nonmarket valuation* in efficient use of natural and environmental resources, the theoretical basis for nonmarket valuation and alternative nonmarket-valuation methods.

Importance of Nonmarket Valuation

Suppose a parcel of public land can be utilized for coal extraction or cattle grazing. The best use of the land from an economic viewpoint is the use that generates the highest net social benefit (total social benefit minus production costs minus environmental damages). Total social benefit of the land in coal and cattle production depends on market prices for coal and cattle. Production costs are based on labor and capital costs, royalty payments and severance taxes on coal, and public grazing fees. Unfortunately, there are no markets that determine the environmental damages caused by coal extraction and cattle grazing. For this reason, it is difficult to determine whether using the land for coal or cattle production is the more socially efficient. Similarly, the net private benefit of commercial timber harvesting in a rain forest should be weighed against the benefit of preserving the forest's biological diversity in determining socially efficient rates of timber harvesting. While net private benefit of commercial timber harvesting can be determined based on market prices for timber and inputs used in timber harvesting, there is much more limited and uncertain information available for estimating the benefit of preserving biological diversity.

Some progress has been made in developing markets for certain natural and environmental resources. Consider air, water and fish or wildlife. The Congress of the United States included a market-based scheme for reducing air pollution in the Clean Air Act Amendments of 1990. This scheme uses tradable emission permits to curtail air pollution from industrial sources.[1] While water markets are being used in parts of the United States to allocate water between competing commercial, residential and agricultural uses, most of the world's free-flowing and stagnant waters are allocated using nonmarket schemes. There are private markets for fishing and hunting.

Unpriced benefits and/or damages associated with the use of natural and environmental resources can be handled three ways. First, they can be ignored. Ignoring benefits and damages is likely to result in socially inefficient resource use. Privately efficient rates of coal production, cattle grazing and timber harvesting are likely to exceed the socially efficient rates when environmental damages are ignored. Second, physical natural resource and environmental accounts can be used to keep track of resource depletion (see Chapter 10). While this approach is better than ignoring depletion, it is difficult to compare the monetary net private benefits of resource use with the physical depletion or overexploitation of resources. Furthermore, when resource impacts are not measured in monetary terms, it is difficult to evaluate the effectiveness and efficiency of resource policies designed to influence production rates and environmental damages, such as royalty rates on energy minerals and public grazing fees for livestock.

Third, the monetary impacts of resource use can be estimated using nonmarket valuation methods. While such methods are not perfect, they provide monetary values for determining the net social benefit of different intertemporal rates of resource use. Basing natural and environmental resource use on net social benefit rather than net private benefit contributes to the achievement of sustainable resource development and use.

The remainder of this section discusses the importance of nonmarket valuation in a) determining efficient use of exhaustible resources, b) implementing natural resource–environmental accounting, c) evaluating resource protection policies, and d) evaluating the feasibility of alternative resource investments.

EFFICIENT USE OF EXHAUSTIBLE RESOURCES. Chapter 7 derives the following criterion for efficient intertemporal use of an exhaustible resource:

$$MB = MEC + MUC + MNC,$$

where MB is marginal benefit of exhaustible resource use, MEC is marginal extraction cost, MUC is marginal user cost and MNC is marginal environmental cost. All four terms in this complete efficiency criterion are monetary. If MNC is measured in physical units, such as the number of fish and wildlife adversely affected by coal extraction, cattle grazing or timber harvesting, rather than in monetary units, then MNC cannot be included in the efficiency criterion.

Excluding MNC from the complete efficiency criterion results in the following private efficiency criterion: MB = MEC + MUC. Figure 12.1 compares the privately efficient extraction rate with the socially efficient extraction rate. If resource extraction is based on the private efficiency criterion, then the firm chooses an extraction rate of Q_r, which exceeds the socially efficient extraction rate of Q_s. Specifically, equality between MB and MEC + MUC occurs at a higher extraction rate than equality between MB and MEC + MUC + MNC. When MNC is measured in monetary terms, the socially efficient extraction rate (Q_s) can be determined and public policies can be developed to reduce the privately efficient extraction rate to the socially efficient extraction rate. While knowing the physical impacts of different resource extraction or use rates and management policies is better than ignoring these impacts, determination of socially efficient extraction rates requires monetary estimates of marginal environmental cost.

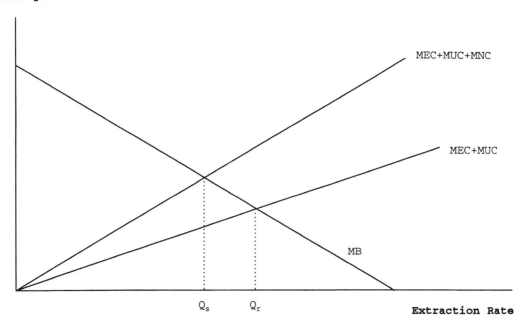

Figure 12.1. Privately efficient (Q_r) and socially efficient (Q_s) resource extraction rates. MB = marginal benefit; MEC = marginal extraction cost; MNC = marginal environmental cost; and MUC = marginal user cost.

NATURAL AND ENVIRONMENTAL RESOURCE ACCOUNTING. Chapter 10 discusses two ways of accounting for resource depletion and environmental degradation. First, it can be reported in a system of accounts that measures depletion and degradation in physical units, such as area in deforestation and extent of ozone depletion. Physical accounting makes it possible to determine whether the timber harvesting rate in a particular rain forest is at or below the rate of regeneration. While choosing a harvesting rate above the rate of regeneration causes long-term yields to decline, it is not clear whether it decreases net social benefit. Second, expressing resource depletion and degradation in monetary terms allows them to be integrated into national income accounts. The monetary approach enables direct comparison of the net social benefit of different rates of resource extraction and use.

Daly and Cobb[2] developed a natural and environmental resource accounting method called the Index of Sustainable Economic Welfare (ISEW). The ISEW adjusts the United States' gross national product (GNP) for monetary losses due to natural resource depletion and environmental damages. Growth in GNP was 0.80 percentage points lower in the 1970–1980 period and 0.42 percentage points lower in the 1980–1986 period after these adjustments. Downward adjustment in GNP for resource depletion and environmental damages is expected to be significant in countries whose economies are highly dependent on extraction of natural resources. For example, Repetto et al.[3] found that annual growth in Indonesia's gross domestic product (GDP) during the 1971–1984 period decreased from 7.1 percent to 4.0 percent after adjusting for natural resource depletion.

RESOURCE PROTECTION POLICIES. Chapter 10 shows that when there is a proportional relationship between the production of a good and associated pollution damages, the socially efficient rate of production occurs where marginal net private benefit (MNPB) equals marginal pollution damage (MPD). Provided that the needed information is available, socially efficient production can be achieved by imposing a Pigouvian-like tax on production equal to MPD at the socially efficient production rate. To determine the socially efficient rate of production, the environmental authority must know both MNPB and MPD. While the authority can require firms to report their MNPBs, it is much more difficult to determine pollution damages.

When the relationship between production and pollution is not proportional, the rate of pollution abatement, not the level of production, needs to be controlled. In this case, pollution abatement is at the socially efficient level when marginal social benefit of abatement (MSBA) equals marginal cost of pollution abatement (MCA). The socially efficient rate can be achieved by levying a Pigouvian tax on pollution equal to the difference between MSBA and marginal private benefit of pollution abatement (MPBA) at the socially efficient rate of abatement. Even if the environmental authority has information on MPBA and MCA, a Pigouvian tax cannot be determined without knowledge of MSBA. Unfortunately, market prices are not available for estimating MSBA.

EFFICIENT RESOURCE INVESTMENT. Chapter 11 discusses how benefit–cost analysis is used to evaluate the economic feasibility of alternative resource investments. Benefits and costs of most resource investments contain elements, such as the value of the extracted resource or capital investment, which can be determined from market prices. Other elements, such as reduction in environmental damages, are more difficult to measure due to nonexistent or incomplete markets. Consider, once again, the benefits and costs of investing in two technologies for removing timber from mountainous areas. The conventional technology employs trucks and roads and the new technology employs helicopters. The new technology results in significantly less soil erosion and sedimentation of streams and rivers than the conventional technology. Reduced sedimentation improves fish and wildlife habitat. Comparing the net social benefit of the two technologies requires estimating soil erosion and sedimentation damages with each technology.

The preceding discussion shows that a) marginal environmental cost is needed to determine the socially efficient use of exhaustible resources, b) monetary damages from resource depletion and environmental degradation are essential for developing monetary natural and environmental resource accounts, and c) either marginal pollution damage or marginal pollution abatement costs are required to determine the socially efficient level of production or pollution abatement. Most of these elements are not subject to market forces. Nonmarket valuation is the primary method for estimating the monetary value of these elements.

Theoretical Basis for Nonmarket Valuation

Chapters 7 and 8 indicate that resources are being used efficiently when net social benefit is maximized. Net social benefit equals the present value of total surplus (consumer surplus plus producer surplus) over some planning horizon. Total surplus in each time period equals the area between the demand and supply curves up to the equilibrium production rate. For natural and environmental resources having well-established markets, such as coal and timber, total surplus can be derived from estimated market demand and supply curves. For this procedure to be reliable, the supply curve should include the marginal environmental cost of production.

WILLINGNESS TO PAY AND ACCEPT COMPENSATION. The primary theoretical construct used in nonmarket valuation of natural and environmental resources is *willingness to pay* (WTP) and *willingness to accept compensation* (WTC). Willingness to pay and WTC for changes in the price, quantity or quality of a resource can be measured by compensating surplus, compensating variation, equivalent surplus and equivalent variation. Because compensating surplus and equivalent surplus assume that households are entitled to their current levels of satisfaction and because public resource management policies typically deal with potential benefits relative to current levels of satisfaction, the surplus measures are more relevant to policy analysis than are the variation measures. Willingness to pay pertains to a) paying a lower price or receiving a higher quantity or quality of the resource or b) avoiding a higher price or lower quantity or quality of the resource. Willingness to accept compensation pertains to a) forgoing a lower price or higher quantity or quality of the resource or b) tolerating a higher price or lower quantity or quality of the resource.

If the entity valuing the changes in price, quantity or quality is a firm, then WTP and WTC are measured by changes in profits. There is no difference between a firm's WTP and WTC for changes in price, quantity and quality. Willingness to pay is typically different than WTC for households because the value of changes in resource price, quantity or quality depends on the assignment of property rights. The remainder of this section examines a household's WTP and WTC for changes in resource price, quantity or quality.

VALUING CHANGES IN RESOURCE PRICE. Consider measuring a household's WTP and WTC for a decrease in the price of water pollution abatement. If the household does not have a property right to unpolluted water, then it is appropriate to ask the household's WTP for a lower price. The gain in consumer surplus from paying a lower price is the maximum WTP. On the other hand, if the household has a property right to unpolluted water, then it is appropriate to ask the WTC for giving up a lower price. The loss in consumer surplus from forgoing the price decrease is the minimum WTC.

Willingness to pay or WTC for a change in resource price is measured using the *Hicksian demand curve*. Along a Hicksian demand curve, resource price is varied while utility (level of satisfaction) and the prices of other resources are held constant. The Hicksian demand curve is used to estimate how changes in resource price influence consumer surplus. Along a *Marshallian demand curve*, money income and the prices of other resources are held constant. The Marshallian demand curve is used to determine how changes in resource price affect quantity demanded. The difference between a Marshallian demand curve (D_M) and a Hicksian demand curve (D_H) for pollution abatement is illustrated in Figure 12.2. Both demand curves indicate the quantity demanded of pollution abatement on the horizontal axis and the price of pollution abatement on the vertical axis.

Demand curves are determined by the household's preferences for pollution abatement, income and the prices of substitutes for and complements to pollution abatement. A negatively sloped demand curve implies that the per unit value of pollution abatement decreases as abatement increases. In other words, the marginal benefit of pollution abatement decreases as pollution abatement increases. For a reduction in price from p_1 to p_2, the increase in quantity demanded of pollution abatement is greater along the Marshallian demand curve (A_1 to A_3) than along the Hicksian demand curve (A_1 to A_2) because the latter is steeper.

Why is the Hicksian demand curve steeper than the Marshallian demand curve? In general, the change in quantity demanded from a change in price can be divided into a *substitution effect* and an *income effect*. The substitution effect is always negative, which means that quantity demanded increases (decreases) when price decreases (increases). The income effect of a price change is the change in quantity demanded with respect to a change in real income. Real income is cash income adjusted for inflation or deflation. As price increases (decreases), real income

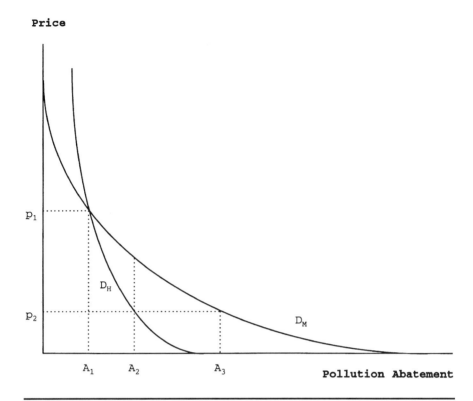

Figure 12.2. Marshallian (D_M) and Hicksian (D_H) demand curves for pollution abatement.

decreases (increases). Normal goods have a positive income effect, which means quantity demanded increases as income increases. Therefore, the income effect of a price change for a normal good causes quantity demanded to increase (decrease) when price decreases (increases). As quantity demanded increases (decreases), utility increases (decreases).

The Marshallian demand curve holds money income constant, which means that the increase in pollution abatement resulting from a price decrease consists of both a substitution effect and an income effect. Real income is constant along the Hicksian demand curve, which has the same effect as holding utility constant. As Figure 12.2 shows, the decrease in price from p_1 to p_2 causes abatement to increase from A_1 to A_3. The substitution effect is the increase from A_1 to A_2 and the income effect is the increase from A_2 to A_3. Because the Hicksian demand curve excludes the income effect, it is steeper than the Marshallian demand curve. The smaller (larger) the proportion of income spent on a good, the smaller (larger) the income effect and the more similar (dissimilar) the Hicksian and Marshallian demand curves. Hicksian and Marshallian demand curves are identical when the income effect is zero.

Suppose a household is asked its WTP for a decrease in the price of abatement from p_1 to p_2. The change in quantity demanded is found by tracing the price decrease along the Marshallian demand curve (D_M) as shown in Figure 12.3. Quantity demanded of abatement increases from A_1 to A_3. The household's WTP for this price decrease is the increase in consumer surplus between p_1 and p_2 along the Hicksian demand curve D_H, namely, $p_1 adp_2$. Note that the increase in Hicksian consumer surplus is less than the increase in Marshallian consumer surplus ($p_1 adp_2 < p_1 aep_2$) because the Hicksian demand curve is steeper than the Marshallian demand curve.

The household's WTC for paying a higher price (p_1 instead of p_2) is the decrease in consumer surplus between p_2 and p_1 along the Hicksian demand curve D'_H, namely, $p_1 bep_2$. Willingness to accept compensation exceeds Marshallian consumer surplus, which exceeds WTP ($p_1 bep_2 > p_1 aep_2 > p_1 adp_2$). The difference between WTP and WTC for price changes is attributed to the income effect. When the income effect is zero, the Marshallian and Hicksian demand curves are identical and WTP and WTC are equal. Willig[4] concluded that a) Marshallian consumer surplus is a good measure of changes in household welfare because it lies between WTP and WTC and b) the difference between Marshallian and Hicksian consumer surplus measures of WTP or WTC is frequently less than 5 percent. Hence, Marshallian consumer surplus is a good approximation of the value of changes in resource prices.

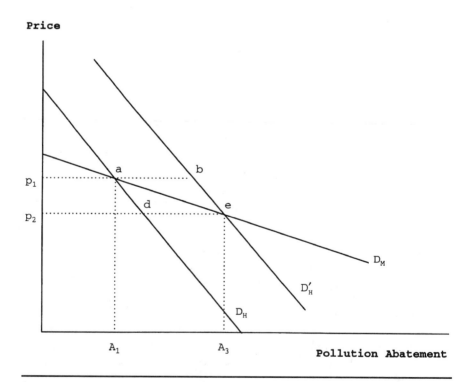

Figure 12.3. Willingness to pay for a decrease in price (p_1 to p_2) and willingness to accept compensation for a higher price (p_1 instead of p_2) for pollution abatement.

VALUING CHANGES IN RESOURCE QUALITY. Consider estimating the value that a household places on an increase in water pollution abatement. Willingness to pay and WTC for water pollution abatement are evaluated in Figure 12.4. The horizontal axis is household expenditures on abatement (EA) and the vertical axis is household expenditures on other goods (EO). Because prices of abatement and other goods are being held constant, changes in expenditures translate directly into changes in consumption. All combinations of expenditures on abatement and expenditures on other goods along an indifference curve provide the same level of total utility or satisfaction to the household. Expenditure combinations along U_2 provide greater utility than expenditure combinations along U_1. The indifference curves are convex (∪). This means that the household is willing to give up smaller and smaller amounts of the other goods to gain additional units of abatement when utility is held constant. Conversely, the household is willing to give up less and less abatement to gain another unit of other goods when utility is held constant.

If the household does not have a property right to clean water, then WTP for an increase in water pollution abatement is the appropriate measure of the change in consumer surplus. The maximum amount the household is willing to pay for an increase in abatement from EA_1 to EA_2 while maintaining the original level of util-

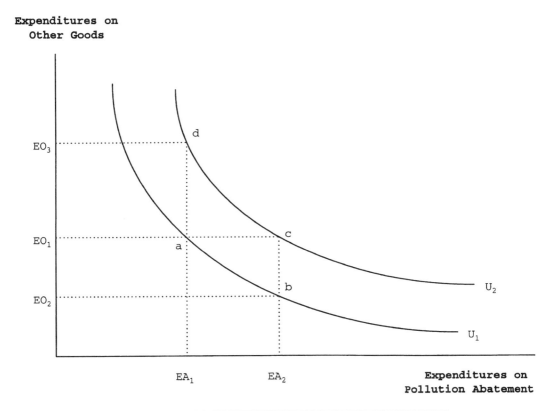

Figure 12.4. Willingness to pay for an increase in pollution abatement and willingness to accept compensation for a lower level of pollution abatement holding prices constant. U = utility.

ity (U_1) is the reduction in expenditures on other goods, namely, $EO_1 - EO_2$. In other words, WTP = $EO_1 - EO_2$. Why? Increasing abatement from EA_1 to EA_2 causes utility to increase unless consumption of other goods is decreased. Therefore, maintaining utility at U_1 when abatement increases requires the household to decrease consumption of other goods. Therefore, between a and b, the household must offset the increase in abatement from EA_1 to EA_2 by decreasing consumption of other goods from EO_1 to EO_2. Due to the convexity of the indifference curve, the offset amount decreases (increases) as abatement increases (decreases). Therefore, WTP for water pollution abatement decreases (increases) as abatement increases (decreases).

When the household has a property right to clean water, WTC for a lower level of abatement is the appropriate measure of the change in consumer surplus. If the household has a right to EA_2, then the starting point for valuing changes in environmental quality is at c where pollution abatement is EA_2 and expenditures on other goods is EO_1. Because point c provides higher pollution abatement ($EA_2 > EA_1$) but the same expenditure on other goods (EO_1) as point a, it must lie on a higher indifference curve, namely, U_2. The smallest amount of compensation the household is willing to accept for EA_1 instead of EA_2 is the additional expenditure on other goods required to maintain the original level of utility (U_2). Therefore, WTC = $EO_3 - EO_1$. When expenditure on other goods is increased from EO_1 to EO_3 and abatement is EA_1, the household is at point d on U_2.

Comparing WTP to WTC for a change in water pollution abatement shows that WTC > WTP ($EO_3 - EO_1 > EO_1 - EO_2$). This relationship holds when the indifference curves are convex ($\cup$). The magnitude of the difference between WTC and WTP depends on the size of the income effect and the degree of substitutability between pollution abatement and other goods. If the income effect is zero or there is another good that is a perfect substitute for the resource, then WTP and WTC are identical.[5] Income effects are negligible whenever a small proportion of household income is spent on the good. A similar procedure is used to determine the value of changes in resource quantity.

INEQUALITY BETWEEN WILLINGNESS TO PAY AND WILLINGNESS TO ACCEPT COMPENSATION.

Not surprisingly, empirical estimates of WTC are generally higher than estimates of WTP. Hammack and Brown[6] found that WTC was over four times greater than WTP for changes in waterfowl benefits. In addition to the theoretical argument just presented, there are several other reasons why WTP is less than WTC. First, WTP is constrained, whereas WTC is unconstrained, by income. Asking a household their WTP for pollution abatement is appropriate when the household does not have a property right to an unpolluted resource and the reduction in pollution represents a potential gain in welfare. Like the purchases of any other good, WTP for environmental improvements is constrained by the household's income. This is not the case for WTC. Willingness to accept compensation represents the household's willingness to accept compensation for the loss in welfare from forgoing an environmental improvement. Compensation required by the household would be paid by someone else, perhaps the entity whose actions deny the improvement in environmental quality. Because WTC is not constrained by the household's income, it is likely to be greater than WTP.

Second, WTC is likely to exceed WTP when households view the welfare loss from a reduction in environmental quality as more serious than the welfare gain from an improvement in environmental quality. When the household has the right to environmental quality, as implied by WTC, denying that right reduces welfare. When the household does not have the right to environmental quality, as implied by WTP, improving the environment provides a welfare gain.

Third, WTC is likely to exceed WTP when households believe they should not have to give up their rights to environmental quality. Households might express their unwillingness to give up such rights by placing high values on WTC. Fourth, high WTC values could result from cautious behavior on the part of households. True WTC is likely to be overstated by households that are not sure how to respond, are adverse to risk, or have limited experience with WTC questions.

Inequality between WTP and WTC suggests that care needs to be taken when these values are used as a basis for allocating resources. As shown in Chapter 11, resource investment and policy decisions involve a comparison of benefits and costs. It is quite possible for benefits to exceed costs when benefits are measured by WTC and for costs to exceed benefits when benefits are measured by WTP. In this case, the best strategy is to postpone a decision regarding the investment or policy until a more detailed assessment of benefits and costs is undertaken.

Use and Nonuse Values

The preceding section explains how WTP or WTC are used to measure changes in consumer surplus brought about by changes in resource price, quantity or quality. Total social benefit of changes in resource price, quantity or quality in a given time period is determined by adding up WTP or WTC for all households who place a value on the resource. If the present value of the stream of total social benefits minus total social costs is positive (negative), then the change is socially efficient (inefficient). Overestimation of social benefits and/or underestimation of costs can lead to acceptance of resource price, quantity or quality changes that are socially inefficient. Conversely, underestimation of benefits and/or overestimation of costs can lead to rejection of changes that are socially inefficient. Both types of errors are reduced by properly measuring all relevant benefits and costs. While measurement of benefits and costs is equally important, most nonmarket valuation studies concentrate on social benefits.

Four resource values comprise total social benefit *use value*, *option value*, *existence value* and *bequest value*. Option, existence and bequest values are nonuse values that are independent of resource use. This section discusses how use and nonuse values are used to evaluate a policy for reducing the frequency and acreage of federal timber sales in a national forest. The policy is expected to reduce both the acreage available for timber harvesting and the rate of timber harvesting and to enhance recreational opportunities.

USE VALUE. *Use value* is the value of the forest to those who use it for timber harvesting, fishing, hunting, hiking, camping and wildlife viewing. Timber harvesting, fishing and hunting are consumptive uses. Hiking, camping and wildlife viewing are nonconsumptive uses of the forest. In addition, fishing, hunting, camping and wildlife viewing are recreational uses of the forest. Suppose timber harvesting and recreation are competitive uses of the forest at current timber harvesting rates. Then, reducing timber harvesting rates decreases timber values and increases recreational value. Specifically, the proposed policy lowers income, employment and local taxes from timber harvesting and raises income, employment and local taxes from recreation. One way to evaluate a reduction in timber harvesting is in terms of an increase in the supply of recreational opportunities, which lowers the price of recreation. The effect on consumer surplus of a lower price for recreation is illustrated in Figure 12.5. When the price of recreation decreases from p_1 to p_2, consumer surplus increases by a + b.

OPTION VALUE. *Option value* is a household's WTP to preserve the option of having an irreplaceable resource available for future use. Option value is a legitimate resource value when resource supply and/or demand are uncertain, use is infrequent, and converting the resource to an alternative use makes it very costly to restore the resource to its original use (irreplaceability).[7]

Consider the option value for a unique natural resource such as Grand Canyon National Park or Everglades National Park. Demand uncertainty arises when a household is not sure whether he or she or others (children, grandchildren) will ever

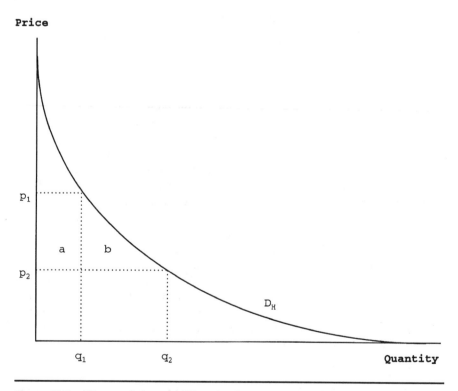

Figure 12.5. Changes in consumer surplus from a decrease in the price of forest recreation from p_1 to p_2. D = demand.

visit the park. Supply uncertainty arises when there is a possibility that the park might be developed for other purposes or that land uses outside the park might reduce the recreational and ecological benefits of the park. Infrequency of use is a source of option value.

Irreplaceability becomes an issue when alternative uses of the park make it very costly to return it to its original use or value. Irreplaceability was the basis for rejecting a proposal to build a commercial jetport in Everglades National Park. Opponents argued that the proposed jetport would cause severe and irreversible ecological damage to an irreplaceable natural asset.

Option value is applicable to private as well as public resources. For example, high rates of commercial timber harvesting in rain forests have the potential for destroying potentially valuable plant and animal species. This is especially true when there are few, if any, restrictions on the location and rate of harvesting. There is supply uncertainty because, quite often, no inventory is made of potentially valuable plant and animal species before harvesting. Demand uncertainty exists because relatively little is known about the potential pharmaceutical, medicinal and ecological value of affected species.

When the conditions necessary for option value are present, households might be willing to pay a fee to ensure the option of gaining access to the resource. Weisbrod (1964) argued that the most socially beneficial use of Sequoia National Park should be determined by comparing the recreational value (including use and option values) to the value of harvesting the park's redwood trees. Ignoring option value for the park leads to underestimation of its recreational value and increases the likelihood of choosing the harvesting alternative.

Option value becomes relevant in decision making when there is a threat that the unique services provided by a resource are in danger of being lost. For example, the option value of Everglades National Park was relevant to the decision of whether or not to construct a commercial jetport in the park. The jetport would have done irreversible damage to the Everglades' delicate ecosystems. Furthermore, there are no good substitutes for the Everglades. After the decision was made not to build the jetport, option value was no longer a relevant part of the total social benefit of the park. As long as the ecological services of the park are preserved by current management practices, provision of the option to preserve the park is automatically provided at no additional cost. Of course, option value could become a relevant part of total social benefit in the event another ecologically threatening activity is proposed.

Weisbrod's concept of option value has been expanded by Henry.[8] They define option value as a household's WTP for the right to have an irreplaceable natural resource preserved until such time that enough information is available to make a decision about how best to manage the resource. This argument is quite similar to the no-regrets management scheme proposed for environmental problems such as global warming and ozone depletion. Applied to these environmental problems, option value is WTP for the right to preserve climate resources by limiting the increase in global temperature and ozone concentration until such time as better information is available regarding the economic and ecological consequences of these events. Option value is relevant to climate resources because a) it applies to areas and time periods other than the one in which the household resides, b) damages are irreversible, at least on a human time scale, and c) the resource is irreplaceable because there are no substitutes.

EXISTENCE VALUE. *Existence value* is the amount a household is WTP for the assurance that a resource is preserved in a particular state. Existence value is based on the knowledge that the resource exists and is completely independent of use or option value. People derive satisfaction from knowing that humpback whales are not harvested to extinction; Australia's Great Barrier Reef is preserved in perpetuity; and lion, elephant, rhinoceros and giraffe are protected in South Africa's Kruger National Park. The memberships of many organizations, such as the National Wildlife Federation, Audubon Society, Sierra Club and Wilderness Society, are made up of households who value wild animals and wild places. The closest many of the members come to wild animals and places is a magazine article or documentary film. Yet, they are willing to support educational, scientific research, habitat and breeding programs that protect a wide range of wild places and animals. These households derive satisfaction from the mere knowledge that wildness is being preserved. Protection of certain species is sometimes controversial, such as the reduction in timber harvesting rates in old growth forests of the Pacific Northwest to protect the habitat of the northern spotted owl, introduction of timber wolf in Yellowstone National Park and the preservation of grizzly bears in Idaho, Montana and Wyoming.

BEQUEST VALUE. Krutilla[9] identifies another nonuse value called *bequest value*. Bequest value is the satisfaction that a household derives from knowing that a resource will be available to future generations. Krutilla uses bequest value to argue for public protection of unique natural areas. Bequest value is like option value in that it represents a willingness to pay for future availability of a resource. It differs from option value because what is being preserved is future use, not by the person holding the value, but by someone else. Bequest value is similar to existence value because both values pertain to satisfaction derived from simply knowing that a resource exists. Both users and nonusers of a resource can have bequest value, although it is likely to be higher for households who have used the resource. Option, existence and bequest values are relevant only when the resource is being threatened or damaged.

In summary, the effect of a change in resource price, quantity or quality on total social benefit includes use and nonuse values. While the value of many uses can be estimated from market prices and quantities, use and nonuse values are usually estimated with nonmarket valuation methods.

Estimation of Nonmarket Values

There are two major approaches to estimating the total social benefit of changes in resource price, quantity or quality: *indirect market methods* and *direct questioning methods*. Indirect market methods impute values to natural and environmental resources based on the market value of related resources. For example, the recreational value of reducing timber harvesting rates in a national forest can be estimated by comparing the recreational demands for forests having different timber harvesting rates. Indirect market methods include *averting behavior*, *weak complementarity* (which includes the *travel cost method*) and *hedonic pricing*. The direct questioning method asks households their WTP or WTC for specific changes in price, quality and/or quantity of a resource. The primary direct questioning technique is the *contingent valuation method*. This section discusses the estimation of the value of changes in resource price, quantity and quality using indirect market and direct questioning methods.

INDIRECT MARKET METHODS. Economists prefer to estimate changes in consumer surplus from statistically estimated demand curves. Because estimation of demand curves requires market data on prices and quantities, it works well for marketed resources. Indirect market methods allow the estimation of use and nonuse values of nonmarketed goods based on the market for related goods. Indirect market methods exploit the fact that changes in the price, quantity and quality of nonmarketed natural and environmental resources influence the value of marketed goods. This section examines three indirect market methods: cost of averting behavior, weak complementarity and hedonic pricing.

Cost of Averting Behavior. This approach recognizes that households respond to changes in the price, quantity or quality of a nonmarketed resource by modifying their purchases of related marketed goods. Smith and Desvousges[10] point out that households can avert some of the damages from polluted drinking water by purchasing bottled water. Dickie and Gerking[11] indicate that households install air conditioners to reduce the adverse effects of poor air quality. Consider averting behavior for a rural community whose drinking water has nitrate–nitrogen concentrations that exceed the safe drinking water standard. The standard is exceeded in the spring when nitrogen fertilizer is applied to nearby farmland. A possible averting behavior is for households in the community to switch from well water to bottled water in the spring. The cost of bottled water is an estimate of the maximum total willingness to pay for improving well water quality.

The cost of averting behavior has certain drawbacks. First, expenditures are likely to underestimate the surplus value of an improvement in resource quality. If, in addition to reducing their exposure to polluted tap water, a household switches to bottled water because it tastes better, then the cost of bottled water is likely to overestimate the household's WTP to reduce pollution of tap water. Second, the approach only works when there is an appropriate averting behavior that can be valued in the marketplace. For example, an appropriate averting behavior for reducing the negative impacts of high timber harvesting rates on recreation in a national forest is to increase recreation in a state forest. Unfortunately, there is no market for estimating the value of recreation in a state forest.

Weak Complementarity. This approach estimates the value of a resource attribute in terms of the difference in values of the resource with and without the attribute. The weak complementarity approach is commonly used to estimate the value of altering the attributes of a recreational site, such as a forest or park. Validity of the approach requires that access to the site is not free and that weak complementarity exists.

How well do these conditions apply to the timber harvesting–recreation example? Access to national forests is not free. There is a public access fee as well as the out-of-pocket and time costs of traveling to the site. The relationship between the reduction in timber harvesting and the recreational value of the forest satisfies the two conditions for weak complementarity given by Maler and by Bockstael and McConnell.[12] Specifically a) recreationists' marginal utility of reducing timber harvesting in national forests is zero when no visits are made to the forest, and b) there is a choke price above which consumption of forest recreation is zero.

The weak complementarity method for measuring the recreational value of reduced timber harvesting in a national forest is illustrated in Figure 12.6. In this figure, the vertical axis measures the cost per visit and the horizontal axis measures the number of visits in a particular national forest. The demand curves represent the sum of the Hicksian demand curves for all users of the site.

With policy 1, timber harvesting is maintained at the current high rate and the Hicksian recreational demand curve for the forest is D_1. If use of the forest requires payment of a daily fee of p_1, total consumer surplus for the forest with policy 1 is $p_1 ab$. With policy 2, timber harvesting is lowered, which enhances recreational opportunities causing recreational demand for the forest to increase from D_1 to D_2. If the access fee remains at p_1, then total consumer surplus with policy 2 is $p_1 dc$. Therefore, reducing the harvesting rates in the forest increases total consumer sur-

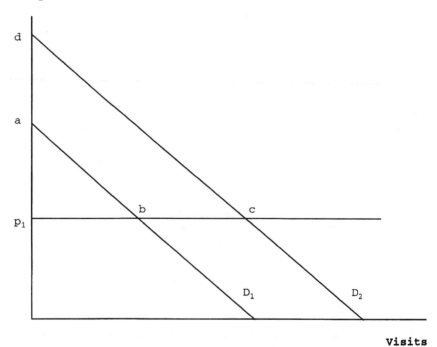

Figure 12.6. Increased value of forest recreation from a reduction in timber harvesting. D = demand.

plus from recreation by an amount equal to the area above p_1 between D_1 and D_2, namely, abcd. The net social benefit of reducing timber harvesting equals the increase in consumer surplus due to enhanced recreational opportunities minus the decrease in the value of commercial timber harvesting.

Travel Cost. Application of the weak complementarity method of valuing recreational sites requires estimation of the demand curve for the site. Demand is estimated by relating the number of visits to cost per visit. While number of visits varies across users, the access fee is the same for all users of a particular site. Hence, the price variation needed to estimate the recreational demand curve must come from another source. The *travel cost method* recognizes that visitors who reside farther away from the site incur a higher travel cost than visitors who reside near the site. In other words, the travel cost method uses the variation in the cost of traveling to the site (travel cost) as a proxy for variation in the cost per visit.

The theoretical basis for the travel cost method is the household production function theory developed by Becker, Lancaster, Hori, and Bockstael and McConnell.[13] This theory states that households generate utility by combining external inputs (purchased goods) with internal labor to produce desired goods. In terms of recreation, households produce recreational experiences by combining time, transportation services, recreational equipment and recreational sites.

The basic travel cost method has its origins in the writings of Hotelling.[14] He theorized that the demand curve for outdoor recreation at a particular site can be estimated statistically by relating participation rates to the corresponding travel costs for different user groups. Recreational value of the site is then estimated by the total area under the demand curve. Clawson[15] utilized the basic travel cost method to estimate demand functions for several national parks.

Hotelling, and Clawson and Knetsch[16] used surveys to determine the number of visits to a recreation site by users who reside in various concentric distance zones around the site. Total visits are divided by the population of users in each distance zone to obtain average visits per person for each zone. Per capita visits to a site account for the two major components of a recreation decision, namely, whether or not to recreate at the site and, when recreation occurs, the number of visits. Average round-trip distance between a zone and the site is multiplied by the average travel cost per mile to obtain an average travel cost for the zone.

The demand curve for a site is estimated using a two-step procedure. First, a statistical technique called regression analysis is used to estimate the travel cost demand function for the site. A typical travel cost demand function is as follows:

$$V_{jk} = f_j(p_{jk}, p_{js}, Y_j, S_j),$$

where:

V_{jk} = average per capita visits from the jth distance zone to the kth site,

p_{jk} = average travel cost from the jth distance zone to the kth site,

p_{js} = average travel cost from the jth distance zone to the sth substitute site,

Y_j = average income of households from the jth distance zone, and

S_j = average socioeconomic characteristics of households from the jth distance zone.

Second, the travel cost demand curve for the site is derived from the travel cost demand function by holding all variables except V_{jk} and p_{jk} constant at their means and taking the inverse of the resulting function, namely:

$$p_{jk} = f_j^{-1}(V_{jk}),$$

where f_j^{-1} stands for the inverse of f_j. A typical travel cost demand curve (D_e) is depicted in Figure 12.7.

Several refinements have been made to the basic travel cost method to improve accuracy of consumer surplus estimates. These include adding income and other socioeconomic factors as demand shifters,[17] accounting for the quality of the site,[18] improving statistical accuracy of estimated demand curves by using data for individual recreationists,[19] treatment of on-site costs as a fixed cost,[20] separating the price variable into transportation costs, travel time costs and on-site costs,[21] accounting for substitute sites,[22] and others. These and other aspects of the travel cost method are reviewed by Ward and Loomis.[23]

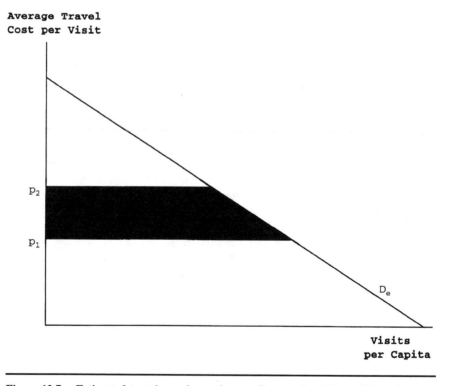

Figure 12.7. Estimated travel cost demand curve for a national forest (D_e) and decrease in consumer surplus (*shaded area*) from an increase in access fee from p_1 to p_2.

Suppose the recreational demand curve for a national forest is D_e in Figure 12.7 and that D_e has been estimated with the travel cost method. Recreational value of the forest equals consumer surplus, which is the total area under the demand curve. Because the travel cost method uses actual visits and travel costs, the resulting demand curve and consumer surplus are Marshallian. A typical use of a recreational demand curve is to estimate the change in consumer surplus from an increase or decrease in the access fee for the forest. For example, suppose the access fee is increased from p_1 to p_2. The estimated decrease in consumer surplus from the price increase equals the shaded area.

As shown in Figure 12.7, changes in recreational value due to price changes are easy to evaluate with the travel cost method once the recreational demand function has been estimated. It is much more difficult to estimate the value of changes in recreational quality that shift the demand curve for the site. The basic travel cost method estimates the demand curve for a particular site at a given point in time when the recreational qualities of the site are fixed. One way to overcome this difficulty is to estimate simultaneously the demand curves for multiple sites having different recreational qualities. Consider once again Figure 12.6. Let D_1 be the Hicksian demand curve for a national forest with high harvest rates and D_2 the Hicksian demand curve for a different national forest with low harvest rates. If other qualities of the two forests are similar, then area abcd is an estimate of the increased recreational value of the site from reducing the harvest rate.

The travel cost method has certain limitations and ambiguities, some of which have been remedied. These include exclusion of nonuse values such as option, existence and bequest values, selection of the variable for measuring use rates (number of visits or visitor days), downward bias in estimates of consumer surplus from omitting time cost and the price of substitute sites from the demand function, valuing the opportunity cost of travel time and on-site time, accounting for multiple characteristics of a site and multiple reasons for visiting a site, and choice of the mathematical form of the travel cost demand function.

Travel Cost Studies. Reiling et al.[24] used the travel cost method to estimate the recreational benefits of improving water quality in Upper Klamath Lake, Oregon. Travel cost data for four lakes, including Upper Klamath Lake, were used to estimate a regression equation for the four sites. Separate equations were estimated for visits per capita and number of days per visit. Results showed that improving water quality in the lake increases the value of recreational visits by $10 per visit or $3.50 per user day for visits that averaged 2.8 days (in 1968 dollars). The improvement in water quality was projected to increase total visits by 158 percent and total benefits by $3.8 million per year.

Smith et al.[25] used several travel cost models to estimate the increase in consumer surplus from improving water quality on the Monongahela River. Survey data used in their evaluation covered 13 sites along the river. Estimated Marshallian consumer surplus for improving water quality from boatable to fishable ranged from $0 to $8.74 per household in 1982 dollars.

Hedonic Pricing Method. The hedonic pricing method is based on the premise that the observed market price of a marketed good is a function of the prices of the numerous attributes of that good. For example, the incremental value of owning a house with a scenic view is the price of a house with a scenic view minus the price of a house without a scenic view, provided all other attributes of the two houses are similar. Likewise, the monetary value of the additional risk that iron workers take in installing the steel understructure of buildings is the average wage received by iron workers minus the average wage received by construction workers who perform tasks involving less risk, such as plumbers.

A typical hedonic price function is as follows:

$$p = g(a_1, a_2, ..., a_n, S)$$

where p is the price of a marketed good (price of a house or wage rate), g is the functional relationship between price and attributes, $a_1, a_2, ..., a_n$ are the n attributes of the good and S represents socioeconomic characteristics of the household. When g is an inverse function:

$$p = c_0 - c_1(1/a_1) - c_2(1/a_2) - ... - c_n(1/a_n),$$

where c_0 is the intercept of the hedonic price function and c_i is the change in the price of the good for a unit change in attribute i, holding all other attributes constant. c_i is the *marginal price* of attribute i. It equals the change in the price of the good with respect to a unit change in the value of attribute i when all other attributes and socioeconomic characteristics of the household are held constant. If p is the price of houses and attribute i is a scenic view, then c_i is the marginal price of having a house with a scenic view.

The hedonic price function is typically estimated using regression analysis. If a_i is scenic view and the corresponding regression coefficient (c_i) is $5,000, then a scenic view adds $5,000 to the value of a house when all other attributes are held constant. To ensure accuracy in estimating marginal prices, all relevant attributes of the good must be included in the hedonic price function. In equilibrium, households equate their marginal willingness to pay for an attribute to its marginal price. Therefore, marginal prices are estimates of the marginal willingness to pay for attributes. The hedonic pricing method estimates the value of relatively small changes in the attributes of a good. For large changes in attributes, the estimates of c_i represent upper limits on long-run changes in attribute values.[26]

Mitchell and Carson[27] identify several drawbacks of the hedonic pricing method. First, marginal prices are not accurate unless all the attributes of the good are included in the hedonic price function. Second, data for estimating the hedonic price function are often difficult to acquire, especially for goods that have slow turnover rates. For example, it is difficult to acquire comparable prices for houses that have a scenic view because turnover rates for such houses are low. Third, even though different functional forms result in different marginal prices, there is no theoretical basis for selecting a particular functional form (g) for the hedonic price function. Fourth, households may not be aware of differences in the attribute being valued. This is especially true of attributes that are not so obvious, such as the concentration of chemical contaminants in drinking water or air. Other attributes, such as a scenic view, are very noticeable.

Fifth, expectations about how an attribute changes over time can affect the

12. Nonmarket Valuation of Natural and Environmental Resources

value that households place on that attribute. For example, if residents of a community believe that the odor caused by a nearby animal feedlot is going to decrease because of new regulations on air pollution, then their WTP for a reduction in air pollution is likely to be low. While such expectations influence WTP, they are not directly observable. Sixth, it is difficult to estimate simultaneously changes in marginal prices of an attribute at substitute sites. Finally, the assumptions commonly used to identify demand and supply functions in the hedonic pricing model are generally incorrect.

Hedonic Travel Cost. The hedonic pricing method was originally used to estimate the marginal prices of attributes of marketed goods such as houses and labor. When the good under consideration is not marketed, there are no market prices and the hedonic price function cannot be estimated. Public recreation is a prime example of a good that does not have a market price. Unlike a house with a view, a public recreation site cannot be purchased. Hence, the marginal prices of attributes of recreational sites cannot be estimated with the hedonic pricing method.

The *hedonic travel cost* method, which is a variant of the hedonic pricing method, was developed to estimate the marginal prices of attributes of recreational sites and other nonmarketed goods.[28] Because the attributes of a given site are fixed, the hedonic travel cost method must be applied to multiple sites with different levels of attributes. The demand equations for site attributes are:

$$a_i = g(p_1, p_2, \ldots, p_m, Y, S)(i = 1, \ldots, n)$$

where a_i is the level of attribute i, n is the number of attributes, $p_1, p_2, \ldots, p_m$ are the respective travel costs per visit for the m sites, Y is household income and S represents the socioeconomic characteristics of the household. When g is an inverse function, travel cost to site k is:

$$p_k = C - b/a_i (k = 1, \ldots, m)$$

where C is travel cost with an infinite amount of attribute i, b is a positive constant and a_i is the level of attribute i. C and b are influenced by other site attributes and socioeconomic characteristics of the user. The inverse travel cost function illustrated in Figure 12.8 is asymptotic to C, which means that p_k approaches C as a_i increases. Travel cost functions for an attribute can be estimated using survey data on travel costs to various sites incurred by different households.

The marginal price of attribute i (p_i), also known as the marginal price function for attribute i, equals the change in travel cost associated with a change in the level of the attribute. The marginal price function is:

$$p_i = b/a^2_i.$$

This function is illustrated in Figure 12.9. Marginal price is inversely related to the level of the attribute, which implies diminishing marginal utility for the attribute. In equilibrium, the marginal price of an attribute equals the marginal WTP for the attribute. Therefore, the marginal price function is equivalent to the demand function for the attribute. Because the marginal price of an attribute in the hedonic travel cost

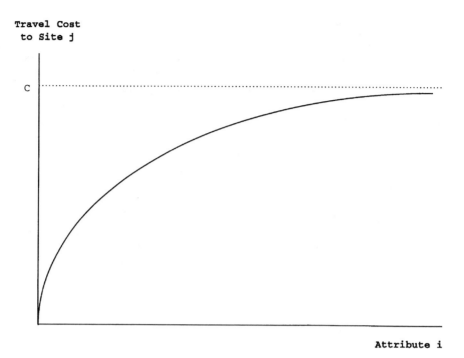

Figure 12.8. Inverse travel cost function for site j.

method is not derived from market prices for recreation, there is no guarantee that it is positive. For example, ease of access to recreation sites is likely to be better the closer the site is to an urban area. Travel cost to sites near urban areas is likely to be less than travel cost to more remotely located sites. In this case, travel cost is inversely related to ease of access, which means that the marginal price of access is negative.

Due to the equivalence of the marginal price function and demand function, the marginal price is an estimate of the value of changes in the level of an attribute. Suppose the recreational attribute of interest is the degree of solitude experienced by backpackers in a wilderness area. Solitude is inversely related to the number and size of other hiking parties encountered while backpacking. In terms of Figure 12.9, the marginal price of solitude decreases from p_{i1} to p_{i2} when solitude increases from a_{i1} to a_{i2}. The value of more solitude equals the increase in consumer surplus, namely, $p_{i1}cdp_{i2}$.

DIRECT VALUATION METHODS. Indirect valuation methods have the limitations mentioned above and require estimation of Marshallian and Hicksian demand curves. Direct valuation methods do not have these limitations. The primary direct valuation technique is the *contingent valuation method* (CVM). With the CVM, the value of a good is estimated by multiplying the average WTP or WTC for that good in a sample of households by the number of households in the relevant population. Willingness to pay and/or WTC for sample households are estimated

12. Nonmarket Valuation of Natural and Environmental Resources

based on their responses to a set of structured questions asked by phone, mail or in person.

The CVM is applicable to both private and public goods. In the case of private goods, market data are available to estimate demand and supply curves and consumer surplus. This is not the case for public goods. For this reason, most applications of the CVM deal with valuation of changes in the price, quantity and quality of public goods, especially changes in environmental quality. The next section describes the theoretical basis and application of the CVM.

Theoretical Basis for Contingent Valuation Method. As explained earlier, WTP and WTC are the main theoretical constructs underlying nonmarket valuation of public goods. Consider the following Hicksian demand function (D_H) for a public good:

$$p_u = D_H(q_u, q_v, F, U),$$

where p_u is the price of the good, q_u is the quantity consumed of the good, q_v is a vector of quantities consumed of private goods, F represents household preferences for all goods and U is a constant level of utility. Hicksian demand functions of this type are depicted in Figures 12.2, 12.3, 12.5 and 12.6.

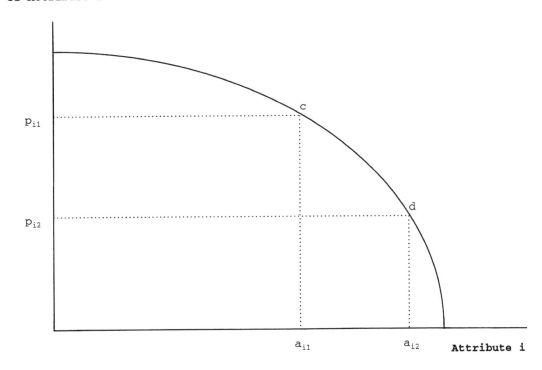

Figure 12.9. Marginal price function for attribute i.

Suppose the quantity of a public good is increased from q_1 to q_2. The Hicksian compensating surplus (CS) and Hicksian equivalent surplus (ES) values for the increased quantity are depicted in Figure 12.10. D_{H1} (D_{H2}) is the Hicksian demand function when utility is constant at U_1 (U_2). Compensating surplus is the area under D_{H1} between q_1 and q_2, namely, q_1abq_2. Holding utility constant at U_1 implies that the household is entitled to the endowment of property rights associated with q_1.

The Hicksian equivalent surplus (ES) from the increased provision of the public good is the area under D_{H2} between q_1 and q_2, namely, q_1cbq_2. Holding utility constant at U_2 implies that the household is entitled to the endowment of property rights associated with q_2. In general, $U_1 \neq U_2$. Compensating surplus and ES measures can also be derived for a change in the price and/or quality of a public good.

A second and theoretically equivalent way to define consumer surplus is in terms of the *expenditure function*. An expenditure function indicates the minimum expenditure or minimum amount of income needed by a household to achieve a particular level of utility when prices and quantities are at certain levels. The expenditure function is defined as follows:

$$Y = e(p, q, U),$$

where Y is expenditure, p is a vector of prices for private and public goods, q is the quantity and quality of the public good and U is utility. Based on the expenditure function, CS for an increase in the amount of the public good from q_1 to q_2 is:

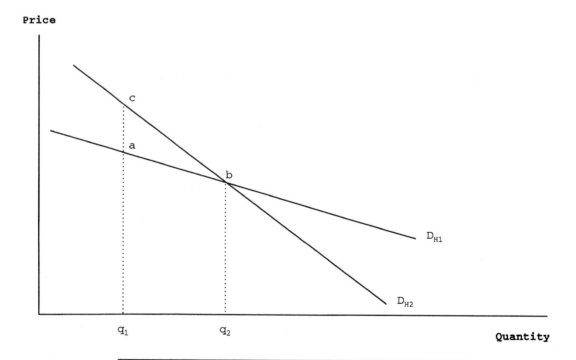

Figure 12.10. Hicksian compensation surplus (q_1abq_2) and equivalent surplus (q_1cbq_2) from an increase in the amount supplied of a public good from q_1 to q_2. D = demand.

$$CS = e(p, q_1, U_1) - e(p, q_2, U_1) = Y_1 - Y_2.$$

Using CS implies the household is entitled to q_1. Because an increase from q_1 to q_2 makes the household better off, the only way utility can be held constant at U_1 is for the household to reduce total expenditures. Therefore, $Y_1 > Y_2$ which implies CS > 0. $Y_1 - Y_2$ equals the maximum willingness to pay for the increased quantity of the public good.

The corresponding ES that makes the household willing to take q_1 instead of q_2 is:

$$ES = e(p, q_1, U_2) - e(p, q_2, U_2) = Y'_1 - Y'_2.$$

Using ES implies the household is entitled to q_2. Making the household take q_1 when there is an entitlement to q_2 reduces utility. The only way utility can be held constant at U_2 is for the household to increase total expenditure. Therefore, $Y'_1 < Y'_2$, which implies ES < 0. $Y'_1 - Y'_2$ equals the smallest amount the household is willing to accept to consume q_1 instead of q_2. In general, $Y_1 - Y_2 \neq Y'_1 - Y'_2$, which implies CS ≠ ES.

Two definitions have been given for CS and ES. The first definition, based on the area under the Hicksian demand function, is used in indirect valuation methods such as travel cost and hedonic pricing. The second definition, based on the difference between two expenditures, is the basis for direct valuation, notably the CVM. While both definitions are mathematically equivalent, the CVM approach has certain advantages. First, the CVM does not require estimation of Hicksian demand functions. Second, the CVM only requires that households be asked their WTP or WTC for changes in the price, quantity or quality of a good, which avoids some of the difficulties encountered in applying direct valuation methods. The CVM has its share of problems as discussed in the next section.

Application of CVM. The CVM method has been used to estimate the value of improving water quality, increasing visibility by reducing air pollution, protecting groundwater, preserving endangered species, reducing congestion and increasing the harvest success in big game hunting, reducing the likelihood of oil spills, and enhancing fish, wildlife and wilderness resources. A major concern with the CVM is the reliability of WTP and WTC values. Efforts to enhance reliability have focused on improving the design and execution of CVM surveys and the statistical analysis of the results.

A major criticism is that the WTP and WTC values from CVM surveys are derived from responses to hypothetical questions. This criticism is often stated as: Ask a hypothetical question, get a hypothetical answer. Hypothetical questioning runs the risk of giving unreliable results when respondents are not well informed enough to state their WTP or WTC because of inexperience with the good or misunderstanding regarding the information being requested in the survey. Surplus values estimated using indirect procedures are less vulnerable to the reliability criticism because they are derived from observed market behavior. The travel cost method uses information on how far recreationists actually travel to reach a site and the average cost of traveling. The hedonic pricing method employs actual differences in market-determined values of a related good to estimate the value of some attribute of that good.

Proponents of the CVM argue that reliability of CVM-based surplus values can be achieved by proper design of the survey questionnaire and careful analysis of the data. Bishop and Heberlein[29] identify six design elements for maintaining the reliability of CVM surveys and results. A brief description of each element is given below.

1. Know whose and which values will be estimated. One advantage of the CVM is that it can elicit both use and nonuse values of a good. The use and nonuse values cannot be differentiated in surveys, however, unless the corresponding populations are sampled in a statistically valid manner.

2. Provide respondents with a clear and meaningful description of the good. The ability of survey respondents to provide accurate and reliable information is directly related to their understanding of the good being valued. Giving respondents a clear, concise and meaningful description of the good increases the reliability of their responses. For example, suppose a sample of recreationists are being asked their willingness to pay for a reduction in harvest rate that improves recreational opportunities in a national forest. Asking them to state their WTP for reducing the harvest rate by 10,000 board feet per year is meaningless to most respondents. However, showing respondents several well-labeled, high-quality, color pictures depicting the visual, aesthetic and recreational impacts of successively lower harvest rates improves their ability to respond.

3. Use a realistic and neutral payment method for asking WTP and WTC questions. The reliability of WTP and WTC values is increased by choosing a realistic and neutral payment method. For example, if respondents are being asked their WTP for specific reductions in harvest rates, then it makes little sense to ask them to respond in terms of how much more they would be willing to pay in the form of higher taxes on cigarettes and liquor. Such taxes are not a realistic payment method for this particular good. It is more realistic to elicit their WTP in terms of a higher access fee for national forests. Sometimes, it is difficult to select a realistic and neutral payment method. For example, if the good being valued is clean air and the payment method is higher property taxes, respondents who believe property taxes are already too high may understate their WTP for clean air simply as a protest against further increases in property taxes.

4. Select a question format that gives reliable values. Several question formats are possible, including a) bidding games, b) open-ended questions, c) payment cards, d) dichotomous choice and e) contingent ranking. A bidding game asks respondents whether or not they are willing to pay successively higher amounts above some starting point amount until their maximum WTP is reached. An open-ended question directly asks respondents their maximum WTP. A bidding game is appropriate when respondents have difficulty handling open-ended WTP questions.

The payment card is a refinement of the open-ended question. Each respondent is shown a list of average expenditures made on related goods by households in the same income bracket as the respondent. After the card is shown, the respondent is asked the maximum WTP for the good. The payment card provides respondents with a reference point for making WTP bids. Both the bidding game and payment card can introduce bias into the WTP estimates by virtue of the choice of starting values.

Dichotomous choice requires respondents to answer yes or no regarding their

12. Nonmarket Valuation of Natural and Environmental Resources

WTP specified amounts. This format for eliciting WTP is thought to be easier for respondents because it does not require them to state their WTP in absolute terms. On the negative side, it is more difficult to estimate maximum WTP from dichotomous data than from continuous data provided by bidding games, open-ended questions and payment cards. Contingent ranking derives the value that a household places on a nonmarketed good from the way that household ranks various combinations of environmental quality and expenditures. Comparisons of value estimates derived using these five payment methods indicate that no one method is superior to all others. Consequently, selection of the payment method is subjective.

5. Collect information on other factors that affect values. Because WTP and WTC are derived from Hicksian demand functions or expenditure functions, they are affected by the same variables that influence demand and expenditure. These variables, lumped together under the heading of socioeconomic characteristics, include income, age, sex, education, household size, preferences, previous experience and involvement with the good and location of residence. Willingness to pay and/or WTC are typically expressed as a function of these socioeconomic variables, and the estimated WTP and/or WTC functions are used to predict surplus values for other resources and/or time periods than the ones used in the survey.

6. Analyze the data using valid statistical procedures. Most of the statistical techniques for estimating WTP or WTC functions are used in other studies. These include regular regression analysis and probit or logit regression analysis. A few statistical problems are encountered, however, in estimating WTP and/or WTC functions, such as the handling of extremely low or extremely high WTP or WTC values.

Even if these six elements are incorporated in the design and execution of a CVM survey, the method is susceptible to *strategic bias* and *hypothetical bias*. Strategic bias results when respondents intentionally overstate or understate the true value that they place on a good. For example, upward strategic bias in WTP is likely when a respondent believes that stating a high WTP increases the likelihood of an improvement in the good but does not affect what, if anything, he or she will be required to pay for the improvement in the good. On the other hand, downward strategic bias of WTP is likely when a respondent thinks the amount he or she will have to pay for an improvement in the good depends on his or her stated WTP.

Hypothetical bias typically occurs when a respondent is unable to assess WTP or WTC for a good accurately. It can occur even when the respondent has no intention of engaging in strategic behavior. Hypothetical bias is possible even in well-designed surveys, particularly when respondents have very limited prior experience with the good. Contingent valuation method evaluation of a public good is especially vulnerable to hypothetical bias because lack of a market makes it difficult for respondents to become familiar with their own preferences for the good. As a result, respondents may be unable to place a value on changes in price, quantity or quality. Unfortunately, trying to reduce hypothetical bias by providing reference points for the respondent, such as starting points in bidding games and comparable expenditures on payment cards, can introduce another source of bias in the results.

Hypothetical bias is different from the criticism that CVM surveys are unreliable because they involve hypothetical questions. For example, asking respondents to value changes in a private good is subject to the hypothetical question criticism but not to hypothetical bias because prior market experience is likely to exist for private goods. Like the hypothetical question criticism, strategic bias and/or hypothetical bias reduce the reliability of WTP and WTC values derived from CVM surveys.

The CVM is vulnerable to two other measurement problems: the *order effect* and *embedding*. When several items are being valued in the same survey, the order of the questions can influence the magnitude of the WTP or WTC values given by respondents. Tolley and Randall[30] found that respondents' WTP for improved visibility in the Grand Canyon fell by a factor of 3 when the order of this question was changed from first to third. The embedding problem occurs when the value given by respondents varies depending upon whether the good is being valued separately or as part of an inclusive good. For example, the CVM value of improved water quality at a lake is likely to be different depending on whether it is valued by itself or as part of a more inclusive good such as the recreational site where the lake is located or recreational sites in a region. If embedding is present, then WTP or WTC values are higher when the good is valued separately than when it is valued as part of an inclusive good. It is inappropriate to estimate the value of an inclusive good by simply adding up the values of the goods that comprise it.

Empirical Validity of CVM. The validity of CVM-based WTP and WTC values have been examined by comparing them to values obtained using other methods. Generally speaking, the divergence between CVM and actual values depends on a) whether the good in question is public or private, b) the method used to elicit values and c) whether the value being compared is WTP or WTC. Studies show that differences between CVM and actual market values of WTP for strawberries and hunting permits are not statistically different.[31] The CVM WTC for hunting permits (private good) are statistically and significantly higher than actual WTP in two out of three cases.[32] The CVM WTP and actual WTP values appear to be different for public goods.[33] Several studies indicate that CVM-based values are similar to values obtained with the travel cost[34] and hedonic pricing[35] methods.

Bishop and Heberlein[36] developed three simulated market experiments to compare CVM values to actual market values of WTP and WTC for hunting permits. These comparisons evaluated the extent of hypothetical bias and strategic bias. Simulated market values were determined by observing how hunters exchanged a specific amount of money (provided by the experiment) for deer and goose hunting permits. Based on their analysis, Bishop and Heberlein concluded there is a considerable upward bias in CVM-based WTC values but not in CVM-based WTP values. Other studies have concluded that strategic bias is not significant.

CVM Studies. In 1989 the *Exxon Valdez* oil tanker hit a reef in Prince William Sound, Alaska causing nearly 11 million gallons of oil to spill into the Sound. About 1,500 miles (2,415 km) of Alaska coast were polluted as a result of the spill. The state of Alaska surveyed 1,043 households in 50 states to determine their willingness to pay for an escort ship program, which would virtually eliminate another large oil spill in Prince William Sound for the next 10 years. By the end of the 10-year period, single-hull tankers would be replaced by double-hull tankers and the program would no longer be needed. The program requires oil companies removing oil from Alaska to pay a one-time tax and households to pay a one-time charge in the first year of the program. Only households who said they would never visit

Prince William Sound were included in the survey. Hence, the resulting values are existence and/or bequest values.

Sampled households supporting the program were asked if they would be willing to pay a larger amount ($30, $60, $120, $250). Households not supporting the program were asked if they would be willing to pay a smaller amount ($5, $10, $30, $60) for the program. The average WTP was $94 and the median willingness to pay was $31 per household. Extrapolating to all U.S. households resulted in a total willingness to pay for the program of $2.8 billion based on median WTP and $8.6 billion based on average WTP. Because these are existence and bequest values, the state of Alaska argued that Exxon Corporation should be held liable for the median total existence value of $2.8 billion. This amount is above and beyond the billions of dollars spent by Exxon and other groups to clean up the oil spill. Exxon settled a liability suit for $1 billion before the survey was introduced as evidence.[37]

Summary

Efficient use of natural and environmental resources and economic evaluation of resource management polices require knowledge of the private and social values of these resources in alternative uses. For natural or environmental resources that are subject to market forces, market prices should be used to evaluate private benefits, and where possible, social benefits of alternative patterns of resource use. When markets are nonexistent or incomplete, however, which is the case for public goods, nonmarket-valuation methods are often used to determine the social value of changes in resource price, quantity and quality. Nonmarket values are useful in a) determining efficient use of exhaustible resources, b) implementing natural and environmental resource accounting, c) evaluating resource protection policies, and d) evaluating the economic feasibility of alternative resource investments.

There are two primary measures of the surplus value of changes in the price, quantity or quality of natural and environmental resources, namely, willingness to pay (WTP) and willingness to accept compensation (WTC). For firms, WTP equals WTC and surplus value can be measured in terms of profit. Willingness to pay and WTC are generally different for households and the correct value to use depends on the assignment of property rights. Willingness to pay is the correct consumer surplus measure when the household does not have a property right to the good. Willingness to accept compensation is the correct consumer surplus measure when the household does have a property right to the good. Willingness to pay or WTC for changes in resource prices are measured by changes in consumer surplus along the Hicksian demand curve. The latter indicates the change in quantity demanded associated with a change in price when real income is held constant. Whenever the income effect of a price change is not zero, the Hicksian demand curve is steeper than the Marshallian demand curve and WTC exceeds WTP.

Willingness to pay and WTC for improvements in environmental quality are measured with reference to the indifference curve, which shows all combinations of expenditures on other goods and expenditures on improvement in environmental quality that provide the same level of utility. When the income effect is not zero, WTC exceeds WTP for improvements in environmental quality. Estimates of WTC generally exceed estimates of WTP.

Total social benefit of a change in resource price, quantity or quality is determined by summing WTP or WTC across all households who place a value on the resource. Total social benefit has four elements: use value, option value, existence value and bequest value. Use value is the value of the resource by those who use it. Option value is a household's WTP to preserve the option of keeping an irreplaceable resource available for future use. Existence value is the amount a household is willingness to pay for the assurance that a resource is preserved in a particular state. Bequest value is the satisfaction that a household derives from knowing that a resource will be available to future generations.

Nonmarket values can be measured using indirect methods and/or direct methods. Indirect market methods include averting behavior, weak complementarity (including the travel cost method), and hedonic pricing. Averting behavior determines nonmarket values in terms of changes in the purchases of related marketed goods. Weak complementarity estimates the value of changing an attribute of a good in terms of changes in the total value of the good.

The most prominent weak complementarity method is the travel cost method. In this method, the value of a recreational site is measured by the area under the estimated travel cost demand curve for the site. Demand curves are estimated by relating average use of the site to the average cost of traveling to the site plus income and other socioeconomic characteristics for households residing in unique distance zones around the site. The hedonic pricing method determines the value of an attribute of a good by variation in the market price of a good that has that attribute. The hedonic travel cost method is based on the hedonic pricing method.

The primary indirect valuation method is the contingent valuation method (CVM). The CVM determines the value of a good by multiplying the average WTP or WTC for that good in a sample of respondents by the number of respondents. Willingness to pay and/or WTC values are estimated from survey responses to a set of structured questions. The theoretical basis for the CVM approach is the expenditure function. Unlike direct methods, the CVM does not require estimation of Hicksian demand curves.

Indirect and direct methods of nonmarket valuation have certain advantages and disadvantages. The cost of averting behavior works when there is an appropriate averting behavior that can be valued in the marketplace. It can overestimate the value of a good when the averting behavior involves more than one good. Limitations of the travel cost method include its exclusion of nonuse values, ambiguity about how to measure use rates, bias due to omission of time cost and the price of substitutes from the demand function, differences of opinion regarding the value of time spent traveling to the site and the value of time spent at the site, difficulty in accounting for multiple characteristics of a site and multiple reasons for visiting a site, and choosing a correct mathematical form for the travel cost demand function.

Limitations of the hedonic pricing method are inaccurate marginal prices when some attributes of the good are not considered, difficulty in obtaining needed data, lack of theoretical justification for the functional form of the hedonic price function, inability of households to detect the attribute being valued, varying expectations about changes in attributes over time, difficulty in simultaneously estimating marginal prices of attributes that exist at substitute sites, and incorrect assumptions about underlying demand and supply functions. Contingent valuation, which is the primary direct valuation method, has been criticized because it involves asking respondents hypothetical questions. In addition, CVM-based estimates of WTP and WTC are subject to hypothetical bias, strategic bias, the order effect and embedding. Many of these pitfalls can be avoided by proper design and execution of CVM surveys.

Questions for Discussion

1. What role does nonmarket valuation play in the management of natural and environmental resources?

2. Use the Marshallian and Hicksian demand curves to explain the difference between WTP and WTC for an increase in the price of a nonmarketed good.

3. What is the appropriate measure of an improvement in air quality when a household has the right to clean air? When the household does not have the right to clean air?

4. The cost of new safety features for oil tankers is $2 million per year. The safety features are expected to reduce damages to fishery resources by $1.7 million per year. Are the safety features socially efficient? Explain.

5. A moderate oil spill causes authorities to close down Tarmin Beach, located 25 miles (40.3 km) from the city of Elmwood. The next best alternative to Tarmin Beach is Raymond Beach, which is located 40 miles (64.4 km) from Elmwood. How would you estimate the loss in the recreational value of Tarmin Beach from the spill?

6. What are the strengths and weaknesses of the travel cost method and contingent valuation method?

Further Readings

Adamowicz, W. L. 1991. "Valuation of Environmental Amenities." *Canadian Journal of Agricultural Economics* 39:609–618.

Cropper, Maureen L. and Wallace E. Oats. 1992. "Environmental Economics: A Survey." *Journal of Economic Literature* 30(2):675–740.

Cummings, R. G., D. S. Brookshire and W. D. Schultze. 1986. *Valuing Environmental Goods: An Assessment of the Contingent Valuation Method.* Totowa, New Jersey: Rowman and Allanheld.

Fletcher, J. J., W. L. Adamowicz and T. Grahm-Tomasi. 1990. "The Travel Cost Model of Recreation Demand: Theoretical and Empirical Issues." *Leisure Sciences* 12:119–147.

Freeman, A. Myrick III. 1993. *The Measurement of Environmental and Resource Values.* Washington, D.C.: Resources for the Future.

Greenley, Douglas A., Richard G. Walsh and Robert A. Young. 1982. *Economic Benefits of Improved Water Quality: Public Perceptions of Option and Preservation Values*, Studies in Water Policy Management, No. 3. Boulder, Colorado: Westview Press.

Hanemann, W. Michael. 1994. "Valuing the Environment through Contingent Valuation." *Journal of Economic Perspectives* 8:19–43.

Mendelsohn, Robert. 1987. "Modeling the Demand for Outdoor Recreation." *Water Resources Research* 23:961–967.

Randall, A., J. P. Hoen and D. S. Brookshire. 1983. "Contingent Valuation Surveys for Evaluating Environmental Assets." *Natural Resources Journal* 23:635-648.

Walsh, R. G., J. B. Loomis and R. S. Gillman. 1985. "Valuing Option, Existence and Bequest Demands for Wilderness." *Land Economics* 60:14–29.

Notes

1. U.S. Environmental Protection Agency, Office of Air and Radiation. *Implementation Strategy for the Clean Air Act Amendments of 1990* (Update, 1992), ANR-443, 400-K-92-004 (Washington, D.C.: July 1992).

2. Herman E. Daly and John B. Cobb, Jr., Appendix: The Index of Sustainable Economic Welfare, *For the Common Good: Redirecting the Economy Toward Community, the Environment, and a Sustainable Future* (Boston, Massachusetts: Beacon Press, 1989), pp. 401–455.

3. Robert Repetto et al. "The Need for Natural Resource Accounting," in *Wasting Assets, Natural Resources in National Income Accounts* (Washington, D.C.: World Resources Institute, 1989), pp. 1–25.

4. Robert D. Willig, "Consumer's Surplus Without Apology," *American Economic Review* 66(1976):589–597.

5. Michael W. Hanemann, "Willingness to Pay and Willingness to Accept: How Much Can They Differ?" *American Economic Review* 81(1993):635–647.

6. Judd Hammack and Gardner Mallard Brown, Jr., *Waterfowl and Wetlands: Toward Bioeconomic Analysis* (Baltimore, Maryland: The Johns Hopkins University Press for Resources for the Future, 1974).

7. Burton A. Weisbrod, "Collective-Consumption Services of Individualized-Consumption Goods," *Quarterly Journal of Economics* 78(1964):471–477.

8. Claude Henry, "Option Values in the Economics of Irreplaceable Assets," in *The Review of Economic Studies: Symposium on the Economics of Exhaustible Resources* (1974):89–104; Kenneth J. Arrow and Anthony C. Fisher, "Environmental Preservation, Uncertainty and Irreversibility," *Quarterly Journal of Economics* 88(1974):312–319.

9. John V. Krutilla, "Conservation Reconsidered," *American Economic Review* 57(1967):777–786.

10. V. Kerry Smith and William H. Desvousges, "Averting Behavior: Does It Exist?" *Economics Letters* 20(1986):291–296.

11. Mark Dickie and Shelby Gerking, "Willingness to Pay for Ozone Control: Inferences from the Demand for Medical Care," *Journal of Environmental Economics and Management* 21(1991):1–16.

12. Karl-Goran Maler, *Environmental Economics: A Theoretical Inquiry* (Baltimore, Maryland: Johns Hopkins University Press, 1974); Nancy E. Bockstael and Kenneth E. McConnell, "Welfare Measurement in the Household Production Framework," *American Economic Review* 73(1983):806–814.

13. Gary S. Becker, "A Theory of the Allocation of Time," *Economic Journal* 75(1965):493–517; K. Lancaster, "A New Approach to Consumer Theory," *Journal of Polit-*

12. Nonmarket Valuation of Natural and Environmental Resources

ical Economy 74(1966):132–157; H. Hori, "Revealed Preference for Pubic Goods," *American Economic Review* 65(1975):978–991; Nancy E. Bockstael and Kenneth E. McConnell, "Welfare Measurement in the Household Production Function Framework," *American Economic Review* 73(1983):806–814.

14. H. Hotelling, Letter to Director Newton B. Drury. *The Economics of Public Recreation, An Economic Study of the Monetary Evaluation of Recreation in the National Parks* (Washington, DC: National Park Service, 1947).

15. M. Clawson, *Methods of Measuring the Demand for and Value of Outdoor Recreation* (Washington, D.C.: Resources for the Future, 1959).

16. Hotelling (1947); Marion Clawson and Jack Knetsch, *The Economics of Outdoor Recreation* (Baltimore, Maryland: Johns Hopkins University Press, 1966).

17. W. G. Brown, A. Singh and E. N. Castle, "An Economic Evaluation of the Oregon Salmon and Steelhead Sport Fishery," Technical Bulletin No. 78, (Corvallis, Oregon: Oregon Agricultural Experiment Station, 1964); Cooperative Regional Research Technical Committee WM59, "An Economic Study of the Demand for Outdoor Recreation," collection of papers presented at Annual Meeting, Report No. 1, San Francisco, California and Report No. 2, Reno, Nevada.

18. J. B. Stevens, "Recreation Benefits form Water Pollution Control," *Water Resources Research* 2(1966):167–182; S.D. Reiling, D.C. Gibbs and H.H. Stoevener, "Economic Benefits from and Improvement in Water Quality" (Washington, D.C.: U.S. Environmental Protection Agency, 1973).

19. W. G. Brown, F. H. Nawas and J. B. Stevens, "The Oregon Big Game Resource: An Economic Evaluation," Special Report No. 379 (Corvallis, Oregon: Oregon Agricultural Experiment Station, 1973); Russell L. Gum and William E. Martin, "Problems and Solutions in Estimating the Demand for and Value of Outdoor Recreation," *American Journal of Agricultural Economics* 56(1974):558–566.

20. J. L. Knetsch, "Outdoor Recreation Demand and Benefits," *Land Economics* 39(1963):387–396.

21. P. H. Pease, "A New Approach to the Evaluation of Non-Priced Recreational Resources," *Land Economics* 44(1968):87-99; Reiling, Gibbs and Stoevener (1973); F. J. Cesario and J.L. Knetsch, "Time Bias in Recreation Benefit," *Water Resources Research* 6(1970):700–704; N. E. Bockstael, I. E. Strand and W. M. Hanemann, "Time and the Recreational Demand Model," *American Journal of Agricultural Economics* 69(1987):293–302; Frank Ward, "Specification Considerations for the Price Variable in Travel Cost Demand Models," *Land Economics* 69(1984): 301–305.

22. P. P. Caulkins, R. C. Bishop and N. W. Bouwes, "The Travel Cost Model for Lake Recreation: A Comparison of Two Methods for Incorporating Site Quality and Substitution Effects," *American Journal of Agricultural Economics* 68(1986):291–297; J. Knetsch, R. Brown and W. Hansen, "Estimating Expected Use and Value of Recreation Sites," *Planning for Tourism Development: Quantitative Approaches,* C. Gearing, W. Swart and T. Var, eds. (New York: Praeger Publishers, 1976); Cindy Sorg, John Loomis and Dennis Donnelly, "Net Economic Value of Cold and Warm Water Fishing in Idaho," Technical Report RM-107 (Fort Collins, Colorado: U.S. Forest Service, 1984).

23. Frank A. Ward and John B. Loomis, "The Travel Cost Demand Models in Environmental Policy Assessment: A Review of Literature," *Western Journal of Agricultural Economics* 11(1986):164–178.

24. Reiling, Gibbs and Stoevener (1973).

25. V. Kerry Smith, William H. Desvousges and Ann Fisher, "A Comparison of Direct and Indirect Methods for Estimating Environmental Benefits," *American Journal of Agricultural Economics* 68(1986):280–289.

26. Yoshitsugu Kanemoto, "Hedonic Prices and the Benefits of Public Prices," *Econometrica* 56(1988):981–990.

27. Robert Cameron Mitchell and Richard T. Carson, *Using Survey to Value Public Goods: The Contingent Valuation Method* (Washington, D.C.: Resources for the Future, 1989), pp. 80–81.

28. G. Brown and R. Mendelsohn, "The Hedonic Travel Cost Method," *Review of Economics and Statistics* 66(1984):427–433; Sherwin Rosen, "Hedonic Prices and Implicit Markets: Product Differentiation in Pure Competition," *Journal of Political Economy* 82(1974):34–55.

29. Richard C. Bishop and Thomas A. Heberlein, "The Contingent Valuation Method," *Economic Valuation of Natural Resources: Issues, Theory and Applications,* Rebecca L. Johnson and Gary V. Johnson, eds. (Boulder, Colorado: Westview Press, 1990).

30. George S. Tolley and A. Randall, "Establishing and Valuing the Effects of Improved Visibility in the Eastern United States," Report to the U.S. Environmental Protection Agency, Washington, D.C., 1983.

31. Mark Dickie, Ann Fisher and Shelby Gerking, "Market Transactions and Hypothetical Demand Data: A Comparative Study," *Journal of American Statistical Association* 82(1987):69–75; Richard C. Bishop and Thomas A. Heberlein, "Measuring Values of Extramarket Goods: Are Indirect Measures Biased?" *American Journal of Agricultural Economics* 61(1989):926–930; Richard C. Bishop, Thomas A. Heberlein and Mary Jo Kealy, "Contingent Valuation of Environmental Assets: Comparisons with a Simulated Market," *Natural Resources Journal* 23(1983):619–634.

32. Bishop and Heberlein (1989); Bishop, Heberlein and Kealy (1983).

33. Mary Jo Kealy, Jack Dovidio and Mark L. Rockel, "Willingness to Pay to Avoid Additional Damages to the Adirondacks from Acid Rain," *Regional Science Review* 15(1987):118–141.

34. J. L. Knetsch and R. L. Davis, "Comparison of Methods for Recreation Valuation," *Water Research,* A.V. Kneese and S.C. Smith, eds. (Baltimore, Maryland: Johns Hopkins University Press, 1966); W.H. Desvousges, V.K. Smith and M.P. McGivney, "A Comparison of Alternative Approaches for Estimating Recreation and Related Benefits of Water Quality Improvements," *Report to the Environmental Protection Agency by Research Triangle Institute*; C. Sellar, R. Stoll and J.P. Chavas, "Validation of Empirical Measures of Welfare Change: A Comparison of Nonmarket Techniques," *Land Economics* 61(1985):156–175.

35. D. S. Brookshire, M. A. Thayer, W. D. Schultze and R. C. D'Arge, "Valuing Public Goods: A Comparison of Survey and Hedonic Approaches," *American Economic Review* 72(1982):165–177.

36. Bishop and Heberlein (1990).

37. "Polls May Help Government Decide the Worth of Nature," *New York Times,* September 6, 1993.

Index

Access fee, 184
Accounting, resource, 243–62
   deficiencies, 244–50
      defensive expenditures, 245
      residual pollution damages, 249–50
      resource capacity, 246–49
   methods, 250–57
      monetary accounts, 252–57
      physical accounts, 250–52
      satellite accounts, 257
   nonmarket valuation, 304
   resource-specific, 257–59
   sustainable development and, 259–60
Acid deposition, 7–8, 157, 194
Adjusted GNP (Gross National Product), 252–57
   defined, 252–53
   Index of sustainable economic welfare, 255–57
   resource capacity accounting, 253–55
Agricultural production
   capital expenses and, 249
   deforestation, 15
   genetic diversity and, 8
   global warming and, 4, 15, 190
   ozone depletion and, 6
   pesticide use, 10–11
   scarcity, 67
   soil erosion, 257–59
   technological change, 15–17
Air pollution, 302. *See also* Acid deposition
   health costs, 15
   open access resources and, 99
Aluminum recycling, 149–50, 191
Amenity value
   circular flow economy, 75
   conservationist view of, 58
   material balances economy model, 75
Amortization expenses, 279
Annual net benefit, 286
Averting behavior, 196, 315

Balance sheet method, 253
Benefit-cost analysis of resource investments. *See* Investment analysis
Benefit-cost ratio, 286–87, 293
Benefits, 49–50
   annual net, 286
   marginal private of recycling, 150–51
   primary and secondary, 281–82
   private, 97–98, 266
   social, 120–22
      divergence from private benefit, 266
   efficient intertemporal extraction, 127–40
   erosion damage and, 259
   feasibility, 279
   investment analysis, 266–70
   pollution abatement, 201, 205–6, 209
   primary and secondary, 281–82
   renewable resources, 173–75, 178, 182–83
   valuation of resources, nonmarket, 301–31
Bequest value, 314
Best management practices (BMPs), 234, 247
Bias, survey, 327–28
Biological diversity, 8–9, 82
Biological resources
   harvest rate, 160–86
   natural growth, 159–62
Biomass
   global warming and, 4
   natural growth, 159–62
   ozone depletion and, 6
Bribes and externality reduction, 104–6
Brundtland Commission, 81
Budget set (line), 30–32, 275
Bureau of Land Management, 159, 167

Cancer and ozone depletion, 6, 15, 193
Capacity, resource, 55–70
   accounting, 246–49
      balance sheet method, 253
      net price method, 254
      value of depletion, 254
   conservativism, 60–62
   contemporary views, 62–63
   economic views
      classical economics, 55–59
      neoclassical economics, 59–60
   indicators
      economic, 66–68
      environmental, 68
      physical, 65–66
   material balances model, 75
   non-economic factors, 63–65
Capital, 244–49
   constraint, 277
   depreciation, 244, 246–47
   Index of sustainable economic welfare, 255–57
   investment, 248–49
   opportunity cost, 274, 276–78, 291
   rationing, 281, 291, 295
   sustainability and, 245
   technological change and, 247
Capital cost, 279

Carbon dioxide, 190
  Earth Summit, 82
  emissions, cumulative, 6
  global warming and, 5–6
  residual assimilation, 193
Cattle grazing. *See* Grazing
Centrally planned economy, 91–92
CFCs. *See* Chlorofluorocarbons (CFCs)
Chesapeake Bay, 9, 15
Chlorofluorocarbons (CFCs), 82, 193
  global warming, 4
  ozone depletion, 6–7
Circular flow economy, 73–75, 111–12, 114
Classical economics, 55–59
Clean Air Act, 99
  acid deposition, 8
  sulfur dioxide emissions, 68, 99, 115, 213–14
  tradable emission permits, 223, 302
Clean Water Act, 68, 285–86
Climate. *See* Global warming
Closed-loop recycling, 149
Coal, 4, 115
  acid deposition and, 7
  production, 65
  sulfur dioxide and, 201
Coase, Ronald, 104, 109
Coastal flooding, 5
Commodities
  circular flow economy, 74
  consumption and demand
    commodity bundle, 28–31
    elasticity of demand, 33
    indifference curve, 28–33
    utility, 28–33
  marginal rate of substitution, 274–75
  market equilibrium, 43–49
  material balances economy model, 75–76
  natural price, 56
  production and supply, 33–43
Commodity bundle, 28–31
Common property resources, 97–99
Complementarity, weak, 315–17
Complement commodities, 33, 47
Compounding, 49
Comprehensive Environmental Response, Compensation and Liability Act of 1980. *See* Superfund
Conditionally renewable resources, 157
Conservation, energy, 218
Conservationism, 60–62
Conservation Reserve Program (CRP), 94
Conservation tillage, 234, 235
Consumer surplus, 118, 128, 267–68, 305–9, 396. *See also* Valuation, nonmarket
Consumption
  circular flow economy, 74–75
  and demand theory, 27–33
  commodity bundle, 28–31
  constraints, 29–30
  residuals, 27
  utility, 28–33
efficient intertemporal, 274–75
marginal rate of time preference, 274–75
material balances economy model, 75–77
sustainable income and, 244
Contingent valuation method (CVM), 322–29
  application of, 325–28
  studies, 328–29
  theoretical basis, 323–25
  validity of, 328
Continuous compound interest factor, 50
Continuous discount factor, 51
Continuous vs discrete time, 271
Cornucopian view, 59, 69, 80
Costs
  benefit ratio, 286–87, 293
  budget, 36–40
  external, 144–47
  long-run marginal, 43
  marginal extraction, 128–48
  marginal user, 136–38, 140–48, 179–81
  opportunity, 118–20, 274, 276–78, 291
  primary and secondary, 281–82
  replacement, 247
  short-run marginal, 40–42
  travel, 317–19, 321–22
  variable and fixed, 40–41
Cost sharing, 234
Crop rotation, 258–59

DDT, 11
Decision hierarchy, resource, 111–14
Decisions, natural resource. *See* Management, natural resource
Defensive expenditures, 245
Deficit, 47
Deforestation, 9, 14–15, 116
Demand
  conditionally renewable resource, 174–77
  dynamic market model, 126–27
  efficient intertemporal extraction of exhaustible resource, 127–42, 144–45
  efficient production and, 206–8
  elasticity, 33, 38
  income effect, 306–7
  substitution effect, 306–7
Demand curve
  bribes and externality reduction, 104–6
  Hicksian, 306–8, 316, 323–25
  household, 31–33
  input, 38–40
  land, 118–21
  market, 43–45, 47–49
  Marshallian, 33, 306–8

monopsony, 96
pure competition, 92
travel cost, 317–19
Depletion, resource
 circular flow economy and, 74–75
 contributing factors, 11–17
  environmental damages, 14–15
  population growth, 11–13
  resource use per capita, 13–14
  technology, 15–17
 material balances economy model, 75–77
 public attitudes, 10
Depletion allowance, 248
Depreciation, 244, 246–47
 benefit-cost analysis, 279
 NNP and, 246
Dioxin, 249
Discounted value, 50
Discount rate, 50
 efficiency and equity, 272–73
 efficient intertemporal extraction of exhaustible resource, 147–49
 intergenerational equity, 82
 selection of, 273–78
 sensitivity analysis, 278
 stock effect, 272
 water resource developments and, 64
Discrete compound interest factor, 49
Discrete discount factor, 50
Discrete time, 271
Dolphins, 171
Drinking water, 11, 15
Dynamic efficiency, 173–81
 necessary conditions, 175–81
 objectives and constraints, 174–75
 pollution and, 207–10

Earth Summit, 81, 82
Ecological economics, 20, 78–80, 112–14
 material balances model compared, 80
 neoclassical economics compared, 75, 78–79
 optimal scale of the economy, 79–80
Ecology, 78
Economic feasibility, 279
Economic rent, 57, 163
Economics, 19–20
 classical, 55–59
 ecological, 20, 75, 78–80, 112–14
 environmental, 19–20, 78
 exhaustion of a resource, 116
 metaeconomics, 20
 natural resource, 78
 neoclassical, 59–60, 75, 78–79
 resource capacity
  classical, 55–59
  indicators of, 65–68
  institutional, 60

 limits to growth, 62–63
 neoclassical, 59–60
Economic stagnation, 56, 58
Economic welfare, 244–45, 255–57, 304
Economy
 centralized, 91–92
 environment and, 19–20, 73–87
  circular flow economy, 73–75
  ecological economics, 78–80
  ecosystem support of economy, 78
  material balances model, 75–77
  sustainable development, 80–86
 optimum scale, 79, 112–13
Efficiency
 defined, 19
 dynamic, 173–81
  necessary conditions, 175–81
  objectives and constraints, 174–75
 energy use, 65
 extraction of exhaustible resources, 125–53
 frontier, 83–86
 intertemporal use of renewable resources, 174–81
 local vs global, 280
 Pareto, 121
 pollution reduction, 195–210
  dynamic, 207–10
  static, 195–207
 private, 266
 privately efficient production level, 92–93
 social, 266
 static, 118–22
  common property renewable resources, 167–70
  multiple-use resources, 171–73
  privately owned renewable resources, 162–67
Effort restrictions, 184
Elasticity of demand, 33
Embedding, survey, 328
Emissions
 charges
  nonpoint source pollution abatement, 231–32
  point source pollution abatement, 213–18
 physical accounts, 250–51
 restrictions, 201–6
 standards, 218–21
 tradable permits, 302
  nonpoint source pollution abatement, 235
  point source pollution abatement, 221–26
Endangered Species Act, 68, 171, 173
 economic criteria prohibition, 64
 pesticides and, 11
Energy. *See also specific energy resources*
 per capita consumption, 13
 recycling, 149

Energy (*continued*)
  return on investment, 65–66
Entropy, 74, 77
Environmental economics, 19–20, 78
Environmental liability. *See* Liability, environmental
Environmental problems
  acid deposition, 7–8
  biological diversity, 8–9
  contributing factors, 11–17
    environmental damages, 14–15
    population growth, 11–13
    resource use per capita, 13–14
    technology, 15–17
  global warming, 4–5
  ozone depletion, 5–7
  policy aspects, 10–11
Environmental Protection Agency. *See* EPA (Environmental Protection Agency)
Environment and the economy, 19–20, 73–87
  circular flow economy, 73–75
  ecological economics, 19–20, 78–80
  material balances model, 75–77
  reductionist vs holistic approach, 17–19, 20–21
  sustainable development, 80–86
EPA (Environmental Protection Agency)
  CFC ban, 6
  creation of, 10
  emission standards, 219
  mandated production methods, 234
  pesticide restrictions, 11
  Superfund and, 229
  water quality, 10, 191
Equilibrium
  household, 31–33
  long-run supply, 43
  market, 43–49
    imperfect competition, 122
    pure competition, 120–21
Equity, intergenerational, 82, 272–73
Erosion Productivity Impact Calculator, 258
Eutrophication, 94
Everglades National Park, 313
Exhaustible resources, 115–16. *See also* Depletion, resource
  capacity loss estimation, 247–48
  discount rate and, 272–73
  economic exhaustion, 116
  economics of, 19
  extraction, efficient intertemporal, 127–53, 272–73, 302–3
    multiple-period efficiency, 139–42
    two-period efficiency, 128–39
  global warming and, 4–5
  influencing factors
    discount rate, 147–49

  exploration and development, 153
  external costs, 144–47
  imperfect competition, 143–44
  recycling, 149–52
  technological progress, 143
  market dynamics, 125–27
  nonmarket valuation, 302–3
  per capita use, 13–14
  recycling, 116
  reserves-to-use ratio, 115–16
Existence value, 314
Expansion path, 37–38
Expenditure function, 324–25
External costs, 144–47
External diseconomy, 101–8
  common property resources, 182
  extraction of exhaustible resources and, 144–45
  feasibility and, 279
  pollution damages, 189, 194
External economies, 101–3
Externalities, 100–108
  examples, 102–4
  Pareto relevant, 101, 201, 205–6, 210
  property rights and, 104–8
  reduction, 104–8
  types of, 101
  undepletable, 211
Extinction, 159–60
  renewable resource, 159–60, 168–70, 182
  species, 9, 15, 159–60
Extraction, efficient intertemporal
  influencing factors
    discount rate, 147–49
    external costs, 144–47
    imperfect competition, 143–44
    recycling, 149–52
    technological progress, 143
  multiple-period efficiency, 139–42
  net social benefit and, 127–40
  two-period dynamic efficiency, 128–39
    constant demand, 129–33
    variable demand, 133–39
Exxon Valdez oil spill, 68
  contingent valuation method, 322–29
  environmental liability, 226

Feasibility
  economic vs financial, 279
  evaluation criteria, 283–90
Fertilizers
  input restrictions, 232
  residual assimilation, 193
  sources of, 16
  taxes, 230
Financial feasibility, 279
Fisheries

Index

access control, 184
biodiversity, 8–9
harvest rate, 15, 179
hydroelectric dams, 272
resource accounting, 159
scarcity of resource, 67
Flow resource, 117
Food safety, 10
Forests
   access control, 184
   acid deposition and, 7
   biodiversity, 9
   as common property resources, 167
   deforestation, 9, 14–15
   efficient harvest rate, 171–72
   Endangered Species Act and, 64, 171
   global warming and, 4
   investment analysis example, 287–90
   net social benefit example, 267–70
   nonmarket valuation example, 311–14
   productivity, 14
   as renewable resource, 116–17
   sediment pollution and, 194
Fossil fuels
   acid deposition and, 7–8
   entropy energy, 77
   global warming and, 4–6
   per capita use, 13
   prices, 5
   production capacity, 65–66
France, resource accounts and, 251
Free rider problem, 106
Fund resource, 117

General Agreement on Tariffs and Trade (GATT), 171
Geothermal energy, 5
Global efficiency, 280
Global warming, 4–5, 190–91
   cost of control, 82
   no-regrets public policy, 78–79
GNP (Gross National Product), 244
   adjusted, 252–57
      defined, 252–53
      Index of sustainable economic welfare, 255–57
      resource capacity accounting, 253–55
   defensive expenditures, 245
   depreciation and, 246–47
   Index of sustainable economic welfare, 304
   investment in capital, 248–49
   net national welfare, 252
Government. *See also* Taxation
   centralized economy, 91–92
   common property resources and, 98–99, 167
   defensive expenditures, 245
   discount rate, 272, 278
   externality reduction, 107–8
   physical accounts of resources, 250–51
   public goods, 100
   renewable resource policies, 182–85
   water resource development projects, 272
Grazing
   common property resource, 97–98, 167
   fee, 167, 246
   resource capacity, 246
Greenhouse gases, 4–5, 15, 190–91. *See also specific emissions*
Gross National Product. *See* GNP (Gross National Product)
Groundwater
   mining, 17
   pollution, 235

Habitat destruction, 8–9
Harvest rate, 160–86
   access control, 184–85
   conditional function, 162
   efficient intertemporal, 175–81
   general function, 160–62
   taxation, 182–84
Hazardous waste, 15, 245
Health care, property rights, 95
Hedonic pricing method, 319–21
Hedonic travel costs, 321–22
Hicksian demand curve, 306–8, 316, 323–25
Holistic approach, 18–19, 20–21
Hotelling's condition, 140–41, 147
Households
   consumption, 27–33, 47
      constraints, 29–30
      equilibrium, 31–33
      market demand curve, 43–45, 47
   demand curve, 31–33
   externalities, 103–4
   production function theory, 317
Hubbert's curve, 65
Hunting, 185
Hypothetical bias, 327–28

Imperfect competition
   efficient intertemporal extraction of exhaustible resources, 143–44
   efficient resource use, 122
   pollution and, 206–7
Income, household and consumption, 29–33, 47
Independent investments, 280–81, 290–91
Index of sustainable economic welfare, 255–57, 304
Indifference map (curve), 28–33, 83, 274
Inferior good, 33–34
Inflation, 244
Input, 34–40
   demand curve, 38–40

Input (*continued*)
  efficient use, 37–40
  elasticity of demand, 38
  restrictions, 232–34
  tax, 230–31
  variable and fixed, 40
Institutional economics, 60
Interdependent investments, 280–81, 291–94
  type I, 291–92
  type II, 292–94
Interest rate, 49–50
Intergenerational equity, 82, 272–73
Internal rate of return, 287
  independent investments, 291
  interdependent investments, 293–94
  multiple resource investments, 294–95
Investment analysis, 265–97
  capital and operating costs, 279
  capital rationing, 281
  continuous vs discrete time, 271
  discount rate, 272–78
  efficiency, local vs global, 280
  efficient resource investment, 304–5
  evaluation criteria, 283–90
    annual net benefit, 286
    benefit-cost ratio, 286–87
    payback period, 283–84
    present value, net, 285–86
    rate of return, 284–85, 287
  evaluation period, 271–72
  feasibility, economic vs financial, 279
  independent vs interdependent, 280–81, 290–95
  multiple resource, 294–95
  mutually exclusive investments, 292–94
  net social benefit, 266–70
  primary and secondary benefits and costs, 281–82
  risk, 282–83
  social efficiency, 266–70
Irreplaceability, 312–13
Irrigation, 194
Isoquants, 34–38

Jevons, William Stanley, 59

Keynesian macroeconomic model, 244

Labor
  natural price and, 56
  value theory, 58, 59–60
  as variable input, 40
Land. *See also* Soil
  demand and pure competition, 118–21
  economics, 60
  ethic of Alfred Leopold, 61–62
  scarcity, 56–59
  absolute, 56–58
  relative, 57–58
Landfills and recycling, 149–50
Leopold, Aldo, 61–62
Liability, environmental, 226–29, 237
Local efficiency, 280
Logistic growth curve, 160–61

Malthus, Thomas, 56–58
Management, natural resource, 111–23. *See also* Exhaustible resources; Renewable resources
  decision hierarchy, 111–14
    circular flow model, 111–12, 114
    ecological economics model, 112–14
    material balances model, 112, 114
    sustainable development model, 112–14
  renewable resource, 157–86
  types of resources
    exhaustible resources, 115–16
    renewable resources, 116–18
Mandated production methods, 234
Marginal costs, 92–93
  food production, 58
  long-run, 43
  social, 93
Marine Mammals Protection Act, 171, 173
Market dynamics for exhaustible resources, 125–27
Market equilibrium, 43–49
Market failures
  common property, 97–99
  externalities, 100–108
  imperfect competition, 96–97
  open access resources, 99
Markets, 91–93
Marsh, George P., 61
Marshall, Alfred, 59
Marshallian demand curve, 306–8
Marx, Karl, 59–60
Material balances, 19–20, 75–77, 80, 112, 114, 191–94. *See also* Pollution, economics of
Maximum sustainable yield, 160
Metaeconomics, 20
Mill, John Stuart, 58–59
Monetary accounts, 252–57
Monopsony, 122
Montreal Protocol, 6–7, 82
Moral hazard, 229
Motor vehicles subsidies, 64
Muir, John, 61
Multiple-period efficiency, 139–42
Multiple-use management, 159, 171–73
Mutually exclusive investments, 292–94

National income accounts, 244–50
  defensive expenditures, 245
  residual pollution damage, 249–50

Index

resource capacity, 246–49
National Oceanic and Atmospheric Administration (NOAA), 7
Natural and environmental resource accounting. *See* Accounting, resource
Natural growth, biomass, 159–62
Natural price, 56
Natural resource economics, 78
Neoclassical economics, 59–60
    ecological economics compared, 75, 78–79
Net National Welfare (NNW), 252–53
Net present value, 285–86
    independent investments, 290–91
    interdependent investments, 291–94
    multiple resource investments, 295
Net price method, 254
Net social benefit. *See* Benefits, social
Nitrogen
    acid deposition and, 7–8
    global warming and, 4
    residual assimilation, 193
NNP (Net National Product), 244, 246, 253
Nondegradable renewable resources, 157–58
Nonmarket valuation. *See* Valuation, nonmarket
Nonpoint source pollution, 191
    abatement policies, 229–36
        cost sharing, 234
        emission charges, 231–32
        input restrictions, 232–34
        input tax, 230–31
        mandated production methods, 234
        tradable emissions permits, 235
    property rights and, 210–11
No-regrets public policy, 78–79
Normal good, 33–34
North American Free Trade Agreement (NAFTA), 171
Norway, resource accounts and, 250–51
No-till farming, 11

Offset system, pollution, 221–22
Oil
    as common property resource, 98–99
    efficient intertemporal extraction, 125–42
    Exxon Valdez spill, 68, 226, 322–29
    global warming and, 4–5
    market dynamics, 125–27
    production, 65–66
    reductionist approach example, 17–18
    scarcity, 67–68
    stock uncertainty, 115
Open access resources, 99, 107
Open-loop recycling, 149
Operating cost, 279
Opportunity cost, 118–20, 274, 276–78, 291
Optimal scale, 79–80, 112–13
Optimal stock, renewable resource, 178–81

Option value, 312–13
Order effect, survey, 328
Output, 34, 40–43. *See also* Production
Overgrazing, 97–98, 246
Overpopulation. *See* Population growth
Ownership. *See* Property rights
Ozone depletion, 5–7, 82, 193
    Montreal Protocol, 6–7, 82
    skin cancer, 6, 15, 193

Pareto efficiency, 121
Pareto relevant externalities, 101, 201, 205–6, 210
Payback period, 283–84
Pecuniary externality, 101
Pecuniary spillovers, 282
Per capita resource use, 13–14
Persistence, pollutant, 193
Pesticides, 10–11
Petroleum. *See* Oil
Physical accounts, 250–52
Pigouvian tax, 184, 211–13, 237, 304
Point source pollution
    abatement policies, 211–29
        emission charges, 213–18
        emission standards, 218–21
        environmental liability, 226–29
        taxes, 211–13
        tradable emission permits, 221–26
    defined, 191
Politics and resource development, 63–65. *See also* Government
Pollution, economics of, 189–238
    abatement policies, 211–36
        nonpoint source, 229–36
        point source, 211–29
    averting behavior, 196
    defensive expenditures, 245
    dynamic efficiency, 207–10
    material balances economy model, 75–76
    nonpoint source, 191, 210–11, 229
    persistence, 193
    point source, 191, 210–29
    potency, 193
    property rights, 94, 210–11
    static efficiency with imperfect competition, 206–7
    static efficiency with pure competition, 195–206
        emission restrictions, 201–6
        production restrictions, 195–201
Pollution abatement
    dynamic efficiency, 207–10
    efficiency in pure competition, 203–6
    efficiency with imperfect competition, 206–7
    nonpoint source policies, 229–36
        cost sharing, 234

Pollution abatement (*continued*)
    emissions charges, 231–32
    input restrictions, 232–34
    input tax, 230–31
    mandated production methods, 234
    tradable emissions permits, 235
  point source policies, 211–29
    emission charges, 213–18
    emission standards, 218–21
    environmental liability, 226–29
    taxes, 211–13
    tradable emission permits, 221–26
Population growth, 11–13
  Malthusian trap, 56–57
  stabilizing, 79–80
Potency, pollutant, 193
Preference, marginal rate of time, 274–78
Preferences, individual, 83, 86
Present value, 49–51
  net, 285–86
    independent investments, 290–91
    interdependent investments, 291–94
    multiple resource investments, 295
Preservationists, 60–61
Preservation value, 173–74
Prices
  as constraints on consumption, 29–33
  market equilibrium, 47–49
  natural, 56
  production and supply effects of, 35–43
Prior appropriation doctrine, 64
Producer surplus, 118–20, 128, 267–68
Production
  efficient intertemporal, 276–77
  equilibrium, 277
  function, 34–43
  mandated methods, 234
  rate and efficiency
    dynamic, 207–10
    imperfect competition, 206–7
    pure competition, 199–206
  restrictions and pollution reduction, 195–201
  supply theory and, 33–43
    constraints, 35–36
    input use, efficient, 37–40
    output, efficient, 40–43
Profit maximization
  common property resources, 97
  long-run, 42–43
  privately efficient level of production, 92–93
  rate of harvest, 163–68
  short-run, 40–42
Property rights, 64, 92–108
  assignment of, 105–7
  common property, 97–99
  defined, 92
  efficient, 93, 104

enforceability, 94
environmental liability and, 228–29
environmental resources, 194, 210–11
externalities, 104–8
government intervention, 107–8
open access resources, 99, 107
ownership, 93
public goods, 100
specificity, 93–94
transaction costs, 95
transferability, 94
willingness to pay and accept compensation, 309–11
Protection policies, resource, 304
Public goods, 100
Public policy, 10–11. *See also* Government
Pure competition, 92
  efficient resource use, 118–21
  pollution and, 195–206

Rate of return
  average, 284–85
  internal, 287, 291, 293–94, 294–95
Rationing, capital, 281, 291, 295
Recession, 244
Recycling, 16, 75–77, 116, 149–52, 191
Reductionist method, 17–18, 20
Renewable resources, 116–18. *See also* Depletion, resource
  conditionally renewable, 157
  discount rate and, 272–73
  dynamic efficiency, 173–81
    necessary conditions, 175–81
    objectives and constraints, 174–75
  economics of, 19
  energy, 5
  intertemporal use, 272–73
  natural growth, 159–62
  nondegradable, 157–58
  public policies, 182–85
  replacement cost, 247
  static efficiency, 162–73
    common property, 167–70
    multiple objective management, 171–73
    private property, 162–67
Rent-seeking behavior, 213
Replacement cost, 247
Res communis, 93
Reserves-to-use ratio, 115–16, 248
Residual pollution damages, 249–50
Residuals, 33, 189–238
  accumulated, 192
  assimilated, 193
  markets for, 112
  recycling, 75–77, 149–51, 191
Res nullius, 93, 97
Resource economics, 19

Index

Ricardo, David, 57–58
Risk, 282–83
Roosevelt, Theodore, 61

Salinization, 194
Salvage value, 279
Satellite accounts, 257
Scarcity, 67
    absolute, 56–58
    relative, 57–58
Scientific conservationists, 60
Sea-level changes, 4
Sensitivity analysis, 278
Shrimp production, 266
Skin cancer and ozone depletion, 6, 15, 193
Smith, Adam, 56
Social benefits. *See* Benefits, social
Socially efficient investment, 266–70
Soil
    depletion, 252
    erosion
        cropland losses, 15
        natural resource accounting, 257–59
        no-till farming, 11
        offsite damages, 257–59
        replacement cost, 247
    as renewable resource, 116
    salinization, 194
    sediment pollution, 194
Solar energy, 5, 16, 117, 157–58
Specificity, property rights, 93–94
Spillovers, 282
Static efficiency, 118–22
    common property renewable resources, 167–70
    multiple-use resources, 171–73
    pollution with imperfect competition, 206–7
    pollution with pure competition, 195–206
        emission restrictions, 201–6
        production restrictions, 195–201
    privately owned renewable resources, 162–67
Strategic bias, 327
Strip mining, 4
Substitution
    capital
        for labor, 58
        manufactured and natural, 245
    effect on demand, 33, 47, 306–7
    irreplaceability, 312–13
    marginal rate of commodity, 274–75
    marginal rate of technical, 276–77
    price increases and, 63
    recycling and, 150–52
Sulfur dioxide, 193–94, 201
    acid deposition and, 7–8
    Clean Air Act, 68, 99, 115, 213–14
    tradable emission permits, 223

Superfund, 228–29, 245, 249
Supply
    curve
        land, 118–22
        long-run, 43
        market, 43, 46–49
        short-run, 40–42
    dynamic market model, 126–27
Surplus. *See* Consumer surplus; Producer surplus
Surveys, contingent valuation method, 325–28
Sustainability, 62–63, 79, 245
Sustainability contour, 85–86
Sustainable development, 80–86, 112–14
    Brundtland Commission, 81
    obstacles to, 82
    sustainable income and, 259–60
    sustainable resource use, 83–86
Sustainable income. *See also* Accounting, resource
    defined, 244
    sustainable development and, 259–60
    weak and strong sustainability, 245
Sustainable yield, maximum, 164–70, 178–79
System of National Accounts (SNAs), 244, 257
Systems analysis, 18

Taxation
    externality reduction, 107
    on harvest of renewable resources, 182–84
    input tax, 230–31
    marginal cost effects, 93
    Pigouvian, 184, 211–13, 237, 304
    pollution abatement and, 211–13
    recession and, 244
Technological externality, 101
Technological spillovers, 282
Technology, 15–17
    ecological economics conservative assumption, 78–79
    harvest rate and, 159
    marginal productivity effects of, 58
    pollution abatement, 215, 220
    product and supply influences of, 34, 43, 48
    resource capacity and, 247
    resource extraction and, 143
    socially efficient investment and, 267–70
Temperature. *See* Global warming
Thailand, 266
Thermodynamics, law of, 77, 80
Tillage, conservation, 234, 235
Timber Harvesting. *See* Forests
Time, continuous vs discrete, 271
Tradable emission permits, 302
    nonpoint source pollution abatement, 235
    point source pollution abatement, 221–26
Tragedy of the commons, 97
Transaction costs, 95, 105–6

Transfer payments, 259
Transformation curve, 276–77
Travel costs, 317–19, 321–22
Tuna harvesting, 171
Two-period dynamic efficiency, 128–39
  constant demand, 129–33
  variable demand, 133–39

Ultraviolet radiation, 5–6
Uncertainty, 282–83
Undepletable externality, 211
United Nations System of National Accounts (SNAs), 244, 257
User cost, 136–38, 140–48, 179–81, 248, 273
Use value, 311–12
Utility, 28–33
  Hicksian demand curve, 306
  pollution and, 195, 203
  sustainable resource use, 83–85

Valuation, nonmarket, 301–31
  importance, 301–5
    exhaustible resources, efficient use of, 302–3
    investment, efficient resource, 304–5
    protection policies, 304
    resource accounting, 304
  nonuse value, 311–12
    bequest value, 314
    existence value, 314
    option value, 312–13
  theoretical basis, 305–11
    price changes, valuing, 306–8
    quality changes, valuing, 309–10
    willingness to accept compensation, 305–11
    willingness to pay, 305–11

use value, 311–12
value estimation, 314–29
  direct valuation methods, 322–29
  indirect market methods, 315–22
Value of depletion, 254

Water
  allocation, 64
  pollution
    oyster production, 9, 15
    strip mining and, 4
  quality, 10–11
    enforcement, 94
    nonpoint source pollution, 191
    standards, 213, 217, 232
    total watershed management, 191
Weak complementarity, 315–17
Wetlands
  agricultural expansion and, 17
  biodiversity and, 9
  industrial activities in, 68
Willingness to pay and accept compensation, 305–29
  bequest value, 314
  estimation of nonmarket values, 314–29
    averting behavior, 315
    CVM, 322–29
  existence value, 314
  option value, 312–13
  theoretical basis for nonmarket valuation, 305–11
Wood, 4
World3 model, 63
World Resources Institute, 251

Yield function, 162–63, 166–67, 169–70